Graduate Texts in Mathematics 87

Springer
New York
Berlin
Heidelberg
Barcelona
Budapest
Hong Kong
London
Milan
Paris
Santa Clara
Singapore
Tokyo

Graduate Texts in Mathematics

1 TAKEUTI/ZARING. Introduction to Axiomatic Set Theory. 2nd ed.
2 OXTOBY. Measure and Category. 2nd ed.
3 SCHAEFER. Topological Vector Spaces.
4 HILTON/STAMMBACH. A Course in Homological Algebra.
5 MAC LANE. Categories for the Working Mathematician.
6 HUGHES/PIPER. Projective Planes.
7 SERRE. A Course in Arithmetic.
8 TAKEUTI/ZARING. Axiomatic Set Theory.
9 HUMPHREYS. Introduction to Lie Algebras and Representation Theory.
10 COHEN. A Course in Simple Homotopy Theory.
11 CONWAY. Functions of One Complex Variable I. 2nd ed.
12 BEALS. Advanced Mathematical Analysis.
13 ANDERSON/FULLER. Rings and Categories of Modules. 2nd ed.
14 GOLUBITSKY/GUILLEMIN. Stable Mappings and Their Singularities.
15 BERBERIAN. Lectures in Functional Analysis and Operator Theory.
16 WINTER. The Structure of Fields.
17 ROSENBLATT. Random Processes. 2nd ed.
18 HALMOS. Measure Theory.
19 HALMOS. A Hilbert Space Problem Book. 2nd ed.
20 HUSEMOLLER. Fibre Bundles. 3rd ed.
21 HUMPHREYS. Linear Algebraic Groups.
22 BARNES/MACK. An Algebraic Introduction to Mathematical Logic.
23 GREUB. Linear Algebra. 4th ed.
24 HOLMES. Geometric Functional Analysis and Its Applications.
25 HEWITT/STROMBERG. Real and Abstract Analysis.
26 MANES. Algebraic Theories.
27 KELLEY. General Topology.
28 ZARISKI/SAMUEL. Commutative Algebra. Vol.I.
29 ZARISKI/SAMUEL. Commutative Algebra. Vol.II.
30 JACOBSON. Lectures in Abstract Algebra I. Basic Concepts.
31 JACOBSON. Lectures in Abstract Algebra II. Linear Algebra.
32 JACOBSON. Lectures in Abstract Algebra III. Theory of Fields and Galois Theory.
33 HIRSCH. Differential Topology.
34 SPITZER. Principles of Random Walk. 2nd ed.
35 WERMER. Banach Algebras and Several Complex Variables. 2nd ed.
36 KELLEY/NAMIOKA et al. Linear Topological Spaces.
37 MONK. Mathematical Logic.
38 GRAUERT/FRITZSCHE. Several Complex Variables.
39 ARVESON. An Invitation to C^*-Algebras.
40 KEMENY/SNELL/KNAPP. Denumerable Markov Chains. 2nd ed.
41 APOSTOL. Modular Functions and Dirichlet Series in Number Theory. 2nd ed.
42 SERRE. Linear Representations of Finite Groups.
43 GILLMAN/JERISON. Rings of Continuous Functions.
44 KENDIG. Elementary Algebraic Geometry.
45 LOÈVE. Probability Theory I. 4th ed.
46 LOÈVE. Probability Theory II. 4th ed.
47 MOISE. Geometric Topology in Dimensions 2 and 3.
48 SACHS/WU. General Relativity for Mathematicians.
49 GRUENBERG/WEIR. Linear Geometry. 2nd ed.
50 EDWARDS. Fermat's Last Theorem.
51 KLINGENBERG. A Course in Differential Geometry.
52 HARTSHORNE. Algebraic Geometry.
53 MANIN. A Course in Mathematical Logic.
54 GRAVER/WATKINS. Combinatorics with Emphasis on the Theory of Graphs.
55 BROWN/PEARCY. Introduction to Operator Theory I: Elements of Functional Analysis.
56 MASSEY. Algebraic Topology: An Introduction.
57 CROWELL/FOX. Introduction to Knot Theory.
58 KOBLITZ. p-adic Numbers, p-adic Analysis, and Zeta-Functions. 2nd ed.
59 LANG. Cyclotomic Fields.
60 ARNOLD. Mathematical Methods in Classical Mechanics. 2nd ed.

continued after index

Kenneth S. Brown

Cohomology of Groups

Springer

Kenneth S. Brown
Department of Mathematics
White Hall
Cornell University
Ithaca, NY 14853
U.S.A.

Library of Congress Cataloging in Publication Data
Brown, Kenneth S.
Cohomology of groups.
(Graduate texts in mathematics; 87)
Bibliography: p.
Includes index.
1. Groups, Theory of. 2. Homology theory.
I. Title. II. Series.
QA171.B876 512′.22 82-733
AACR2

Printed and bound by Braun-Brumfield, Inc., Ann Arbor, MI.
Printed in the United States of America

9 8 7 6 5 4 3

ISBN 0-387-90688-6 Springer-Verlag New York Berlin Heidelberg
ISBN 3-540-90688-6 Springer-Verlag Berlin Heidelberg New York SPIN 10536744

Preface

This book is based on a course given at Cornell University and intended primarily for second-year graduate students. The purpose of the course was to introduce students who knew a little algebra and topology to a subject in which there is a very rich interplay between the two. Thus I take neither a purely algebraic nor a purely topological approach, but rather I use both algebraic and topological techniques as they seem appropriate.

The first six chapters contain what I consider to be the basics of the subject. The remaining four chapters are somewhat more specialized and reflect my own research interests. For the most part, the only prerequisites for reading the book are the elements of algebra (groups, rings, and modules, including tensor products over non-commutative rings) and the elements of algebraic topology (fundamental group, covering spaces, simplicial and CW-complexes, and homology). There are, however, a few theorems, especially in the later chapters, whose proofs use slightly more topology (such as the Hurewicz theorem or Poincaré duality). The reader who does not have the required background in topology can simply take these theorems on faith.

There are a number of exercises, some of which contain results which are referred to in the text. A few of the exercises are marked with an asterisk to warn the reader that they are more difficult than the others or that they require more background.

I am very grateful to R. Bieri, J-P. Serre, U. Stammbach, R. Strebel, and C. T. C. Wall for helpful comments on a preliminary version of this book.

Notational Conventions

All rings (including graded rings) are assumed to be associative and to have an identity. The latter is required to be preserved by ring homomorphisms. Modules are understood to be *left* modules, unless the contrary is explicitly stated. Similarly, group actions are generally understood to be left actions.

If a group G acts on a set X, I will usually write X/G instead of $G\backslash X$ for the orbit set, even if G is acting on the left. One exception to this concerns the notation for the set of cosets of a subgroup H in a group G. Here we are talking about the left or right translation action of H on G, and I will always be careful to put the H on the appropriate side. Thus $G/H = \{gH : g \in G\}$ and $H\backslash G = \{Hg : g \in G\}$.

A symbol such as

$$\sum_{g \in G/H} f(g)$$

indicates that f is a function on G such that $f(g)$ depends only on the coset gH of g; the sum is then taken over a set of coset representatives.

Finally, I use the "topologists' notation"

$$\mathbb{Z}_n = \mathbb{Z}/n\mathbb{Z};$$

in particular, $\mathbb{Z}_p$ denotes the integers mod p, not the p-adic integers.

Contents

Introduction

The cohomology theory of groups arose from both topological and algebraic sources. The starting point for the topological aspect of the theory was the work of Hurewicz [1936] on "aspherical spaces." About a year earlier, Hurewicz had introduced the higher homotopy groups $\pi_n X$ of a space X ($n \geq 2$). He now singled out for study those path-connected spaces X whose higher homotopy groups are all trivial, but whose fundamental group $\pi = \pi_1 X$ need not be trivial. Such spaces are called *aspherical.*

Hurewicz proved, among other things, that the homotopy type of an aspherical space X is completely determined by its fundamental group π. In particular, the homology groups of X depend only on π; it is therefore reasonable to think of them as *homology groups of* π. [This terminology, however, was not introduced until the 1940's.] Throughout the remainder of this introduction, then, we will write $H_* \pi$ for the homology of any aspherical space with fundamental group π. (Similarly, we could define homology and cohomology groups of π with arbitrary coefficients.) As a simple example, note that $H_2(\mathbb{Z} \oplus \mathbb{Z}) = \mathbb{Z}$. [Take X to be the torus.] Although Hurewicz considered only the *uniqueness* and not the *existence* of aspherical spaces, there does in fact exist an aspherical space with any given group as fundamental group. Thus our topological definition of group homology applies to all groups

For any group π we obviously have $H_0 \pi = \mathbb{Z}$ and $H_1 \pi = \pi_{ab}$, the latter being the abelianization of π, i.e., π modulo its commutator subgroup. For $n \geq 2$, however, it is by no means clear how to describe $H_n \pi$ algebraically. The first progress in this direction was made by Hopf [1942], who expressed $H_2 \pi$ in purely algebraic terms, and who gave further evidence of its importance in topology by proving the following theorem: for any path-connected space X with fundamental group π, there is an exact sequence

(0.1) $$\pi_2 X \to H_2 X \to H_2 \pi \to 0.$$

[To put this result in perspective, one should recall that Hurewicz had introduced homomorphisms $h_n\colon \pi_n X \to H_n X$ $(n \geq 2)$ and had shown that h_n is an isomorphism if $\pi_i X = 0$ for $i < n$. In particular, h_2 is an isomorphism if $\pi = \pi_1 X = 0$. When π is non-trivial, however, h_2 is in general neither injective nor surjective, and Hopf's theorem gives a precise description, in terms of π, of the extent to which surjectivity fails.]

Hopf's description of $H_2\pi$, incidentally, went as follows: Choose a presentation of π as F/R, where F is free and $R \lhd F$; then

$$(0.2) \qquad H_2\pi = R \cap [F, F]/[R, F],$$

where $[A, B]$ for $A, B \subseteq F$ denotes the subgroup generated by the commutators $[a, b] = aba^{-1}b^{-1}$ $(a \in A, b \in B)$.

Following Hopf's paper there was a rapid development of the subject by Eckmann, Eilenberg–MacLane, Freudenthal, and Hopf. (See MacLane [1978] for some comments about this development.) In particular, one had by the mid-1940's a purely algebraic definition of group homology and cohomology, from which it became clear that the subject was of interest to algebraists as well as topologists. Indeed, the low-dimensional cohomology groups were seen to coincide with groups which had been introduced much earlier in connection with various algebraic problems. H^1, for instance, consists of equivalence classes of "crossed homomorphisms" or "derivations." And H^2 consists of equivalence classes of "factor sets," the study of which goes back to Schur [1904], Schreier [1926], and Brauer [1926]. Even H^3 had appeared in an algebraic context (Teichmüller [1940]). These are the algebraic sources of group cohomology referred to at the beginning of this introduction. (Of course, there had been nothing in this algebraic work to suggest that there was an underlying "homology theory"; this had to wait for the impetus from topology.)

It is not surprising, in view of this history, that the subject of group cohomology offers possibilities for a great deal of interaction between algebra and topology. For instance a "transfer map," motivated by a classical group-theoretic construction due to Schur [1902], was introduced into group cohomology (Eckmann [1953], Artin–Tate [unpublished]) and from there into topology, where it has become an important tool. Another example is the theory of Euler characteristics of groups. This theory was motivated by topology, but it has applications to group theory and number theory.

Our approach to the subject will be as follows: We begin in Chapters I and II by defining $H_*\pi$ from the point of view of "homological algebra." This is the point of view which had evolved by the end of the 1940's. The topological motivation, however, will always be kept in sight, and we will immediately obtain the topological interpretation of $H_*\pi$ in terms of aspherical spaces. In particular, we will prove 0.1 and 0.2.

Chapter III contains more homological algebra, involving homology and cohomology with coefficients. These arise naturally in applications, both in algebra and topology. They are also an important technical tool, since they make it possible to prove theorems by "dimension-shifting." In Chapter IV we treat the theory of group extensions, which involves the crossed homomorphisms and factor sets mentioned above.

Chapter V introduces cup and cap products (motivated by topology), and these are then used in Chapter VI in the study of the cohomology of finite groups. Much of the material in Chapter VI (such as the "Tate cohomology theory") was originally motivated by algebra (class field theory), but it turns out to be related to topological questions as well, such as the study of groups acting freely on spheres.

In Chapter VII we introduce spectral sequence techniques, which are used extensively in the remaining chapters. The reader is not expected to have previously seen spectral sequences; I give a reasonably self-contained treatment, omitting only some routine (but tedious) verifications.

Beginning with Chapter VIII the emphasis is on infinite groups, with the most interesting examples being groups of integral matrices. In Chapter VIII we discuss various finiteness conditions which can be imposed on such a group to guarantee that the homology has nice properties. Chapter IX treats Euler characteristics, which can be defined under suitable finiteness hypotheses. This theory, as we mentioned above, has interesting connections with number theory. Finally, Chapter X develops the "Farrell cohomology theory," which is a generalization to infinite groups of the Tate cohomology theory for finite groups.

CHAPTER I

Some Homological Algebra

0 Review of Chain Complexes

We collect here for ease of reference some terminology and results concerning chain complexes. Much of this will be well-known to anyone who has studied algebraic topology. The reader is advised to skip this section (or skim it lightly) and refer back to it as necessary. We will omit some of the proofs; these are either easy or else can be found in standard texts, such as Dold [1972], Spanier [1966], or MacLane [1963].

Let R be an arbitrary ring. By a *graded R-module* we mean a sequence $C = (C_n)_{n \in \mathbb{Z}}$ of R-modules. If $x \in C_n$, then we say x has *degree n* and we write $\deg x = n$. A *map of degree p* from a graded R-module C to a graded R-module C' is a family $f = (f_n: C_n \to C'_{n+p})_{n \in \mathbb{Z}}$ of R-module homomorphisms; thus $\deg(f(x)) = \deg f + \deg x$. A *chain complex* over R is a pair (C, d) where C is a graded R-module and $d: C \to C$ is a map of degree -1 such that $d^2 = 0$. The map d is called the *differential* or *boundary operator* of C. We often suppress d from the notation and simply say that C is a chain complex. We define the *cycles* $Z(C)$, *boundaries* $B(C)$, and *homology* $H(C)$ by $Z(C) = \ker d$, $B(C) = \operatorname{im} d$, and $H(C) = Z(C)/B(C)$. These are all graded modules.

One often comes across graded modules C with an endomorphism d of square zero such that d has degree $+1$ instead of -1. In this case it is customary to use superscripts instead of subscripts to denote the grading, so that $C = (C^n)_{n \in \mathbb{Z}}$ and $d = (d^n: C^n \to C^{n+1})$. Such a pair (C, d) is called a *cochain complex*. There is no essential difference between chain complexes and cochain complexes, since we can always convert one to the other by setting $C_n = C^{-n}$. We will therefore confine ourselves, for the most part, to discussing chain complexes, it being understood that everything applies to cochain complexes by reindexing as above. ⌈Note, however, that there *is* a difference when we consider *non-negative* complexes, i.e., complexes such that C_n

[or C^n] = 0 for $n < 0$; if the differential is thought of as going from left to right, then a non-negative chain complex extends indefinitely to the left, whereas a non-negative cochain complex extends indefinitely to the right.] In discussing cochain complexes, one often prefixes "co" to much of the terminology; thus d may be called a coboundary operator, and we have cocycles $Z(C)$, coboundaries $B(C)$, and cohomology $H(C) = (H^n(C))_{n \in \mathbb{Z}}$.

If (C, d) and (C', d') are chain complexes, then a *chain map* from C to C' is a graded module homomorphism $f: C \to C'$ of degree 0 such that $d'f = fd$. A *homotopy* h from a chain map f to a chain map g is a graded module homomorphism $h: C \to C'$ of degree 1 such that $d'h + hd = f - g$. We write $f \simeq g$ and say that f is *homotopic* to g if there is a homotopy from f to g.

(0.1) Proposition. *A chain map $f: C \to C'$ induces a map $H(f): H(C) \to H(C')$, and $H(f) = H(g)$ iff $f \simeq g$.* □

The abelian group of homotopy classes of chain maps $C \to C'$ will be denoted $[C, C']$. It is often useful to interpret $[C, C']$ as the 0-th homology group of a certain "function complex" $\mathcal{H}om_R(C, C')$, defined as follows: $\mathcal{H}om_R(C, C')_n$ is the set of graded module homomorphisms of degree n from C to C' [thus $\mathcal{H}om_R(C, C')_n = \prod_{q \in \mathbb{Z}} \operatorname{Hom}_R(C_q, C'_{q+n})$], and the boundary operator $D_n: \mathcal{H}om_R(C, C')_n \to \mathcal{H}om_R(C, C')_{n-1}$ is defined by $D_n(f) = d'f - (-1)^n fd$. [The sign here makes $D^2 = 0$. It is also consistent with other standard sign conventions, cf. exercise 3 below.] Note that the 0-cycles are precisely the chain maps $C \to C'$, and the 0-boundaries are the null-homotopic chain maps. Thus $H_0(\mathcal{H}om_R(C, C')) = [C, C']$. More generally, there is an interpretation of $H_n(\mathcal{H}om_R(C, C'))$ in terms of chain maps. Consider the complex $(\Sigma^n C, \Sigma^n d)$ defined by $(\Sigma^n C)_p = C_{p-n}$, $\Sigma^n d = (-1)^n d$; this complex is called the *n-fold suspension* of C. [If $n = 1$, we write ΣC instead of $\Sigma^1 C$.] Let $[C, C']_n = [\Sigma^n C, C']$. Then we have $H_n(\mathcal{H}om_R(C, C')) = [C, C']_n$. The elements of $[\ \ ,\ \ ]_n$ are called *homotopy classes of chain maps of degree n.*

A chain map $f: C \to C'$ is called a *homotopy equivalence* if there is a chain map $f': C' \to C$ such that $f'f \simeq \mathrm{id}_C$ and $ff' \simeq \mathrm{id}_{C'}$. And a chain map f is called a *weak equivalence* if $H(f): H(C) \to H(C')$ is an isomorphism.

(0.2) Proposition. *Any homotopy equivalence is a weak equivalence.* □

A chain complex C is called *contractible* if it is homotopy equivalent to the zero complex, or, equivalently, if $\mathrm{id}_C \simeq 0$. A homotopy from id_C to 0 is called a *contracting homotopy*. Any contractible chain complex is *acyclic*, i.e., $H(C) = 0$.

(0.3) Proposition. *C is contractible if and only if it is acyclic and each short exact sequence $0 \to Z_{n+1} \hookrightarrow C_{n+1} \xrightarrow{\bar{d}} Z_n \to 0$ splits, where $\bar{d}$ is induced by d.*

PROOF. If h is a contracting homotopy, then $(h|Z): Z \to C$ splits the surjection $\bar{d}: C \to Z$. Conversely, suppose we have a splitting $s: Z \to C$, whence a graded module decomposition $C = \ker \bar{d} \oplus \operatorname{im} s = Z \oplus \operatorname{im} s$. We then get a contracting homotopy $h: C \to C$ by setting $h|Z = s$ and $h|\operatorname{im} s = 0$. □

(0.4) Proposition. *A short exact sequence* $0 \to C' \xrightarrow{i} C \xrightarrow{\pi} C'' \to 0$ *of chain complexes gives rise to a long exact sequence in homology:*

$$\cdots \to H_n(C') \xrightarrow{H(i)} H_n(C) \xrightarrow{H(\pi)} H_n(C'') \xrightarrow{\partial} H_{n-1}(C') \to \cdots.$$

The "connecting homomorphism" ∂ is natural, in the sense that a commutative diagram

$$\begin{array}{ccccccccc} 0 & \longrightarrow & C' & \longrightarrow & C & \longrightarrow & C'' & \longrightarrow & 0 \\ & & \downarrow & & \downarrow & & \downarrow & & \\ 0 & \longrightarrow & E' & \longrightarrow & E & \longrightarrow & E'' & \longrightarrow & 0 \end{array}$$

with exact rows yields a commutative square

$$\begin{array}{ccc} H_n(C'') & \longrightarrow & H_{n-1}(C') \\ \downarrow & & \downarrow \\ H_n(E'') & \longrightarrow & H_{n-1}(E'). \end{array}$$

□

(0.5) Corollary. *The inclusion $i: C' \to C$ is a weak equivalence if and only if C'' is acyclic.* □

This shows that the cokernel C'' of i is the appropriate object to consider if we want to measure the "difference" between $H(C)$ and $H(C')$. We now wish to define a "homotopy-theoretic" cokernel for an arbitrary chain map $f: C' \to C$, which plays the same role as the cokernel in the case of an inclusion: The *mapping cone* of $f: (C', d') \to (C, d)$ is defined to be the complex (C'', d'') with $C'' = C \oplus \Sigma C'$ (as a graded module) and $d''(c, c') = (dc + fc', -d'c')$. In matrix notation, we have

$$d'' = \begin{pmatrix} d & f \\ 0 & \Sigma d' \end{pmatrix}.$$

See exercise 2 below for the motivation for this definition.

(0.6) Proposition. *Let $f: C' \to C$ be a chain map with mapping cone C''. There is a long exact homology sequence*

$$\cdots \to H_n(C') \xrightarrow{H(f)} H_n(C) \to H_n(C'') \to H_{n-1}(C') \to \cdots.$$

In particular, f is a weak equivalence if and only if C'' is acyclic.

Proof. There is a short exact sequence $0 \to C \to C'' \to \Sigma C' \to 0$; now apply (0.4). By checking the definition of the connecting homomorphism $H_n(\Sigma C') \to H_{n-1}(C)$, one finds that it equals $H_{n-1}(f): H_{n-1}(C') \to H_{n-1}(C)$. □

The mapping cone is also useful for studying homotopy equivalences, not just weak equivalences:

(0.7) Proposition. *A chain map $f: C' \to C$ is a homotopy equivalence if and only if its mapping cone C'' is contractible.*

PROOF. A straightforward computational proof can be found in the standard references (or can be supplied by the reader). For the sake of variety, we will sketch a conceptual proof. Suppose first that C'' is contractible. One then checks easily that the function complex $\mathscr{H}om_R(D, C'')$ is contractible for any complex D; in particular, it is acyclic. One also checks that $\mathscr{H}om_R(D, C'')$ is isomorphic to the mapping cone of $\mathscr{H}om_R(D,f): \mathscr{H}om_R(D, C') \to \mathscr{H}om_R(D, C)$. It therefore follows from (0.6) that $\mathscr{H}om_R(D,f)$ is a weak equivalence. Looking at H_0, we deduce that f induces an isomorphism $[D, C'] \to [D, C]$ for any D; hence f is a homotopy equivalence by a standard argument. Conversely, suppose f is a homotopy equivalence. Then one shows easily that $\mathscr{H}om_R(D,f)$: $\mathscr{H}om_R(D, C') \to \mathscr{H}om_R(D, C)$ is a homotopy equivalence, so its mapping cone $\mathscr{H}om_R(D, C'')$ is acyclic by 0.6. In particular, $[D, C''] = 0$ for any D, and this implies that C'' is contractible. □

Finally, we recall briefly the Künneth and universal coefficient theorems. If (C, d) (resp. (C', d')) is a chain complex of right (resp. left) R-modules, then we define their *tensor product* $C \otimes_R C'$ by $(C \otimes_R C')_n = \bigoplus_{p+q=n} C_p \otimes_R C'_q$, with differential D given by $D(c \otimes c') = dc \otimes c' + (-1)^{\deg c} c \otimes d'c'$ for $c \in C$, $c' \in C'$. The sign here can be remembered by means of the following *sign convention*: When something of degree p is moved past something of degree q, the sign $(-1)^{pq}$ is introduced. [In the present case, the differential, which is of degree -1, is moved past c, so we get the sign $(-1)^{-\deg c} = (-1)^{\deg c}$.] Note that $C \otimes_R C'$ is simply a complex of abelian groups for general R, but it is a complex of R-modules if R is commutative.

(0.8) Proposition (Künneth Formula). *Let R be a principal ideal domain and let C and C' be chain complexes such that C is dimension-wise free. There are natural exact sequences*

$$0 \to \bigoplus_{p \in \mathbb{Z}} H_p(C) \otimes_R H_{n-p}(C') \to H_n(C \otimes_R C')$$
$$\to \bigoplus_{p \in \mathbb{Z}} \operatorname{Tor}_1^R(H_p(C), H_{n-p-1}(C')) \to 0$$

and

$$0 \to \prod_{p \in \mathbb{Z}} \operatorname{Ext}_R^1(H_p(C), H_{p+n+1}(C')) \to H_n(\mathscr{H}om_R(C, C'))$$
$$\to \prod_{p \in \mathbb{Z}} \operatorname{Hom}_R(H_p(C), H_{p+n}(C')) \to 0,$$

and these sequences split. □

We will not recall the definitions of Tor and Ext at this point, since we will be defining them in much greater generality in §III.2.

An important special case of 0.8 is that where C' consists of a single module M, regarded as a complex concentrated in dimension 0 (i.e. $C'_0 = M$, $C'_n = 0$ for $n \neq 0$). In this case 0.8 is called the *universal coefficient theorem*, and the exact sequences take the following form:

$$0 \to H_n(C) \otimes_R M \to H_n(C \otimes_R M) \to \operatorname{Tor}_1^R(H_{n-1}(C), M) \to 0$$

and

$$0 \to \operatorname{Ext}_R^1(H_{n-1}(C), M) \to H^n(\mathscr{H}om_R(C, M)) \to \operatorname{Hom}_R(H_n(C), M) \to 0.$$

[Here we are following standard conventions in regarding $\mathscr{H}om_R(C, M)$ as a *co*chain complex, with $\mathscr{H}om_R(C, M)^n = \mathscr{H}om_R(C, M)_{-n} = \operatorname{Hom}_R(C_n, M)$; the last equality comes from the fact that the only non-zero component of a graded map $f: C \to M$ of degree $-n$ is $f_n: C_n \to M$.]

Exercises

1. Let T: (R-modules) $\to$ (S-modules) be a covariant functor which takes the zero module to the zero module (or, equivalently, which takes zero maps to zero maps). For any chain complex (C, d) over R, there is then a chain complex (TC, Td) over S. If T is an exact functor, show that $H(TC) \approx TH(C)$. [Recall that T is *exact* if it carries exact sequences to exact sequences. It follows that T preserves injections, surjections, kernels, cokernels, etc.]

2. (Motivation for the definition of the mapping cone) Given a chain map $f: C' \to C$, the "homotopy theoretic cokernel" of f should fit into a diagram $C' \xrightarrow{f} C \xrightarrow{g} C''$ with $gf \simeq 0$. Thus C'' must receive a chain map g [of degree 0] from C and a homotopy h [of degree 1] from C'; this suggests setting $C'' = C \oplus \Sigma C'$ as a graded module, and taking g and h to be inclusions. Show that the definition of the boundary operator d'' is now forced on us by the requirement that g be a chain map and that h be a homotopy from gf to 0. [Using matrix notation, set

$$d'' = \begin{pmatrix} \alpha & \beta \\ \gamma & \delta \end{pmatrix}$$

and note that

$$g = \begin{pmatrix} 1 \\ 0 \end{pmatrix} \quad \text{and} \quad h = \begin{pmatrix} 0 \\ 1 \end{pmatrix}.$$

Now write out the matrix equations $d''g = gd$ and $d''h + hd' = gf$ and solve for $\alpha, \beta, \gamma, \delta$.]

The remaining exercises are designed to illustrate various compatibility properties of our sign conventions in $\mathscr{H}om(-, -)$ and $- \otimes -$. Few readers will have the patience to do all of them, but you should at least do enough to convince yourself that all reasonable identities involving $\mathscr{H}om$ and $\otimes$ are true, provided one follows the sign convention stated above.

3. Given $u \in \mathcal{H}om_R(C, C')_p$ and $x \in C_q$, set $\langle u, x\rangle = u(x) \in C'_{p+q}$. Let d, d', and D be the differentials in C, C', and $\mathcal{H}om_R(C, C')$, respectively. Verify that $d'\langle u, x\rangle = \langle Du, x\rangle + (-1)^p\langle u, dx\rangle$. [In fact, this is nothing but a restatement of the definition of D, but it is a convenient form in which to remember that definition.] In other words, the *evaluation map* $\mathcal{H}om_R(C, C') \otimes C \to C'$, given by $u \otimes x \mapsto \langle u, x\rangle$, is a chain map.

4. Given $v \in \mathcal{H}om_R(C, C')_q$ and $u \in \mathcal{H}om_R(C', C'')_p$, their composite $u \circ v$ is in
$$\mathcal{H}om_R(C, C'')_{p+q}.$$
Verify that $D(u \circ v) = Du \circ v + (-1)^p u \circ Dv$, where D denotes the differential in any of the three $\mathcal{H}om$ complexes; in other words, composition of graded maps defines a chain map $\mathcal{H}om_R(C', C'') \otimes \mathcal{H}om_R(C, C') \to \mathcal{H}om_R(C, C'')$. [Hint: This and the remaining exercises are most conveniently done by taking the definition of D in the form given in exercise 3. From this point of view, one starts with the equation $\langle u \circ v, x\rangle = \langle u, \langle v, x\rangle\rangle$ $(x \in C)$, which is the definition of composition, and one applies d'' to both sides. Applying exercise 3 several times, one obtains
$$\langle D(u \circ v), x\rangle + (-1)^{p+q}\langle u \circ v, dx\rangle = \langle Du, \langle v, x\rangle\rangle + (-1)^p\langle u, \langle Dv, x\rangle\rangle + (-1)^{p+q}\langle u, \langle v, dx\rangle\rangle$$
which simplifies to
$$\langle D(u \circ v), x\rangle = \langle Du \circ v, x\rangle + (-1)^p\langle u \circ Dv, x\rangle,$$
as required.]

5. If C and C' are chain complexes over a commutative ring R, there is an isomorphism of graded modules $C \otimes_R C' \xrightarrow{\approx} C' \otimes_R C$ given by $c \otimes c' \mapsto (-1)^{\deg c \cdot \deg c'} c' \otimes c$. Prove that this isomorphism is a chain map.

6. Let C be a complex of right R-modules, C' a complex of left R-modules, and C'' a complex of $\mathbb{Z}$-modules.

(a) There is a canonical isomorphism of graded abelian groups
$$\varphi: \mathcal{H}om_{\mathbb{Z}}(C \otimes_R C', C'') \xrightarrow{\approx} \mathcal{H}om_R(C, \mathcal{H}om_{\mathbb{Z}}(C', C'')),$$
given by $\langle\langle\varphi(u), c\rangle, c'\rangle = \langle u, c \otimes c'\rangle$ for $u \in \mathcal{H}om_{\mathbb{Z}}(C \otimes_R C', C'')$, $c \in C$, $c' \in C'$. [Sketch of proof: An element of $\mathcal{H}om_{\mathbb{Z}}(C \otimes_R C', C'')_n$ is a family of $\mathbb{Z}$-module maps $C_p \otimes_R C'_q \to C''_{p+q+n}$. In view of the universal mapping property of the tensor product, this is the same as a family of R-module maps $C_p \to \mathrm{Hom}_{\mathbb{Z}}(C'_q, C''_{p+q+n})$, i.e., as a graded R-module map $C \to \mathcal{H}om_{\mathbb{Z}}(C', C'')$ of degree n. Note here that $\mathcal{H}om_{\mathbb{Z}}(C', C'')$ is indeed a complex of right R-modules via the left action of R on C' and the contravariance of Hom in the first variable: $(ur)(c') = u(rc')$ for $u \in \mathcal{H}om_{\mathbb{Z}}(C', C'')$, $r \in R$, $c' \in C'$.] Show that φ is a chain map. [Hint: Let d, d', and d'' be the differentials in C, C', and C'', and let D be the differential in any of the $\mathcal{H}om$ complexes under consideration. Apply d'' to both sides of the equation defining φ; using exercise 3 several times, you should obtain
$$\langle\langle D(\varphi(u)), c\rangle, c'\rangle + (-1)^p\langle\langle\varphi(u), dc\rangle, c'\rangle + (-1)^{p+q}\langle\langle\varphi(u), c\rangle, d'c'\rangle$$
on the left, where $p = \deg u$ and $q = \deg c$, and
$$\langle Du, c \otimes c'\rangle + (-1)^p\langle u, dc \otimes c'\rangle + (-1)^{p+q}\langle u, c \otimes d'c'\rangle$$

on the right. Remembering the definition of φ, you can conclude that

$$\langle\langle D(\varphi(u)), c\rangle, c'\rangle = \langle Du, c \otimes c'\rangle$$

and hence that $\langle\langle D(\varphi(u)), c\rangle, c'\rangle = \langle\langle \varphi(Du), c\rangle, c'\rangle$.]

(b) Deduce from (a) [or check directly] that a map $u: C \otimes_R C' \to C''$ of degree 0 is a chain map iff the corresponding map $C \to \mathscr{H}om_{\mathbb{Z}}(C', C'')$ is a chain map.

7. Let C and C' (resp. E and E') be complexes of right (resp. left) R-modules. Given $u \in \mathscr{H}om_R(C, C')$ and $v \in \mathscr{H}om_R(E, E')$, their *tensor product*

$$u \otimes v \in \mathscr{H}om_{\mathbb{Z}}(C \otimes_R E, C' \otimes_R E')$$

is defined by

$$\langle u \otimes v, c \otimes e\rangle = (-1)^{\deg v \cdot \deg c}\langle u, c\rangle \otimes \langle v, e\rangle \qquad \text{for} \quad c \in C, e \in E.$$

(a) Let D be the differential in any of the three $\mathscr{H}om$ complexes under consideration. Verify that $D(u \otimes v) = Du \otimes v + (-1)^{\deg u} u \otimes Dv$. In other words, the operation "tensor product of graded maps" defines a chain map

$$\mathscr{H}om_R(C, C') \otimes \mathscr{H}om_R(E, E') \to \mathscr{H}om_{\mathbb{Z}}(C \otimes_R E, C' \otimes_R E').$$

[Hint: Once again, you are advised to do this by differentiating both sides of the equation defining $u \otimes v$.]

(b) Deduce from (a) (or check directly) that if $u: C \to C'$ is a chain map (of degree 0), then there is a chain map $\mathscr{H}om_R(E, E') \to \mathscr{H}om_{\mathbb{Z}}(C \otimes_R E, C' \otimes_R E')$ given by $v \mapsto u \otimes v$.

(c) Deduce from (a) (or check directly) that the tensor product of chain maps is compatible with homotopy, in the following sense: Given chain maps and homotopies $u_0 \simeq u_1: C \to C'$ and $v_0 \simeq v_1: E \to E'$, one has $u_0 \otimes v_0 \simeq u_1 \otimes v_1: C \otimes_R E \to C' \otimes_R E'$. [Hint: u_0 and u_1 are homologous 0-cycles in $\mathscr{H}om_R(C, C')$, and similarly for v_0 and v_1; it now follows from the boundary formula in (a) that $u_0 \otimes v_0$ and $u_1 \otimes v_1$ are homologous 0-cycles in $\mathscr{H}om_{\mathbb{Z}}(C \otimes_R E, C' \otimes_R E')$.]

8. Prove that the tensor product operation of the previous exercise is compatible with composition of graded maps, i.e., that

$$(u \otimes v) \circ (u' \otimes v') = (-1)^{\deg v \cdot \deg u'}(u \circ u') \otimes (v \circ v')$$

whenever the composites are defined.

1 Free Resolutions

Let R be a ring (associative, with identity) and M a (left) R-module. A *resolution* of M is an exact sequence of R-modules

$$\cdots \to F_2 \xrightarrow{\partial_2} F_1 \xrightarrow{\partial_1} F_0 \xrightarrow{\varepsilon} M \to 0.$$

If each F_i is free, then this is called a *free resolution*. Free resolutions exist for any module M and can be constructed by an obvious step-by-step procedure: Choose a surjection $\varepsilon: F_0 \twoheadrightarrow M$ with F_0 free, then choose a surjection $F_1 \twoheadrightarrow \ker \varepsilon$ with F_1 free, etc. Note that the initial segment $F_1 \to F_0 \to M \to 0$ of a free resolution is simply a *presentation* of M by generators and relations.

There are two useful ways to interpret a resolution in terms of chain complexes.[1] The first is to regard the resolution itself as an acyclic chain complex, with M in degree -1. We will refer to this as the *augmented chain complex* associated to the resolution. The second way is to consider the non-negative chain complex $F = (F_i, \partial_i)_{i \geq 0}$ and to view $\varepsilon: F \to M$ as a chain map, where M is regarded as a chain complex concentrated in dimension 0. The exactness hypothesis, from this point of view, simply says that ε is a weak equivalence.

In case there is an integer n such that $F_i = 0$ for $i > n$, we will say that the resolution has *length* $\leq n$. In this case we will simply write

$$0 \to F_n \to \cdots \to F_0 \to M \to 0,$$

it being understood that the resolution continues to the left with zeroes.

Examples

1. A free module F admits the free resolution

$$0 \to F \xrightarrow{\text{id}} F \to 0$$

of length 0.

2. If $R = \mathbb{Z}$ (or any principal ideal domain) then submodules of a free module are free, hence any module M admits a free resolution

$$0 \to F_1 \to F_0 \to M \to 0$$

of length ≤ 1. For example, the $\mathbb{Z}$-module $\mathbb{Z}_2 = \mathbb{Z}/2\mathbb{Z}$ admits the resolution

$$0 \to \mathbb{Z} \xrightarrow{2} \mathbb{Z} \to \mathbb{Z}_2 \to 0.$$

3. Let M again be $\mathbb{Z}_2$, regarded now as a module over the polynomial ring $R = \mathbb{Z}[T]$, with T acting as 0 (i.e., $f(T)$ acts as multiplication by $f(0)$). Then M can be regarded as the quotient of R by the ideal generated by T and 2. Using unique factorization in R and the fact that T and 2 are relatively prime, one easily obtains the free resolution of length 2

$$0 \to R \xrightarrow{\partial_2} R \oplus R \xrightarrow{\partial_1} R \xrightarrow{\varepsilon} \mathbb{Z}_2 \to 0$$

where $\varepsilon(f) = f(0) \bmod 2$, and ∂_1 and ∂_2 are given by the matrices $(T\ 2)$ and $\binom{2}{-T}$, respectively.

4. Let $R = \mathbb{Z}[T]/(T^2 - 1)$ and let t be the image of T in R. Let M be the R-module $R/(t - 1)$. (Equivalently, $M = \mathbb{Z}$, with t acting as the identity.)

[1] See §0 for terminology and basic facts concerning chain complexes.

Since $T^2 - 1 = (T - 1)(T + 1)$, it is clear that an element of R is annihilated by $t - 1$ (resp. $t + 1$) if and only if it is divisible by $t + 1$ (resp. $t - 1$). One therefore has a free resolution

$$\cdots R \xrightarrow{t-1} R \xrightarrow{t+1} R \xrightarrow{t-1} R \to M \to 0.$$

Note that, unlike the previous examples, this resolution is of *infinite* length. We will see, in fact, that M admits no free resolution of finite length, cf. §II.3, exercise 2.

In this book we will be primarily interested in the case where R is a group ring, so we digress now to recall what a group ring is.

2 Group Rings

Let G be a group, written multiplicatively. Let $\mathbb{Z}G$ (or $\mathbb{Z}[G]$) be the free $\mathbb{Z}$-module generated by the elements of G. Thus an element of $\mathbb{Z}G$ is uniquely expressible in the form $\sum_{g \in G} a(g)g$, where $a(g) \in \mathbb{Z}$ and $a(g) = 0$ for almost all g. The multiplication in G extends uniquely to a $\mathbb{Z}$-bilinear product $\mathbb{Z}G \times \mathbb{Z}G \to \mathbb{Z}G$; this makes $\mathbb{Z}G$ a ring, called the *integral group ring* of G.

Note that G is a subgroup of the group $(\mathbb{Z}G)^*$ of units of $\mathbb{Z}G$ and that we have the following *universal mapping property*:

(2.1) Given a ring R and a group homomorphism $f: G \to R^*$, there is a unique extension of f to a ring homomorphism $\mathbb{Z}G \to R$. Thus we have the "adjunction formula"

$$\mathrm{Hom}_{(\text{rings})}(\mathbb{Z}G, R) \approx \mathrm{Hom}_{(\text{groups})}(G, R^*).$$

EXAMPLES

1. Let G be cyclic of order n and let t be a generator. Then the powers t^i $(0 \le i \le n - 1)$ form a $\mathbb{Z}$-basis for $\mathbb{Z}G$, and one has $t^n = 1$. It follows that $\mathbb{Z}G \approx \mathbb{Z}[T]/(T^n - 1)$.

2. If G is infinite cyclic with generator t, then $\mathbb{Z}G$ has a $\mathbb{Z}$-basis $\{t^i\}_{i \in \mathbb{Z}}$. Hence $\mathbb{Z}G$ can be identified with the ring (usually denoted $\mathbb{Z}[t, t^{-1}]$) of Laurent polynomials $\sum_{i \in \mathbb{Z}} a_i t^i$ $(a_i \in \mathbb{Z}, a_i = 0$ for almost all $i)$.

EXERCISES

1. For any group G we define the *augmentation map* to be the ring homomorphism $\varepsilon: \mathbb{Z}G \to \mathbb{Z}$ such that $\varepsilon(g) = 1$ for all $g \in G$. The kernel of ε is called the *augmentation ideal* of $\mathbb{Z}G$ and is denoted I or IG.

   (a) Show that the elements $g - 1$ $(g \in G, g \neq 1)$ form a basis for I as a $\mathbb{Z}$-module.

   (b) If S is a set of generators for G, show that the elements $s - 1$ $(s \in S)$ generate I as a left ideal.

(c) Conversely, if S is a set of elements of G such that the elements $s - 1$ ($s \in S$) generate I as a left ideal, show that S generates G. [Hint: The hypothesis on S implies that every element of I is a sum of elements of the form $g - g'$, where $g = g's^{\pm 1}$ for some $s \in S$. Writing $g - 1$ in this way, where $g \in G$ is arbitrary, deduce that there is a sequence $g_1, g_2, \ldots, g_n$ such that $g_1 = g$, $g_n = 1$, and $g_i = g_{i+1}s_i^{\pm 1}$ for some $s_i \in S$. See exercise 2 of §3 below for an alternative proof.]

(d) Show that G is a finitely generated group if and only if I is finitely generated as a left ideal.

2. Let G be cyclic with generator t and let M be the G-module $\mathbb{Z}$, with t acting as the identity. (Equivalently, $M = \mathbb{Z}G/I = \mathbb{Z}G/(t - 1)$.) Write down a free resolution of M. [See §1, example 4, for the case where $|G| = 2$.]

3 G-modules

A (left) $\mathbb{Z}G$-module, also called a *G-module*, consists of an abelian group A together with a homomorphism from $\mathbb{Z}G$ to the ring of endomorphisms of A. In view of 2.1, such a ring homomorphism corresponds to a group homomorphism from G to the group of automorphisms of A. Thus a G-module is simply an abelian group A together with an action of G on A. For example, one has for any A the *trivial* module structure, with $ga = a$ for $g \in G$, $a \in A$. (Thus $ra = \varepsilon(r) \cdot a$ for $r \in \mathbb{Z}G$, where $\varepsilon: \mathbb{Z}G \to \mathbb{Z}$ is the augmentation map discussed in §2, exercise 1.)

One way of constructing G-modules is by linearizing permutation representations. More precisely, if X is a G-set (i.e., a set with G-action), then one forms the free abelian group $\mathbb{Z}X$ (also denoted $\mathbb{Z}[X]$) generated by X and one extends the action of G on X to a $\mathbb{Z}$-linear action of G on $\mathbb{Z}X$. The resulting G-module is called a *permutation module*. In particular, one has a permutation module $\mathbb{Z}[G/H]$ for every subgroup H of G, where G/H is the set of cosets gH and G acts on G/H by left translation.

The operation of disjoint union in the category of G-sets corresponds to the direct sum operation in the category of G-modules:

$$\mathbb{Z}[\coprod X_i] = \bigoplus \mathbb{Z}X_i.$$

It follows that every permutation module $\mathbb{Z}X$ admits a decomposition

$$\mathbb{Z}X \approx \bigoplus \mathbb{Z}[G/G_x],$$

where x ranges over a set of representatives for the G-orbits in X and G_x is the isotropy subgroup of G at x. In particular, if X is a *free* G-set (i.e., if all isotropy groups are trivial), then $G/G_x = G$, and we obtain:

(3.1) Proposition. *Let X be a free G-set and let E be a set of representatives for the G-orbits in X. Then $\mathbb{Z}X$ is a free $\mathbb{Z}G$-module with basis E.*

Remark. If k is an arbitrary commutative ring, then one can form the *group algebra* kG of G over k. This is the free k-module generated by G, with the unique k-bilinear product extending the group multiplication on G. Everything we have done in this section generalizes in an obvious way to this situation. For example, a kG-module is simply a k-module A together with a (k-linear) action of G on A.

EXERCISES

1. Let H be a subgroup of G and let E be a set of representatives for the *right* cosets Hg. Show that $\mathbb{Z}G$, regarded as a left-module over its subring $\mathbb{Z}H$, is free with basis E.

2. Use permutation modules to give a non-computational solution of exercise 1c of §2. [Let H be the subgroup of G generated by S and consider $\mathbb{Z}[G/H]$. It has an element fixed by H and hence annihilated by I. But then the element is fixed by G.]

4 Resolutions of $\mathbb{Z}$ over $\mathbb{Z}G$ via Topology

In this section we will consider $\mathbb{Z}$ as a G-module with trivial G-action, and we will see that free resolutions of this module arise from free actions of G on contractible complexes.

By a *G-complex* we will mean a CW-complex X together with an action of G on X which permutes the cells. Thus we have for each $g \in G$ a homeomorphism $x \mapsto gx$ of X such that the image $g\sigma$ of any cell σ of X is again a cell. For example, if X is a simplicial complex on which G acts simplicially, then X is a G-complex.

If X is a G-complex then the action of G on X induces an action of G on the cellular chain complex $C_*(X)$, which thereby becomes a chain complex of G-modules. Moreover, the canonical augmentation $\varepsilon\colon C_0(X) \to \mathbb{Z}$ (defined by $\varepsilon(v) = 1$ for every 0-cell v of X) is a map of G-modules.

We will say that X is a *free* G-complex if the action of G freely permutes the cells of X (i.e., $g\sigma \neq \sigma$ for all σ if $g \neq 1$). In this case each chain module $C_n(X)$ has a $\mathbb{Z}$-basis which is freely permuted by G, hence by 3.1 $C_n(X)$ is a *free* $\mathbb{Z}G$-module with one basis element for every G-orbit of cells. (Note that to obtain a specific basis we must choose a representative cell from each orbit and we must choose an orientation of each such representative.)

Finally, if X is contractible, then $H_*(X) \approx H_*(\text{pt.})$; in other words, the sequence

$$\cdots \to C_n(X) \xrightarrow{\partial} C_{n-1}(X) \to \cdots \to C_0(X) \xrightarrow{\varepsilon} \mathbb{Z} \to 0$$

is exact. We have, therefore:

(4.1) Proposition. *Let X be a contractible free G-complex. Then the augmented cellular chain complex of X is a free resolution of $\mathbb{Z}$ over $\mathbb{Z}G$.*

The reader who has studied covering spaces has, of course, seen many examples of free G-complexes. Indeed, suppose $p: \tilde{Y} \to Y$ is a regular covering map with G as group of deck transformations. (See the appendix to this chapter for a review of regular covers.) If Y is a CW-complex, then it is an elementary fact that $\tilde{Y}$ inherits a CW-structure such that the G-action permutes the cells, cf. Schubert [1968], III.6.9. Explicitly, the open cells of $\tilde{Y}$ lying over an open cell σ of Y are simply the connected components of $p^{-1}\sigma$; these cells are permuted freely and transitively by G, and each is mapped homeomorphically onto σ under p. Thus $\tilde{Y}$ is a free G-complex and $C_* \tilde{Y}$ is a complex of free $\mathbb{Z}G$-modules with one basis element for each cell of Y.

In view of 4.1, it is natural now to consider CW-complexes Y satisfying the following three conditions:

(i) Y is connected.
(ii) $\pi_1 Y = G$.
(iii) The universal cover X of Y is contractible.

Such a Y is called an *Eilenberg–MacLane complex* of type $(G, 1)$, or simply a *$K(G, 1)$-complex*. Condition (ii) above is to be interpreted as meaning that we are given a basepoint $y \in Y$ and a specific isomorphism $\pi_1(Y, y) \xrightarrow{\approx} G$ by which we identify $\pi_1(Y, y)$ with G.

It is sometimes convenient to note that condition (iii) above can be replaced by:

(iii′) $H_i X = 0$ for $i \geq 2$.

Indeed, we clearly have (iii) $\Rightarrow$ (iii′). Conversely, (iii′) implies that X is *acyclic*, i.e., that $H_* X \approx H_*(\text{pt.})$; and it is shown in elementary homotopy theory that simply-connected acyclic CW-complexes are contractible. [The reader who is not familiar with this fact can simply take (i), (ii), and (iii′) as the *definition* of a $K(G, 1)$-complex, since we will never make serious use of the fact that X is contractible. Note, for instance, that 4.1 remains true if "contractible" is replaced by "acyclic."]

It also follows from elementary homotopy theory that (iii) can be replaced by

(iii″) $\pi_i Y = 0$ for $i \geq 2$.

Thus the Eilenberg–MacLane complexes are precisely the *aspherical* complexes studied by Hurewicz (cf. Introduction). Once again, we will not need to use this fact.

If Y is a $K(G, 1)$ then the universal cover $p: X \to Y$ is a regular cover whose group is isomorphic to $\pi_1 Y = G$. [This depends on a choice of basepoint in X, cf. Appendix.] We therefore obtain from 4.1:

(4.2) Proposition. *If Y is a $K(G, 1)$ then the augmented cellular chain complex of the universal cover of Y is a free resolution of $\mathbb{Z}$ over $\mathbb{Z}G$.*

We now give one example. There will be many other examples in Chapter II and later in the book.

(4.3) EXAMPLE. Let $G = F(S)$, the free group generated by a set S. Let $Y = \bigvee_{s \in S} S_s^1$, a bouquet of circles S_s^1 in 1-1 correspondence with S. Thus Y is a 1-dimensional CW-complex with exactly one vertex and with one 1-cell for each element of S. One knows that $\pi_1 Y \approx F(S)$. Moreover, Y is a $K(F(S), 1)$ since condition (iii′) above holds for trivial reasons. [X is 1-dimensional.] As basepoint in X we take a vertex x_0; this then represents the unique G-orbit of vertices of X and hence generates the free $\mathbb{Z}G$-module $C_0(X)$. As basis for $C_1(X)$ we take, for each $s \in S$, an oriented 1-cell e_s of X which lies over S_s^1 (traversed in the positive sense). Replacing e_s by ge_s for suitable $g \in G$, we can assume that the initial vertex of e_s is the basepoint x_0; the endpoint of e_s is then sx_0 (cf. Appendix, A1) so that $\partial e_s = sx_0 - x_0 = (s - 1)x_0$. We obtain, therefore, the free resolution

$$0 \to \mathbb{Z}G^{(S)} \xrightarrow{\partial} \mathbb{Z}G \xrightarrow{\varepsilon} \mathbb{Z} \to 0 \tag{4.4}$$

where $\mathbb{Z}G^{(S)}$ is a free $\mathbb{Z}G$-module with basis $(e_s)_{s \in S}$, $\partial e_s = s - 1$, and $\varepsilon(g) = 1$ for all $g \in G$.

Remarks

1. Note that ε coincides with the natural augmentation of the group ring $\mathbb{Z}G$, as defined in §2, exercise 1. The exactness of 4.4 says, therefore, that the augmentation ideal of $\mathbb{Z}[F(S)]$ is a free left $\mathbb{Z}[F(S)]$-module with basis $(s - 1)_{s \in S}$. [We will later indicate a purely algebraic proof of this fact, cf. §IV.2, exercise 3b.] The reader who is not used to working with modules over non-commutative rings may find it surprising that $\mathbb{Z}G$, the free module of rank 1, can contain a submodule which is free of rank > 1; this sort of thing cannot happen over a *commutative* ring R, since then any two elements $a, b \in R$ are linearly dependent: $(-b) \cdot a + a \cdot b = 0$.

2. If S contains a single element t (i.e., G is infinite cyclic) then (4.4) reduces to the "obvious" resolution which the reader probably wrote down in §2, exercise 2:

$$0 \to \mathbb{Z}G \xrightarrow{t-1} \mathbb{Z}G \xrightarrow{\varepsilon} \mathbb{Z} \to 0. \tag{4.5}$$

Note that X, in this case, is simply $\mathbb{R}^1$, with t acting by $x \mapsto x + 1$:

$t^{-2}x_0$ $\quad t^{-1}x_0$ $\quad x_0$ $\quad tx_0$ $\quad t^2x_0$

3. The contractible free $F(S)$-complex X, and hence the resolution 4.4, can easily be constructed directly, without appeal to the theory of covering complexes, or to the fact that $\pi_1(\bigvee S^1) \approx F(S)$. Namely, X can be defined as the 1-dimensional simplicial complex whose vertices are the elements of $G = F(S)$ and whose 1-simplices are the pairs $\{g, gs\}$ ($g \in G, s \in S$). The action

of G on itself by left translation induces a simplicial action of G on X, which is easily seen to be free. Finally, one can construct an explicit contracting homotopy which contracts X to the vertex 1 along paths of the form

$$g = g_n \quad g_{n-1} \quad \cdots \quad g_1 \quad g_0 = 1$$

where $g = s_1^{\varepsilon_1} \cdots s_n^{\varepsilon_n}$ is a reduced word of length n ($s_i \in S$, $\varepsilon_i = \pm 1$, $\varepsilon_i = \varepsilon_{i+1}$ if $s_i = s_{i+1}$) and $g_i = s_1^{\varepsilon_1} \cdots s_i^{\varepsilon_i}$ ($0 \le i \le n$). In case S is a two-element set $\{s, t\}$, for example, X is the tree pictured below:

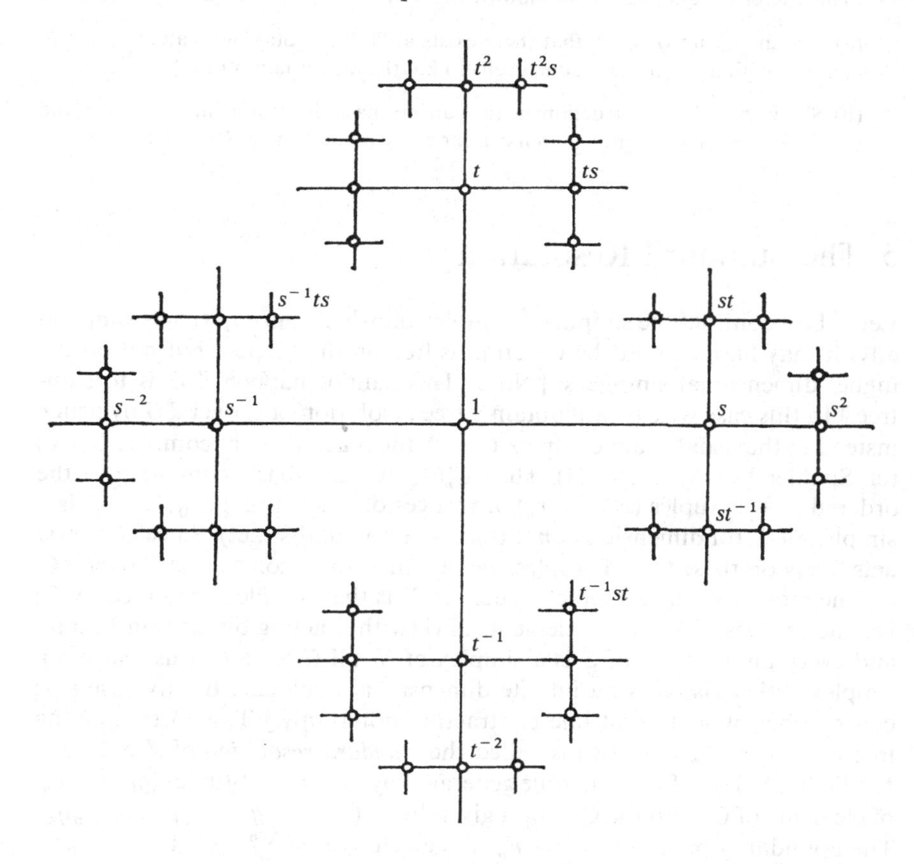

EXERCISES

1. If X is an arbitrary G-complex, is $C_n(X)$ necessarily a permutation module?

*2. Let X be a free G-complex, where G is an arbitrary group. Show that every point of X has a neighborhood U such that $gU \cap U = \varnothing$ for all $g \neq 1$ in G. Deduce that $X \to X/G$ is a regular covering map with G as group of covering transformations. In particular, if X is contractible then X/G is a $K(G, 1)$ and X is its universal cover.

[Hint: A CW-complex X, with explicitly given characteristic maps $(B^n, S^{n-1}) \to (\bar{\sigma}, \partial\sigma)$ for its cells, admits a canonical open cover $\{U_\sigma\}$ indexed by the cells, such that $U_\sigma \cap U_\tau = \emptyset$ if σ and τ are distinct cells of the same dimension. If X is simplicial, for example, we can take U_σ to be the open star of the barycenter of σ in the barycentric subdivision of X.]

3. Let G be a free abelian group of rank 2. Use the methods of this section to construct a free resolution of $\mathbb{Z}$ over $\mathbb{Z}G$.

*4. (This exercise requires some homotopy theory.)

(a) For any group G, show that there exists a $K(G, 1)$. [Start with a connected 2-complex with $\pi_1 = G$; then attach cells to kill the higher homotopy.]

(b) Show that the construction in (a) can be made functorial in G. [Given the $(n-1)$-skeleton Y^{n-1}, attach an n-cell for every possible map $S^{n-1} \to Y^{n-1}$.]

5 The Standard Resolution

Let X be a contractible simplicial complex on which a group G acts simplicially. It may happen that the G-action is free on the vertices but not on the higher-dimensional simplices. [Note: This cannot happen if G is torsion-free.] In this case we can still obtain a free resolution of $\mathbb{Z}$ over $\mathbb{Z}G$ by using, instead of the usual chain complex $C_*(X)$, the *ordered* chain complex $C'_*(X)$ (cf. Spanier [1966], ch. 4, §3). Thus $C'_n(X)$ has a $\mathbb{Z}$-basis consisting of the ordered $(n+1)$-tuples $(v_0, \ldots, v_n)$ of vertices of X such that $\{v_0, \ldots, v_n\}$ is a simplex of X (of dimension $<n$ if the v_i are not all distinct). Since G clearly acts freely on these $(n+1)$-tuples, we obtain a free resolution of $\mathbb{Z}$ over $\mathbb{Z}G$.

The most obvious example of such an X is the "simplex" spanned by G; i.e., the vertices of X are the elements of G (with G acting by left translation), and every finite subset of G is a simplex of X. [If G is finite, this really is a simplex; otherwise it is an infinite dimensional analogue. In any case it is contractible by a straight-line contracting homotopy.] The corresponding free resolution $F_* = C'_*(X)$ is called the *standard resolution* of $\mathbb{Z}$ over $\mathbb{Z}G$. Explicitly, F_n is the free $\mathbb{Z}$-module generated by the $(n+1)$-tuples $(g_0, \ldots, g_n)$ of elements of G, with the G-action given by $g \cdot (g_0, \ldots, g_n) = (gg_0, \ldots, gg_n)$. The boundary operator $\partial: F_n \to F_{n-1}$ is given by $\partial = \sum_{i=0}^{n} (-1)^i d_i$, where

(5.1) $$d_i(g_0, \ldots, g_n) = (g_0, \ldots, \hat{g}_i, \ldots, g_n).$$

The augmentation $\varepsilon: F_0 \to \mathbb{Z}$ is given by $\varepsilon(g_0) = 1$. Note that the acyclicity (i.e., exactness) of this augmented complex has been deduced from the contractibility of X, but one can also verify it directly by writing down a contracting homotopy $h: F_n \to F_{n+1}$ for the underlying augmented complex of $\mathbb{Z}$-modules (i.e., h will *not* be a map of G-modules). Namely, we define h by $h(g_0, \ldots, g_n) = (1, g_0, \ldots, g_n)$ if $n \geq 0$ and $h(1) = (1)$ if $n = -1$.

As basis for the free $\mathbb{Z}G$-module F_n we may take the $(n+1)$-tuples whose first element is 1, since these represent the G-orbits of $(n+1)$-tuples. It is often convenient to write such an $(n+1)$-tuple in the form $(1, g_1, g_1g_2, \ldots, g_1g_2\cdots g_n)$ and to introduce the "bar notation"

$$[g_1|g_2|\cdots|g_n] = (1, g_1, g_1g_2, \ldots, g_1g_2\cdots g_n).$$

(If $n = 0$ there is only one such basis element, denoted $[\ ]$; if we identify F_0 with $\mathbb{Z}G$ in the obvious way, then $[\ ] = 1$.) It is easy to compute $\partial: F_n \to F_{n-1}$ in terms of this $\mathbb{Z}G$-basis $\{[g_1|\cdots|g_n]\}$; one finds $\partial = \sum_{i=0}^n (-1)^i d_i$, where d_i is the $\mathbb{Z}G$-homomorphism given by

$$(5.2)\quad d_i[g_1|\cdots|g_n] = \begin{cases} g_1[g_2|\cdots|g_n] & i = 0 \\ [g_1|\cdots|g_{i-1}|g_ig_{i+1}|g_{i+2}|\cdots|g_n] & 0 < i < n \\ [g_1|\cdots|g_{n-1}] & i = n. \end{cases}$$

This standard resolution F_* is often called the *bar resolution*. In low dimensions it has the form

$$F_2 \xrightarrow{\partial_2} F_1 \xrightarrow{\partial_1} \mathbb{Z}G \xrightarrow{\varepsilon} \mathbb{Z} \to 0,$$

where $\partial_2([g|h]) = g[h] - [gh] + [g]$, $\partial_1([g]) = g[\ ] - [\ ] = g - 1$, and $\varepsilon(1) = 1$.

Finally, we mention the *normalized* standard (or bar) resolution $\bar{F}_* = F_*/D_*$, where D_*, the "degenerate" subcomplex of F_*, is generated by the elements $(g_0, \ldots, g_n)$ such that $g_i = g_{i+1}$ for some i. In terms of the bar notation, D_* can be described as the G-subcomplex of F_* generated by the elements $[g_1|\cdots|g_n]$ such that $g_i = 1$ for some i. Thus $\bar{F}_n$ is a free $\mathbb{Z}G$-module with one basis element (still denoted $[g_1|\cdots|g_n]$) for every n-tuple of *non-trivial* elements of G. The fact that $\bar{F}_*$ is acyclic over $\mathbb{Z}$, and hence a resolution, follows from general facts about normalization (cf. MacLane [1963], VIII.6); alternatively, one can simply observe that the contracting homotopy h defined above carries D_* into itself and hence induces a contracting homotopy of the quotient $\bar{F}_*$.

Exercises

1. Write down the homotopy operator h in terms of the $\mathbb{Z}$-basis $g[g_1|\cdots|g_n]$ for F_n.

2. Write out the normalized bar resolution in case G is the group of order 2; compare it with the resolution given in §1, example 4.

*3. (a) Show that the standard resolution is the cellular chain complex of a free contractible G-complex X. Note that this reproves the result of exercises 4a and 4b of §4. [For each $(n+1)$-tuple $\sigma = (g_0, \ldots, g_n)$, let Δ_σ be a copy of the standard n-simplex with vertices $v_0, \ldots, v_n$. Let $d_i\sigma = (g_0, \ldots, \hat{g}_i, \ldots, g_n)$ and let $\delta_i: \Delta_{d_i\sigma} \to \Delta_\sigma$ be the linear embedding which sends $v_0, \ldots, v_{n-1}$ to $v_0, \ldots, \hat{v}_i, \ldots, v_n$. Form the disjoint union $X_0 = \coprod_\sigma \Delta_\sigma$ and obtain X by identifying $\Delta_{d_i\sigma}$ with its image under δ_i

for all σ and all i. Use the quotient map $C(X_0) \to C(X)$ to compute the boundary operator on $C(X)$ and see that $C(X) \approx F$. To prove X contractible, consider $h\sigma = (1, g_0, \ldots, g_n)$ for each σ, and use the straight-line homotopy between δ_0: $\Delta_\sigma \to \Delta_{h\sigma}$ and the constant map $\Delta_\sigma \to \Delta_{h\sigma}$ at v_0.]

(b) Do the same for the normalized standard resolution. [Use the same method as in (a), but make further identifications in X_0 to collapse degenerate simplices. Explicitly, for each $\sigma = (g_0, \ldots, g_n)$ let $s_i\sigma = (g_0, \ldots, g_i, g_i, \ldots, g_n)$ and collapse $\Delta_{s_i\sigma}$ to Δ_σ via the linear map which sends $v_0, \ldots, v_{n+1}$ to $v_0, \ldots, v_i, v_i, \ldots, v_n$.]

6 Periodic Resolutions via Free Actions on Spheres

Let X be a free G-complex which is homeomorphic to an odd dimensional sphere S^{2k-1}. (G is then necessarily finite since X is compact.) By an easy special case of the Lefschetz fixed-point theorem (cf. exercise 1 below), G acts trivially on $H_{2k-1}X = \mathbb{Z}$. Writing $C_* = C_*(X)$, we have then an exact sequence of G-modules

$$(6.1) \qquad 0 \to \mathbb{Z} \xrightarrow{\eta} C_{2k-1} \to \cdots \to C_1 \to C_0 \xrightarrow{\varepsilon} \mathbb{Z} \to 0,$$

where each C_i is free. Splicing together an infinite number of copies of 6.1, we obtain a free resolution of $\mathbb{Z}$ over $\mathbb{Z}G$ which is periodic of period $2k$:

$$(6.2) \qquad \cdots \to C_1 \to C_0 \xrightarrow{\eta\varepsilon} C_{2k-1} \to \cdots \to C_1 \to C_0 \xrightarrow{\varepsilon} \mathbb{Z} \to 0.$$

EXAMPLE. Let G be a finite cyclic group of order n with generator t. Then G acts freely as a group of rotations of S^1, regarded as a CW-complex with n vertices and n 1-cells:

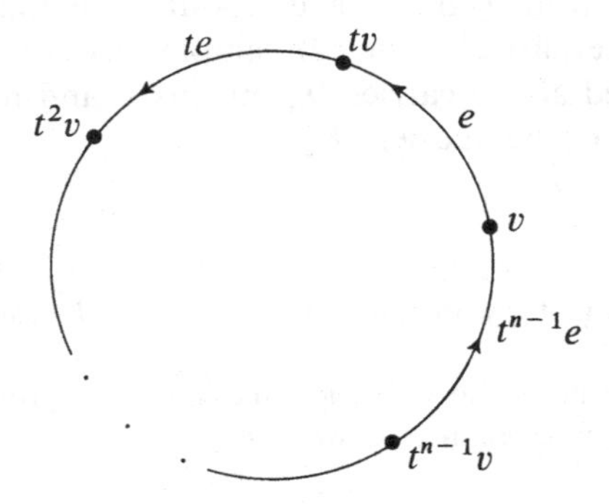

Note that H_1S^1 is generated by the cycle $e + te + t^2e + \cdots + t^{n-1}e = Ne$, where N is the "norm element" $1 + t + \cdots + t^{n-1}$ of $\mathbb{Z}G$, so 6.1 takes the form

$$0 \to \mathbb{Z} \xrightarrow{\eta} \mathbb{Z}G \xrightarrow{t-1} \mathbb{Z}G \xrightarrow{\varepsilon} \mathbb{Z} \to 0,$$

where $\varepsilon(1) = 1$ and $\eta(1) = N$. We therefore obtain the following periodic resolution of period 2, which the reader has probably already seen (§2, exercise 2):

(6.3) $$\cdots \to \mathbb{Z}G \xrightarrow{t-1} \mathbb{Z}G \xrightarrow{N} \mathbb{Z}G \xrightarrow{t-1} \mathbb{Z}G \xrightarrow{\varepsilon} \mathbb{Z} \to 0.$$

We will see other examples of groups with periodic resolutions in Chapter VI.

EXERCISES

1. Prove the following special case of the Lefschetz fixed-point theorem: Let X be a finite CW-complex and $f: X \to X$ a map such that, for every cell σ,
$$f(\sigma) \subseteq \bigcup_{\substack{\tau \neq \sigma \\ \dim \tau \leq \dim \sigma}} \tau.$$
(In particular, f has no fixed-points.) Then the Lefschetz number $\sum (-1)^i \operatorname{trace} f_i$ is zero, where f_i is the endomorphism induced by f on the free abelian group $H_i X/\text{torsion}$. [Hint: The Lefschetz number can be computed on the chain level, where the matrix of f has zeroes on the diagonal.] If $H_*(X) \approx H_*(S^{2n-1})$, deduce that $f_*: H_{2n-1}X \to H_{2n-1}X$ must be the identity.

2. Prove that the group of order 2 is the only non-trivial group which can act freely on an even-dimensional sphere.

7 Uniqueness of Resolutions

We return to the generality of §1, i.e., we work over an arbitrary ring R. It is obvious that a given R-module M admits many different free resolutions; the purpose of this section is to show that all such resolutions are homotopy equivalent. In the course of proving this we will be faced with mapping problems which can be put in the form

(7.1) $$\begin{array}{ccccc} & & P & & \\ & \psi \swarrow & \downarrow \varphi & \searrow 0 & \\ M' & \xrightarrow{i} & M & \xrightarrow{j} & M'', \end{array}$$

where the solid arrows represent given maps (with $j\varphi = 0$) and the dotted arrow represents a map we would like to construct. A solution to this problem, then, consists of a map $\psi: P \to M'$ such that $i\psi = \varphi$. A module P is called *projective* if a solution exists for every mapping problem 7.1 in which the row is exact. More concisely, P is projective if the functor $\operatorname{Hom}_R(P, -)$ is exact.

For our present purposes, the main interest in projectivity is provided by:

(7.2) Lemma. *Free modules are projective.*

PROOF. Let F be free with basis (e_α), and consider a mapping problem

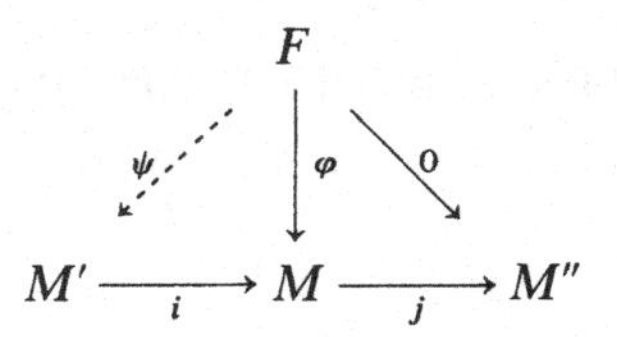

with exact row. Then $\varphi(e_\alpha) \in \ker j = \operatorname{im} i$, so we can choose $x_\alpha \in M'$ with $i(x_\alpha) = \varphi(e_\alpha)$. Now let ψ be the unique R-module map with $\psi(e_\alpha) = x_\alpha$. □

The next lemma treats the particular mapping problems which one has to solve in trying to construct chain maps and homotopies inductively:

(7.3) Lemma. (a) *Suppose given a diagram*

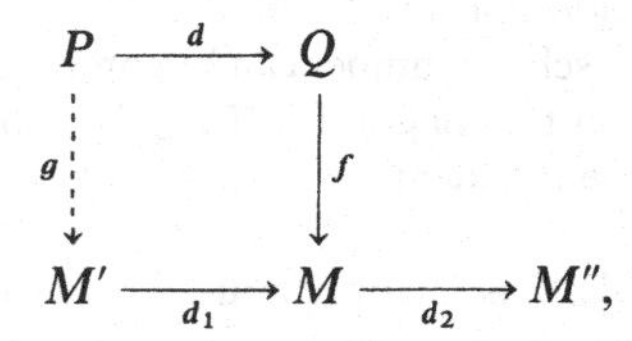

where $d_2 fd = 0$ and it is desired to find a g such that $d_1 g = fd$. If P is projective and the bottom row is exact, then such a g exists.

(b) *Suppose given a diagram (not necessarily commutative)*

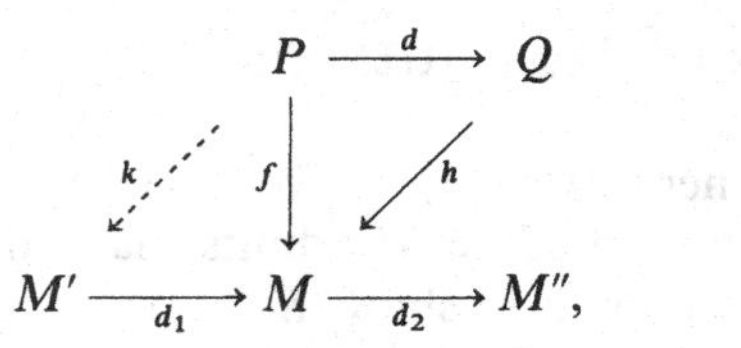

where $d_2 hd = d_2 f$ and it is desired to find a k such that $d_1 k + hd = f$. If P is projective and the bottom row is exact, then such a k exists.

PROOF. (a) is obvious, since the given mapping problem is of the form 7.1 with $\varphi = fd$. Similarly, (b) is a problem of the form 7.1 with $\varphi = f - hd$. □

We can now prove the "fundamental lemma of homological algebra," which says, roughly speaking, that it is easy to construct chain maps and homotopies from a projective complex to an acyclic one:

(7.4) Lemma. *Let (C, ∂) and (C', ∂') be chain complexes and let r be an integer. Let $(f_i: C_i \to C_i')_{i \le r}$ be a family of maps such that $\partial_i' f_i = f_{i-1}\partial_i$ for $i \le r$. If C_i is projective for $i > r$ and $H_i(C') = 0$ for $i \ge r$, then $(f_i)_{i \le r}$ extends to a chain map $f: C \to C'$, and f is unique up to homotopy. More precisely, any two extensions are homotopic by a homotopy h such that $h_i = 0$ for $i \le r$.*

PROOF. Assume inductively that f_i has been defined for $i \leq n$, where $n \geq r$; and that $\partial'_i f_i = f_{i-1}\partial_i$ for $i \leq n$. We then have a mapping problem

$$\begin{array}{ccccc} C_{n+1} & \xrightarrow{\partial} & C_n & \xrightarrow{\partial} & C_{n-1} \\ \vdots f_{n+1} & & \downarrow f_n & & \downarrow f_{n-1} \\ C'_{n+1} & \xrightarrow[\partial']{} & C'_n & \xrightarrow[\partial']{} & C'_{n-1}, \end{array}$$

where $\partial' f_n \partial = f_{n-1}\partial\partial = 0$. The desired f_{n+1} therefore exists by 7.3a.

Suppose now that g is a second extension of $(f_i)_{i \leq r}$. We wish to find a homotopy h between f and g. Assume inductively that $h_i: C_i \to C'_{i+1}$ has been defined for $i \leq n$, where $n \geq r$, and that $\partial' h_i + h_{i-1}\partial = f_i - g_i$. (Note that we can start the induction by setting $h_i = 0$ for $i \leq r$.) Setting $\tau_i = f_i - g_i$, we then have a mapping problem

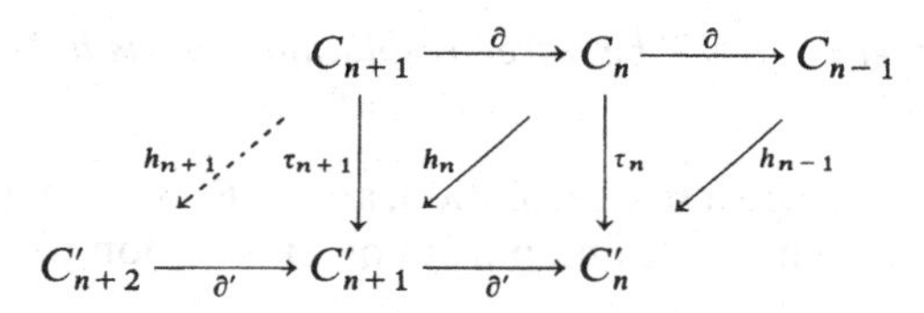

with

$$\begin{aligned} \partial' h_n \partial &= (\tau_n - h_{n-1}\partial)\partial && \text{by the inductive hypothesis} \\ &= \tau_n \partial && \text{since } \partial^2 = 0 \\ &= \partial' \tau_{n+1} && \text{since } \tau \text{ is a chain map.} \end{aligned}$$

The desired h_{n+1} with $\partial' h_{n+1} + h_n \partial = \tau_{n+1}$ therefore exists by 7.3b. □

Remark. The proof of 7.4 should have looked familiar to anyone who has seen the method of acyclic models in algebraic topology. We will explain this "similarity" in exercise 3 below.

Now let $\varepsilon: F \to M$ and $\varepsilon': F' \to M$ be two free (or projective) resolutions of a module M. We can then form the augmented chain complexes with M in dimension -1 and apply 7.4 with $r = -1$:

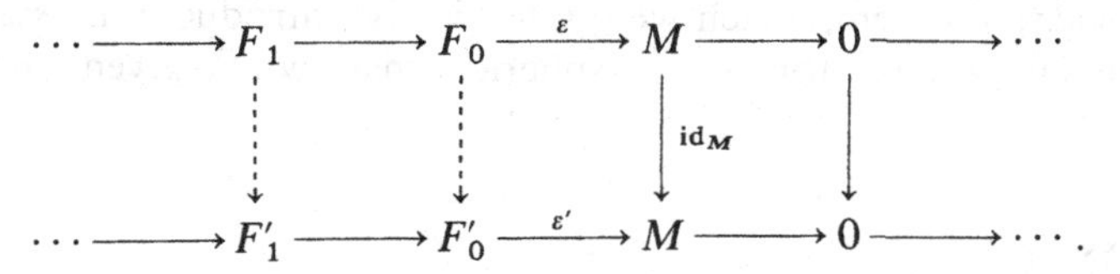

We conclude that there is a chain map $f: F \to F'$ which is *augmentation-preserving*, i.e., which satisfies $\varepsilon' f = \varepsilon$. Moreover, f is unique up to homotopy. (Note that the homotopy h given by 7.4 on the level of *augmented* chain complexes yields a homotopy $F \to F'$, because $h_{-1} = 0$.) Similarly, there is an augmentation-preserving map $f': F' \to F$, and we have $f'f \simeq \mathrm{id}_F$ and $ff' \simeq \mathrm{id}_{F'}$, again by the uniqueness part of 7.4. This proves:

(7.5) Theorem. *Given projective resolutions F and F' of a module M, there is an augmentation-preserving chain map $f: F \to F'$, unique up to homotopy, and f is a homotopy equivalence.*

We will express this informally by saying that projective resolutions are "unique up to canonical homotopy equivalence." At the moment, of course, we are mainly interested in *free* resolutions, but later (beginning in Chapter VIII) we will need to consider projective resolutions which are not known to be free.

For future reference we record two special cases of 7.5:

(7.6) Corollary. *Let $\varepsilon: F \to \mathbb{Z}$ be a free resolution of $\mathbb{Z}$ as a $\mathbb{Z}$-module. Then ε is a homotopy equivalence.*

Proof. $\mathrm{id}_{\mathbb{Z}}: \mathbb{Z} \to \mathbb{Z}$ is also a free resolution, and $\varepsilon: F \to \mathbb{Z}$ can be viewed as an augmentation-preserving chain map from one resolution to another. □

(7.7) Corollary. *If F is a non-negative acyclic chain complex of projective modules (over an arbitrary ring R), then F is contractible.*

Proof. F and the zero complex are projective resolutions of 0, so they are homotopy equivalent. See also exercise 3 of §8 for an alternative proof. □

Remarks

1. In practice, a free resolution usually comes equipped with an explicit basis in each dimension, and one usually proves that it is a resolution by exhibiting a contracting homotopy for the underlying augmented chain complex of $\mathbb{Z}$-modules. In this case, the proof of 7.5 yields a specific map $f: F \to F'$. Namely, if k is a contracting homotopy for the augmented complex associated to F', then f is determined inductively by $f_{n+1}(e_\alpha) = k_n f_n \partial e_\alpha$, where (e_α) is a basis for F_{n+1}.

2. The uniqueness theorem 7.5 can be thought of as the algebraic analogue of Hurewicz's theorem, which we quoted in the introduction, asserting the uniqueness up to homotopy of an aspherical space with a given fundamental group.

Exercises

1. Let G be a finite cyclic group. Let F be the free resolution of $\mathbb{Z}$ over $\mathbb{Z}G$ given in 6.3 and let F' be the bar resolution. Write down an augmentation preserving chain map $f: F \to F'$.

2. The method of proof of 7.4 applies to many situations, some of which will arise later in this book. It is therefore useful to axiomatize 7.4, as follows. By an *additive category* we mean a category $\mathscr{A}$ in which $\operatorname{Hom}(A, B)$ is endowed with an abelian group structure for any two objects A, B, in such a way that the composition law $\operatorname{Hom}(B, C) \times \operatorname{Hom}(A, B) \to \operatorname{Hom}(A, C)$ is $\mathbb{Z}$-bilinear. In particular, since we have a zero element $0 \in \operatorname{Hom}(A, B)$ it is clear what we mean by a *chain complex* in $\mathscr{A}$. The usual definitions of *chain map* and *homotopy* also go through with no change. [Note that additivity is needed to define "homotopy."] Suppose now that we are given a class $\mathscr{E}$ of sequences $M' \to M \to M''$ in $\mathscr{A}$ called *exact sequences.* We then say that an object P of $\mathscr{A}$ is *projective* relative to $\mathscr{E}$ if every mapping problem 7.1 whose row is in $\mathscr{E}$ admits a solution. Verify that the proof of 7.4 is valid in this situation and leads to an analogue of 7.4 for chain complexes in $\mathscr{A}$. [Note: In the statement of this analogue, the "acyclicity" hypothesis on C' should be replaced by the assumption that $C'_{i+1} \to C'_i \to C'_{i-1}$ is in $\mathscr{E}$ for $i \geq r$.] Deduce an analogue of 7.5.

3. Let $\mathscr{C}$ be an arbitrary category and let $\mathscr{A}$ be the additive category whose objects are covariant functors $\mathscr{C} \to$ (abelian groups) and whose maps are natural transformations of functors. Fix a subclass $\mathscr{M}$ of the class of objects of $\mathscr{C}$. A sequence $T' \to T \to T''$ in $\mathscr{A}$ will be called *$\mathscr{M}$-exact* if the resulting sequence $T'(M) \to T(M) \to T''(M)$ of abelian groups is exact for all $M \in \mathscr{M}$. The purpose of this exercise is to show that when exercise 2 is applied to $\mathscr{A}$ (with $\mathscr{E}$ equal to the class of $\mathscr{M}$-exact sequences), the result is the acyclic models theorem as given, for instance, in Spanier [1966], Theorem 4.2.8, or Dold [1972], Proposition VI.11.7. The crucial step is to prove an analogue in $\mathscr{A}$ of Lemma 7.2; this is done in (a) and (b) below.

(a) (Yoneda's lemma) Let A be an object of $\mathscr{C}$ and let $h_A : \mathscr{C} \to$ (sets) be the covariant functor *represented* by A, i.e., $h_A = \operatorname{Hom}_{\mathscr{C}}(A, -)$. Let $u_A \in h_A(A)$ be the identity map $A \to A$. Let $T : \mathscr{C} \to$ (sets) be an arbitrary covariant functor. For any $v \in T(A)$, show that there is a unique natural transformation $\varphi : h_A \to T$ such that $\varphi(u_A) = v$. Thus $\operatorname{Hom}_{\mathscr{F}}(h_A, T) \approx T(A)$, where $\mathscr{F}$ is the category of functors $\mathscr{C} \to$ (sets). [This can be thought of as saying that h_A is "freely generated" by u_A. The proof is straightforward definition-checking. To prove uniqueness, for example, note that we must have $\varphi(f) = T(f)(v)$ for any $f \in h_A(B) = \operatorname{Hom}_{\mathscr{C}}(A, B)$, in view of the diagram

$$\begin{array}{ccc} h_A(A) & \xrightarrow{h_A(f)} & h_A(B) \\ \downarrow{\scriptstyle\varphi} & & \downarrow{\scriptstyle\varphi} \\ T(A) & \xrightarrow{T(f)} & T(B). \end{array}$$

To prove existence, take this equation as a definition and check that it works.]

(b) Let $\mathbb{Z}h_A : \mathscr{C} \to$ (abelian groups) be the composite of h_A with the functor (sets) $\to$ (abelian groups) which associates to a set the free abelian group it generates. A functor $F : \mathscr{C} \to$ (abelian groups) will be called *$\mathscr{M}$-free* if it is isomorphic to a direct sum $\bigoplus_\alpha \mathbb{Z}h_{A_\alpha}$, where $A_\alpha \in \mathscr{M}$ for all α. Deduce from (a) that $\mathscr{M}$-free functors are projective relative to the class of $\mathscr{M}$-exact sequences.

(c) Using (b) and exercise 2, state a theorem about chain maps in $\mathscr{A}$ from $\mathscr{M}$-free complexes to "$\mathscr{M}$-acyclic" complexes. [Note: This theorem is precisely the acyclic models theorem cited above.]

4. Another important example of exercise 2 is obtained by taking $\mathscr{A}$ to be the *dual* of the category of R-modules. Thus $\mathscr{A}$ has one object M° for every R-module M and one map $f^\circ: M_1^\circ \to M_2^\circ$ for every R-module map $f: M_2 \to M_1$. Composition is defined by $f^\circ g^\circ = (gf)^\circ$. As exact sequences in $\mathscr{A}$ we take those sequences $M_1^\circ \to M_2^\circ \to M_3^\circ$ such that the corresponding sequence $M_3 \to M_2 \to M_1$ of R-modules is exact. Applying exercise 2, we get an analogue of 7.4 for chain complexes in $\mathscr{A}$, which can obviously be restated as a theorem about cochain complexes of R-modules. Explicitly state this theorem, as well as the theorem corresponding to 7.5. [Note: An R-module Q is called *injective* if Q° is projective in $\mathscr{A}$, or, equivalently, if the functor $\operatorname{Hom}_R(-, Q)$ is exact. Your theorem should therefore be stated in terms of maps of an acyclic cochain complex into a cochain complex of injectives. To give this theorem substance, of course, we should have an analogue of 7.2, so that we will have examples of injectives. We will provide such an analogue later, in §III.4.]

8 Projective Modules

The reader may be curious at this point to know more about projective modules, other than the fact that free modules are projective. We give in this section, therefore, some miscellaneous results and examples concerning projective modules and complexes. We will not make serious use of these results (except as they apply to free modules) until Chapter VIII. One may therefore skip this section now and return to it later.

The first observation is that to prove a module P is projective one need only consider mapping problems 7.1 in which $M'' = 0$; for 7.1 can be replaced by

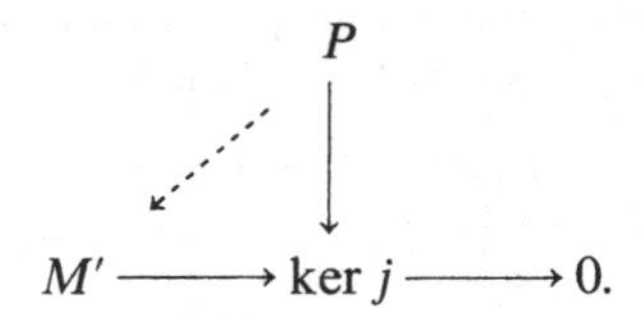

Thus we have:

(8.1) Proposition. *P is projective if and only if for every surjection $\pi: M \to \overline{M}$ and every map $\varphi: P \to \overline{M}$ there is a map $\psi: P \to M$ such that $\varphi = \pi\psi$:*

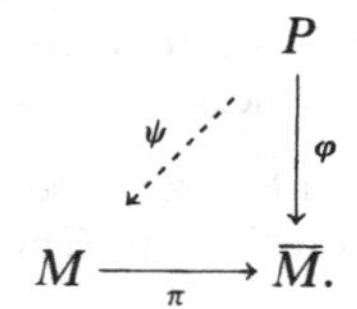

One also has the following characterization of projectivity:

(8.2) Proposition. *The following conditions on an R-module P are equivalent:*

(i) *P is projective.*
(ii) *Every exact sequence $0 \to M' \to M \to P \to 0$ splits.*
(iii) *P is a direct summand of a free module.*
(iv) *There are elements $e_i \in P$ and $f_i \in \mathrm{Hom}_R(P, R)$ (where i ranges over some index set I) such that for every $x \in P$, $f_i(x) = 0$ for almost all i and $x = \sum_{i \in I} f_i(x)e_i$.*

PROOF. If P is projective and we are given an exact sequence as in (ii), then we can split the sequence by lifting $\mathrm{id}: P \to P$ to a map $P \to M$. Hence (i) $\Rightarrow$ (ii). Choosing such an exact sequence with M free, we see that (ii) $\Rightarrow$ (iii). It is immediate from the definitions that any direct summand of a projective is projective; so (iii) $\Rightarrow$ (i). Finally, (iv) is simply a restatement of (iii). □

EXAMPLES

1. Let $e \in R$ be *idempotent*, i.e., $e^2 = e$. Then right multiplication by e is a projection operator of R onto the direct summand Re. So the left ideal Re is projective.

2. Let R be a (commutative) integral domain and I an invertible ideal. [This means that there is an R-submodule J of the field of fractions of R such that $IJ = R$, where IJ is the set of finite sums $\sum a_i b_i$, $a_i \in I$, $b_i \in J$.] Then I is a projective module. For if we write $1 = \sum e_i f_i$ where $e_i \in I$, $f_i \in J$, then the criterion of 8.4(iv) is satisfied. [Each f_i gives rise, by multiplication, to a homomorphism $I \to R$ which plays the role of the f_i in 8.4(iv).] On the other hand, I is free only if it is a principal ideal, since any two elements $a, b \in I$ are linearly dependent.

3. Let $R = \mathbb{Z}[\zeta]$, where ζ is a primitive twenty-third root of unity. It is known from algebraic number theory that R has an ideal I which is not principal and that every non-zero ideal in R is invertible. Hence I is projective but not free.

4. Let G be a cyclic group of prime order p. There is a theorem due to Rim which relates projective modules over $\mathbb{Z}G$ to projective modules over $\mathbb{Z}[\zeta]$, where ζ is a primitive p-th root of unity, cf. Milnor [1971], §3. In particular, if $p = 23$, we deduce from example 3 that $\mathbb{Z}G$ has non-free projectives.

5. If R is the rational group algebra $\mathbb{Q}G$ of a finite group G, then one can show that *every* R-module is projective, cf. exercise 5 below. As an illustration we will prove that $\mathbb{Q}$, with trivial G-action, is projective. Note first that the functor $\mathrm{Hom}_{\mathbb{Q}G}(\mathbb{Q}, -)$ is simply the "invariants" functor $M \mapsto M^G$, where

$$M^G = \{m \in M : gm = m \text{ for all } g \in G\}.$$

Thus we need to show that any surjection $M \to \overline{M}$ of $\mathbb{Q}G$-modules gives rise to a surjection $M^G \to \overline{M}^G$. This is shown by *averaging*: if $\overline{m} \in \overline{M}^G$, lift $\overline{m}$ to $m \in M$; then $(1/|G|) \sum_{g \in G} gm$ is also a lifting of $\overline{m}$ and is in M^G.

We turn now to the *duality theory* for finitely generated projectives, analogous to that for finite dimensional vector spaces over a field. For any left R-module M, let $M^* = \operatorname{Hom}_R(M, R)$. Here R is regarded as a left R-module in forming Hom, but it also has a right R-module structure which we can use to make M^* a right R-module; namely, we set $(ur)(m) = u(m)r$ for $u \in M^*$, $r \in R$, $m \in M$. Similarly, we can define the dual of a right module, and it is a left module.

The main facts about duality are given in the following proposition. Parts (b) and (c) are the most important ones for our purposes; they allow one to use duality to convert Hom to $\otimes$ and vice-versa.

(8.3) Proposition. *Let P be a finitely generated projective left R-module.*

(a) *P^* is a finitely generated projective right R-module.*

(b) *For any left R-module M, there is an isomorphism*

$$\varphi: P^* \otimes_R M \xrightarrow{\approx} \operatorname{Hom}_R(P, M)$$

of abelian groups, given by $\varphi(u \otimes m)(x) = u(x) \cdot m$ for $u \in P^*$, $m \in M$, $x \in P$.

(c) *For any right R-module M, there is an isomorphism*

$$\varphi': M \otimes_R P \xrightarrow{\approx} \operatorname{Hom}_R(P^*, M),$$

given by $\varphi'(m \otimes x)(u) = m \cdot u(x)$ for $m \in M$, $x \in P$, $u \in P^*$.

(d) *There is an isomorphism*

$$\varphi'': P \xrightarrow{\approx} P^{**}$$

of left R-modules, given by $\varphi''(x)(u) = u(x)$ for $x \in P$, $u \in P^*$.

(In connection with (b) and (c), the reader should recall that one can form $M \otimes_R N$ whenever M is a right module and N is a left module; the tensor products written down above therefore make sense.)

Proof. It is clear from the proof of 8.2 that P can be written as a direct summand of a finitely generated free module F. By additivity, then, it suffices to prove the proposition for F. In more detail: If $F = P \oplus Q$, then $F^* = P^* \oplus Q^*$, $F^* \otimes_R M = (P^* \otimes_R M) \oplus (Q^* \otimes_R M)$, etc., and the maps φ, φ', and φ'' preserve these decompositions. So (a)–(d) for P will follow from (a)–(d) for F. [In order to use this argument, of course, one must first check that φ and φ' are well-defined.] By additivity again, it suffices to consider the case where F is free of rank 1, i.e., we may assume $F = R$. In this case $R^* \approx R$, whence (a), and it is easy to verify (b)–(d). To prove (b), for instance, one need only check that φ is the composite of the canonical isomorphisms $R^* \otimes_R M \xrightarrow{\approx} M \xrightarrow{\approx} \operatorname{Hom}_R(R, M)$. □

Next we give some properties of chain complexes of projectives.

(8.4) Theorem. *If $f: P' \to P$ is a weak equivalence between non-negative complexes of projectives, then f is a homotopy equivalence.*

PROOF. The mapping cone of f is non-negative, projective, and acyclic (cf. 0.6). It is therefore contractible by 7.7, so f is a homotopy equivalence by 0.7. □

Using similar methods, we will prove a closely related mapping property of projective complexes, from which we could have deduced 8.4:

(8.5) Theorem. *Let $f: C' \to C$ be a weak equivalence between arbitrary complexes. If P is a non-negative complex of projectives, then*

$$\mathcal{H}om_R(P, f): \mathcal{H}om_R(P, C') \to \mathcal{H}om_R(P, C)$$

is a weak equivalence. In particular, the map $[P, C] \to [P, C']$ induced by f is an isomorphism.

PROOF. Let C'' be the mapping cone of f. It is acyclic, and the mapping cone of $\mathcal{H}om_R(P, f)$ is $\mathcal{H}om_R(P, C'')$; so it suffices to show that $\mathcal{H}om_R(P, C'')$ is acyclic, i.e., that $[P, C'']_n = 0$ for all $n \in \mathbb{Z}$. Now $[P, C'']_n = [P, \Sigma^{-n}C'']$, and the latter is zero by the uniqueness part of the fundamental lemma 7.4, since any map on P is zero in negative dimensions. □

Finally, we prove an analogue of 8.5 for tensor products. Projectivity is unnecessarily strong for this purpose and can be replaced by a "flatness" hypothesis. Recall that a (left) R-module F is *flat* if the functor $- \otimes_R F$ is exact. Free modules are flat, for example, and hence projectives are flat by 8.2(iii).

(8.6) Theorem. *Let $f: C' \to C$ be a weak equivalence between complexes of right R-modules. If P is a non-negative complex of flat left R-modules, then $f \otimes_R P: C' \otimes_R P \to C \otimes_R P$ is a weak equivalence.*

PROOF. Let C'' be the mapping cone of f. It is acyclic, and $C'' \otimes_R P$ is the mapping cone of $f \otimes P$; so it suffices to show that $C'' \otimes_R P$ is acyclic. Let $P^{(n)}$ be the n-skeleton of P, i.e., the truncation $(P_i)_{i \le n}$. We will show inductively that $C'' \otimes_R P^{(n)}$ is acyclic. Note first that $C'' \otimes_R F$ is acyclic for any complex F consisting of a flat module concentrated in a single dimension, since the exact sequences $C''_{i+1} \to C''_i \to C''_{i-1}$ remain exact when tensored with F. But $P^{(n)}/P^{(n-1)}$ is such a complex F. So if we assume inductively that $C'' \otimes_R P^{(n-1)}$ is acyclic, it follows from the exact sequence $0 \to C'' \otimes_R P^{(n-1)} \to C'' \otimes_R P^{(n)} \to C'' \otimes_R (P^{(n)}/P^{(n-1)}) \to 0$ that $C'' \otimes_R P^{(n)}$ is acyclic. Finally, $C'' \otimes_R P$ is the increasing union of the acyclic complexes $C'' \otimes_R P^{(n)}$, hence it is acyclic. □

EXERCISES

1. For what groups G is $\mathbb{Z}$ a projective $\mathbb{Z}G$-module? [Hint: When does $\varepsilon: \mathbb{Z}G \to \mathbb{Z}$ split as a map of G-modules?]

2. If P is a projective $\mathbb{Z}G$-module, show that P is also projective as $\mathbb{Z}H$-module for any $H \subset G$. [Use criterion 8.2(iii).]

3. (a) Use 8.2 to give another proof of 7.7. [According to 0.3, it suffices to show that the surjection $\bar{\partial}_n: P_n \to Z_{n-1}$ induced by ∂_n splits for all n. Assuming inductively that $\bar{\partial}_{n-1}$ splits, $Z_{n-1} = \ker \bar{\partial}_{n-1}$ is a direct summand of P_{n-1}, hence it is projective. Therefore $\bar{\partial}_n$ splits.]

   (b) Use the same method to show that the non-negativity hypothesis in 7.7 can be dropped for certain rings R, e.g., for principal ideal domains. [If R is a principal ideal domain, then submodules of a projective module are projective (in fact, free). So Z_{n-1} above is automatically projective and we do not need to use induction.]

4. If G is a group and X is a G-set such that all isotropy groups G_x are finite, show that the permutation module $\mathbb{Q}X$ is a projective $\mathbb{Q}G$-module.

5. If G is finite and k is a field of characteristic zero, show that every kG-module is projective. [Given an exact sequence as in 8.2(ii), choose a splitting $f: P \to M$ for the underlying sequence of k-vector spaces. Then $x \mapsto (1/|G|) \sum_{g \in G} gf(g^{-1}x)$ is a kG-splitting.]

6. Prove the following converse of 8.3b: If P is a module such that $\varphi: P^* \otimes_R P \to \operatorname{Hom}_R(P, P)$ is surjective, then P is finitely generated and projective. [Write $\mathrm{id}_P = \varphi(\sum f_i \otimes e_i)$ and show that 8.2(iv) is satisfied.]

7. Let P be finitely generated and projective. For any $z \in P^* \otimes_R P$ and any module M there is a map $\psi_z: \operatorname{Hom}_R(P, M) \to P^* \otimes_R M$, defined as follows: $\psi_z(f)$ is the image of z under $P^* \otimes f: P^* \otimes_R P \to P^* \otimes_R M$. Show that the inverse of the canonical isomorphism $\varphi: P^* \otimes_R M \xrightarrow{\approx} \operatorname{Hom}_R(P, M)$ of 8.3b is a map of the form ψ_z for some fixed z (independent of M). [Method 1: View φ^{-1} as a natural transformation $\operatorname{Hom}_R(P, -) \to P^* \otimes_R -$. By Yoneda's lemma (exercise 3a of §7), φ^{-1} is uniquely determined by what it does to id_P. Moreover, the proof of Yoneda's lemma tells you how to describe φ^{-1} in terms of $z = \varphi^{-1}(\mathrm{id}_P)$, and this description says precisely that $\varphi^{-1} = \psi_z$. Method 2: Choose (e_i) and (f_i) as in 8.2(iv) and set $z = \sum f_i \otimes e_i$. By directly checking definitions, verify that $\psi_z \circ \varphi = \mathrm{id}$ and/or that $\varphi \circ \psi_z = \mathrm{id}$.]

8. If P is finitely presented and flat, show that P is projective. [Take a finite presentation $F_1 \to F_0 \to P \to 0$ with F_0 and F_1 free of finite rank. This gives an exact sequence $0 \to P^* \to F_0^* \to F_1^*$ of right R-modules. Tensor with P and deduce that $P^* \otimes_R P \xrightarrow{\approx} \operatorname{Hom}_R(P, P)$. Now apply exercise 6.]

Appendix. Review of Regular Coverings

The material summarized in this appendix can be found in many algebraic topology texts, such as Massey [1967] or Spanier [1966].

Let $p: \tilde{X} \to X$ be a covering map of connected, locally path-connected spaces. A *deck transformation* of p is a homeomorphism $g: \tilde{X} \to \tilde{X}$ such that $pg = p$. The group G of all deck transformations acts *freely* on $\tilde{X}$, in the sense that $g\tilde{x} \neq \tilde{x}$ for all $\tilde{x} \in \tilde{X}$ and $g \neq 1$ in G.

The cover p is said to be *regular* if it satisfies the following conditions, which are equivalent:

(i) G acts transitively on $p^{-1}x$ for all $x \in X$. [Hence $X \approx \tilde{X}/G$.]
(ii) The image of $\pi_1 \tilde{X} \to \pi_1 X$ is normal in $\pi_1 X$ for some (and hence every) choice of basepoints.
(iii) For any closed loop ω in X, if one lift of ω to $\tilde{X}$ is closed, then all lifts of ω are closed.

In this case we have $G \approx \pi_1 X/\pi_1 \tilde{X}$. In particular, if $\tilde{X}$ is simply connected (so that p is the *universal cover* of X), then $G \approx \pi_1 X$.

To get an explicit isomorphism $G \approx \pi_1 X/\pi_1 \tilde{X}$ above, we must choose basepoints $x \in X$ and $\tilde{x} \in p^{-1}x$. We then have a homomorphism $\varphi: \pi_1(X, x) \to G$, defined as follows: Let $\omega: [0, 1] \to X$ represent $[\omega] \in \pi_1(X, x)$ and let $\tilde{\omega}: [0, 1] \to \tilde{X}$ be the lift of ω with $\tilde{\omega}(0) = \tilde{x}$. Then $\tilde{\omega}(1) \in p^{-1}x$, and $\varphi([\omega])$ is defined to be the unique element of G such that

$$\text{(A1)} \qquad \varphi([\omega])\tilde{x} = \tilde{\omega}(1).$$

One verifies that φ is a homomorphism, is surjective, and has kernel $\pi_1 \tilde{X}$, so it induces the required isomorphism.

Finally, we mention a slightly different point of view which is sometimes useful. [For our purposes this will be needed only in exercise 2 of §II.7.] Fix an abstract group G and a connected, locally path-connected space X. By a *regular G-cover* of X (also called a *principal G-bundle* over X) we mean a covering map $p: \tilde{X} \to X$, where $\tilde{X}$ is not necessarily connected, together with a free G-action on $\tilde{X}$ satisfying condition (i) above.

In case $\tilde{X}$ is connected, such a p is simply a regular cover in the usual sense, together with an isomorphism of G with the group of deck transformations.

We will assume that a basepoint $x \in X$ has been chosen and that all covering spaces $\tilde{X}$ come equipped with a basepoint $\tilde{x} \in p^{-1}x$. We can then define a homomorphism $\varphi: \pi_1 X \to G$ exactly as in A1 above, the only difference being that φ will not be surjective if $\tilde{X}$ is disconnected. In fact, one checks easily that $G/\text{im } \varphi \approx \pi_0 \tilde{X}$ (isomorphism of G-sets).

The main theorem on regular G-covers (with basepoint) says that they are completely classified by $\varphi \in \text{Hom}(\pi_1 X, G)$:

(A2) Theorem. *Let $\mathscr{C}_G(X)$ be the set of isomorphism classes of pointed, regular G-covers of X. The assignment of φ to p gives a bijection*

$$\mathscr{C}_G(X) \approx \mathrm{Hom}(\pi_1 X, G).$$

(Note: Isomorphisms are required to preserve basepoints, commute with the G action, and commute with the projection onto X.)

SKETCH OF PROOF. Using the usual classification of connected covering spaces in terms of subgroups of $\pi_1 X$, one easily sees that connected, pointed, regular G-covers correspond to surjections $\varphi: \pi_1 X \to G$. The study of disconnected covers is easily reduced to the connected case by considering the connected components of $\tilde{X}$. □

CHAPTER II

The Homology of a Group

1 Generalities

In homological algebra one constructs homological invariants of algebraic objects by the following process, or some variant of it:

Let R be a ring and T a covariant *additive* functor from R-modules to abelian groups. Thus the map $\text{Hom}_R(M, N) \to \text{Hom}_{\mathbb{Z}}(TM, TN)$ defined by T is a homomorphism of abelian groups for all R-modules M, N. For any R-module M, choose a free (or projective) resolution $\varepsilon: F \to M$ and consider the chain complex TF of abelian groups obtained by applying T to F termwise. Now T, being additive, preserves chain homotopies; so we can apply the uniqueness theorem for resolutions (I.7.5) to deduce that the complex TF is independent, up to canonical homotopy equivalence, of the choice of resolution. Passing to homology, we obtain groups $H_n(TF)$ which depend only on T and M (up to canonical isomorphism).

This construction is of no interest, of course, if T is an *exact* functor; for then the augmented complex

$$\cdots \to TF_1 \to TF_0 \to TM \to 0$$

is acyclic, so that $H_n(TF) = 0$ for $n > 0$ and $H_0(TF) = TM$. Thus we can regard the groups $H_n(TF)$ in the general case as a measure of the failure of T to be exact.

In this chapter we will apply this construction with $R = \mathbb{Z}G$, $M = \mathbb{Z}$, and T equal to the "co-invariants" functor which we will describe in §2 below. This particular choice of R, M, and T is not arbitrary, as we will see, but rather it is a reflection of the topology which motivates the homology theory of groups.

2 Co-invariants

If G is a group and M is a G-module, then the group of *co-invariants* of M, denoted M_G, is defined to be the quotient of M by the additive subgroup generated by the elements of the form $gm - m$ $(g \in G, m \in M)$. Thus M_G is obtained from M by "dividing out" by the G-action. (The name "co-invariants" comes from the fact that M_G is the largest *quotient* of M on which G acts trivially, whereas M^G, the group of invariants, is the largest *submodule* of M on which G acts trivially.) In view of exercise 1a of §I.2, we can also describe M_G as M/IM, where I is the augmentation ideal of $\mathbb{Z}G$ and IM denotes the set of all finite sums $\sum a_i b_i$ $(a_i \in I, b_i \in M)$.

Still another description of M_G is given by:

(2.1)
$$M_G \approx \mathbb{Z} \otimes_{\mathbb{Z}G} M.$$

Here, in order for the tensor product to make sense, we regard $\mathbb{Z}$ as a *right* $\mathbb{Z}G$-module (with trivial G-action). To prove 2.1, note that in $\mathbb{Z} \otimes_{\mathbb{Z}G} M$ we have the identity $1 \otimes gm = 1 \cdot g \otimes m = 1 \otimes m$; hence there is a map $M_G \to \mathbb{Z} \otimes_{\mathbb{Z}G} M$ given by $\bar{m} \mapsto 1 \otimes m$, where $\bar{m}$ denotes the image in M_G of an element $m \in M$. On the other hand, using the universal property of the tensor product, we can define a map $\mathbb{Z} \otimes_{\mathbb{Z}G} M \to M_G$ by $a \otimes m \mapsto a\bar{m}$. These two maps are inverses of one another.

In view of 2.1 and standard properties of the tensor product, we immediately obtain the following two properties of the co-invariants functor:

(2.2) Right-exactness: Given an exact sequence $M' \to M \to M'' \to 0$ of G-modules, the induced sequence $M'_G \to M_G \to M''_G \to 0$ is exact.

(2.3) If F is a free $\mathbb{Z}G$-module with basis (e_i), then F_G is a free $\mathbb{Z}$-module with basis $(\bar{e}_i)$.

Finally, we note that the co-invariants functor arises naturally in the topological setting of §I.4:

(2.4) Proposition. *Let X be a free G-complex and let Y be the orbit complex X/G. Then $C_*(Y) \approx C_*(X)_G$.*

Proof. The projection $C_*(X) \to C_*(Y)$ induces by passage to the quotient a map $\varphi\colon C_*(X)_G \to C_*(Y)$. Now $C_*(X)_G$ has (by 2.3 and the observations in §I.4) a $\mathbb{Z}$-basis with one basis element for each G-orbit of cells of X. But $C_*(Y)$ also has a $\mathbb{Z}$-basis with one element for each G-orbit of cells of X, and it is clear that φ maps a basis element of $C_*(X)_G$ to the corresponding basis element of $C_*(Y)$, hence φ is an isomorphism. □

Exercises

1. If S is an arbitrary G-set, show that $(\mathbb{Z}S)_G \approx \mathbb{Z}[S/G]$.
2. The freeness hypothesis in 2.4 is unnecessarily strong. Find a weaker hypothesis under which the conclusion remains true. [Hint: Use exercise 1.]

3. Let H be a normal subgroup of G and let M be a G-module.

   (a) Show that the action of G on M induces an action of G/H on M_H.

   (b) Show that $M_G \approx (M_H)_{G/H}$.

   (c) Show that $M_H \approx \mathbb{Z}[G/H] \otimes_{\mathbb{Z}G} M$ as G/H-modules. (Here the *right*-translation action of G on $\mathbb{Z}[G/H]$ is used to form the tensor product, and the *left*-translation action of G/H on $\mathbb{Z}[G/H]$ is used to give the tensor product a G/H-module structure.)

3 The Definition of H_*G

Let G be a group and $\varepsilon: F \to \mathbb{Z}$ a projective resolution of $\mathbb{Z}$ over $\mathbb{Z}G$. We define the *homology groups* of G by

$$H_iG = H_i(F_G).$$

As we explained in §1, the right-hand side is independent of the choice of resolution, up to canonical isomorphism. For example, suppose G is a finite cyclic group of order n. Using the resolution

$$\cdots \xrightarrow{t-1} \mathbb{Z}G \xrightarrow{N} \mathbb{Z}G \xrightarrow{t-1} \mathbb{Z}G \to \mathbb{Z} \to 0$$

of I.6.3, we obtain for F_G the complex

$$\cdots \xrightarrow{0} \mathbb{Z} \xrightarrow{n} \mathbb{Z} \xrightarrow{0} \mathbb{Z}.$$

Thus

$$H_iG \approx \begin{cases} \mathbb{Z} & i = 0 \\ \mathbb{Z}_n & i \text{ odd} \\ 0 & i \text{ even}, i > 0. \end{cases} \tag{3.1}$$

The reader is invited to similarly compute the homology of a free group (cf. I.4.3) or a free abelian group of rank 2 (cf. §I.4, exercise 3). We will treat these examples topologically in the next section.

For any group G we can always take F to be the standard resolution (§I.5), in which case we write $C_*(G)$ for the chain complex F_G. Using 2.3 and the formula I.5.1, we can describe $C_*(G)$ explicitly, as follows: Define an equivalence relation on the $(n+1)$-tuples $(g_0, \ldots, g_n)$ $(g_i \in G)$ by setting $(g_0, \ldots, g_n) \sim (gg_0, \ldots, gg_n)$ for all $g \in G$, and let $[g_0, \ldots, g_n]$ denote the equivalence class of $(g_0, \ldots, g_n)$. Then $C_n(G)$ has a $\mathbb{Z}$-basis consisting of the equivalence classes $[g_0, \ldots, g_n]$, and $\partial: C_n(G) \to C_{n-1}(G)$ is given by $\partial = \sum_{i=0}^{n} (-1)^i d_i$, where

$$d_i[g_0, \ldots, g_n] = [g_0, \ldots, \hat{g}_i, \ldots, g_n].$$

$C_*(G)$, when described in this way, is often called the *homogeneous* chain complex of G, because of the analogy with homogeneous coordinates for projective space.

The *non-homogeneous* description of $C_*(G)$ is obtained by using the bar notation and the formula I.5.2. From this point of view $C_n(G)$ has a $\mathbb{Z}$-basis consisting of n-tuples $[g_1|\cdots|g_n]$, and $\partial = \sum_{i=0}^{n}(-1)^i d_i$, where

$$d_i[g_1|\cdots|g_n] = \begin{cases} [g_2|\cdots|g_n] & i = 0 \\ [g_1|\cdots|g_ig_{i+1}|\cdots|g_n] & 0 < i < n \\ [g_1|\cdots|g_{n-1}] & i = n. \end{cases}$$

Note that the symbol $[g_1|\cdots|g_n]$, which previously was used to denote a typical $\mathbb{Z}G$-basis element of F_n, now denotes the image of that basis element in $(F_n)_G = C_n(G)$. This abuse of notation, which is standard, might occasionally cause confusion. The reader is also warned that some authors write $[g_1, \ldots, g_n]$ or $(g_1, \ldots, g_n)$ instead of $[g_1|\cdots|g_n]$.

In low dimensions $C_*(G)$ has the form

$$C_2(G) \xrightarrow{\partial} C_1(G) \xrightarrow{0} \mathbb{Z},$$

where $\partial[g|h] = [h] - [gh] + [g]$. Consequently, $H_0G = \mathbb{Z}$ and H_1G is isomorphic to the *abelianization* $G_{ab} = G/[G, G]$. (Explicitly, if we denote by $\bar{g}$ the homology class of the cycle $[g]$, the reader can easily check that there is an isomorphism $H_1G \to G_{ab}$ such that $\bar{g} \mapsto g \bmod [G, G]$.)

Exercises

1. Let $g_1, \ldots, g_n$ be n elements of G which pairwise commute, and let

$$z = \sum (-1)^{\operatorname{sgn}\sigma}[g_{\sigma(1)}|\cdots|g_{\sigma(n)}]$$

in $C_n(G)$, where σ ranges over all permutations of $\{1, \ldots, n\}$. (If $n = 2$, for example, $z = [g_1|g_2] - [g_2|g_1]$.) Verify that z is a cycle. [Such cycles play an important role in the homology theory of abelian groups, as we will see in Chapter V.]

2. If G is a non-trivial finite cyclic group, show that $\mathbb{Z}$ does not admit a projective resolution of finite length over $\mathbb{Z}G$.

3. If G has torsion (i.e., non-trivial elements of finite order), show that $\mathbb{Z}$ does not admit a projective resolution of finite length over $\mathbb{Z}G$. [Hint: This follows from exercise 2.]

4 Topological Interpretation

If Y is a $K(G, 1)$-complex with universal cover X, then we know (I.4.2) that $C_*(X)$ is a free resolution of $\mathbb{Z}$ over $\mathbb{Z}G$. Since $C_*(X)_G \approx C_*(Y)$ by 2.4 above, we obtain:

(4.1) Proposition. *If Y is a $K(G, 1)$-complex then $H_*G \approx H_*Y$.*

[In some treatments of the homology theory of groups, this result is taken as the *definition* of $H_* G$, as we indicated in the introduction.]

EXAMPLES

1. Let Y be a bouquet of circles, indexed by a set S. Then, as we saw in I.4.3, Y is a $K(F(S), 1)$. Hence

$$H_i(F(S)) = H_i Y = \begin{cases} \mathbb{Z} & i = 0 \\ \mathbb{Z}S\,(=F(S)_{ab}) & i = 1 \\ 0 & i > 1 \end{cases}$$

2. Let g be an integer ≥ 1 and let $G = \langle a_1, \ldots, a_g, b_1, \ldots, b_g; \prod_{i=1}^{g} [a_i, b_i] \rangle$, i.e., G is the group with generators $a_1, \ldots, a_g, b_1, \ldots, b_g$ and the single defining relation $\prod_{i=1}^{g} [a_i, b_i] = 1$. Thus G is the fundamental group of the closed orientable surface Y of genus g (cf. Massey [1967], ch. 4, §5), and I claim that Y is a $K(G, 1)$. To see this, we need only note that the universal cover X of Y is a *non-compact* surface, since G is infinite. Consequently, one knows from the homology theory of manifolds (cf. Dold [1972], ch. VIII, §3) that $H_i X = 0$ for $i \geq 2$, so Y is a $K(G, 1)$. [Alternatively, one can explicitly exhibit X as the hyperbolic plane tiled by $4g$-sided polygons (cf. Siegel [1971], ch. 3, §9), hence X is contractible.] Thus

$$H_i G = H_i Y = \begin{cases} \mathbb{Z} & i = 0, 2 \\ \mathbb{Z}^{2g} & i = 1 \\ 0 & i > 2. \end{cases}$$

The interested reader can similarly treat the group $\langle c_1, \ldots, c_k; \prod_{i=1}^{k} c_i^2 \rangle$ ($k \geq 2$), which is the fundamental group of the non-orientable closed surface with k crosscaps. Finally, we remark that non-compact surfaces and surfaces with boundary are also easily seen to be $K(G, 1)$'s, but we obtain no new examples in this way since the fundamental groups are free.

3. The surface groups just considered are examples of *one-relator* groups. Suppose now that $G = \langle S; r \rangle$ is an arbitrary one-relator group, i.e., G is the quotient of a free group $F(S)$ by the normal closure of a single element r. Let Y be the 2-complex $(\bigvee_{s \in S} S_s^1) \bigcup_r e^2$ obtained from the bouquet of circles $\bigvee S_s^1$ by attaching a 2-cell via the map $S^1 \to \bigvee S_s^1$ corresponding to r. Then $\pi_1 Y = G$ (cf. Massey [1967], ch. 7, §2). If r is not a power u^n ($n > 1$) in $F(S)$, then a deep theorem of Lyndon implies that Y is a $K(G, 1)$; proofs can be found in Lyndon [1950], Dyer–Vasquez [1973], and Lyndon–Schupp [1977] (ch. III, §§9–11). (If r is a power, on the other hand, then one can show that G has torsion, so that there cannot exist a finite-dimensional $K(G, 1)$, cf. exercise 3 of §3 above.) The chain complex $C_* Y$ is easily seen to have the form

$$\mathbb{Z} \xrightarrow{\partial} \mathbb{Z}S \xrightarrow{0} \mathbb{Z},$$

where $\partial(1)$ is the image of r in $\mathbb{Z}S = F(S)_{ab}$. Hence $H_0 G = \mathbb{Z}$, $H_1 G = G_{ab}$,

$$H_2 G = \begin{cases} \mathbb{Z} & \text{if } r \in [F(S), F(S)] \\ 0 & \text{otherwise,} \end{cases}$$

and $H_i G = 0$ for $i > 2$.

4. If $G = \mathbb{Z}^n$, the free abelian group of rank n, then the n-dimensional torus $Y = S^1 \times \cdots \times S^1$ (n factors) is a $K(G, 1)$ since its universal cover $\mathbb{R}^n$ is contractible. Hence $H_i G$ is a free abelian group of rank $\binom{n}{i} = n!/i!(n-i)!$.

Example 4 can be described in terms of the embedding of the group $G = \mathbb{Z}^n$ as a discrete subgroup of the Lie group $L = \mathbb{R}^n$. Indeed, the $K(G, 1)$ $Y = S^1 \times \cdots \times S^1$ is simply the quotient L/G. Our remaining examples will further illustrate this method of constructing $K(G, 1)$'s. These examples will require somewhat more effort to read than the previous ones, and they will not be referred to again until Chapter VIII; the reader may therefore want to simply glance at them now and read them more carefully later.

5. Let G be the $n \times n$ strict upper triangular group over $\mathbb{Z}$, i.e., G is the group of $n \times n$ integral matrices with 1's on the diagonal and 0's below the diagonal. Let L be the $n \times n$ strict upper triangular group over $\mathbb{R}$. Then G is a discrete subgroup of L, and we can form the coset space L/G. Since G is discrete, the projection $L \to L/G$ is a covering map. Indeed, let U be a neighborhood of 1 in L such that $U \cap G = \{1\}$, and let V be a neighborhood of 1 such that $V^{-1}V \subseteq U$. Then for any $l \in L$, the neighborhood $W = lV$ of l has the property that its transforms Wg ($g \in G$) are disjoint; our assertion follows at once. Finally, L is obviously homeomorphic to Euclidean space $\mathbb{R}^d$, $d = n(n-1)/2$, so the manifold L/G is a $K(G, 1)$ and $H_* G \approx H_*(L/G)$. [Strictly speaking, we should verify that L/G admits a CW-structure, since 4.1 was proved only for $K(G, 1)$-*complexes*. This is in fact true by Whitehead's triangulation theorem for smooth manifolds (cf. Munkres [1966], ch. II), but an easier way to deal with the problem is to simply observe that the proof of 4.1 goes through with no difficulty in the context of *singular* homology theory; see exercise 1 below.]

6. Let L be the Lie group $GL_n(\mathbb{R})$. In contrast to the Lie groups considered in examples 4 and 5, L is *not* contractible. Nevertheless, there is a contractible manifold X associated to L which can be used to study the homology of discrete subgroups of L. In order to describe X, we need to recall some elementary linear algebra.

Recall that a *quadratic form* on $\mathbb{R}^n$ is a function $Q: \mathbb{R}^n \to \mathbb{R}$ of the form

$$Q(x_1, \ldots, x_n) = \sum_{i,j=1}^{n} a_{ij} x_i x_j.$$

The matrix $A = (a_{ij})$ can be taken to be symmetric, and it is then uniquely determined by Q. We will often identify Q with A. The form Q and the matrix A are called *positive definite* if $Q(x) > 0$ for all $x \neq 0$ in $\mathbb{R}^n$. The set of positive definite symmetric matrices is a convex open subset of the space of all symmetric $n \times n$ matrices.

We now define X to be the space of positive definite quadratic forms on $\mathbb{R}^n$, topologized as a subspace of the space of symmetric $n \times n$ matrices. It is clear from what we have said above that X is a contractible manifold of dimension $d = n(n+1)/2$. [In fact, $X \approx \mathbb{R}^d$.] There is a right action of $L = GL_n(\mathbb{R})$ on $\mathbb{R}^n$ by right matrix multiplication (where an element of $\mathbb{R}^n$ is thought of as a row vector), and this induces a left action of L on X, said to be given by "change of variable":

$$(gQ)(x) = Q(xg)$$

for $Q \in X, g \in L, x \in \mathbb{R}^n$. [In terms of symmetric matrices, this action takes the form

$$g \cdot A = gAg^t,$$

where g^t is the transpose of g.] It is well-known that any $Q \in X$ is equivalent under change of variable to the standard form $Q_0 = \sum x_i^2$, so the action of L on X is transitive. Moreover, the isotropy group L_{Q_0} is the orthogonal group $K = O_n(\mathbb{R})$. We therefore have a bijection

$$X \approx L/K$$

of left L-spaces, which can be shown to be a homeomorphism.

Since K is compact, it follows that the action of L on X is *proper*, i.e., that the following condition is satisfied: For every compact set $C \subseteq X$, $\{g \in L: gC \cap C \neq \varnothing\}$ is a compact subset of L. [This says, roughly speaking, that the transforms gC of any compact set C tend to ∞ in X as $g \to \infty$ in L.]

Suppose now that G is a *discrete* subgroup of L. Then for any compact $C \subseteq X$, $\{g \in G: gC \cap C \neq \varnothing\}$ is finite. One deduces easily that the isotropy group G_x of any $x \in X$ is finite and that x has a neighborhood U such that $gU \cap U = \varnothing$ for $g \in G - G_x$.

Finally, suppose further that G is *torsion-free*. Then the finite isotropy groups G_x must be trivial, and it follows at once that the projection $X \to X/G$ is a regular covering map with group G. Thus X/G is a $K(G, 1)$ and $H_*G \approx H_*(X/G)$.

This discussion does not apply to $GL_n(\mathbb{Z})$, the most obvious discrete subgroup of $GL_n(\mathbb{R})$, because it is not torsion-free. But $GL_n(\mathbb{Z})$ does have torsion-free subgroups G of finite index (cf. exercise 3 below), and our discussion applies to them. Unfortunately, it is extremely difficult in practice to actually compute $H_*(X/G)$.

7. Consider now the special linear group $SL_n(\mathbb{R}) = \{g \in GL_n(\mathbb{R}): \det g = 1\}$. We can then replace the space X of example 6 by a contractible manifold X_0 on which $SL_n(\mathbb{R})$ acts properly, with $\dim X_0 = \dim X - 1$. Namely, we take X_0 to be the quotient space of X obtained by identifying two quadratic forms which are (positive) scalar multiples of one another. One can verify that

$$X_0 \approx SL_n(\mathbb{R})/SO_n(\mathbb{R}),$$

so that the $SL_n(\mathbb{R})$-action on X_0 is proper, $SO_n(\mathbb{R})$ being compact. Moreover, X_0 is a contractible manifold; in fact, the interested reader can verify that X_0 is diffeomorphic to Euclidean space of dimension $n(n+1)/2 - 1$. As in example 6, then, we have $H_* G \approx H_*(X_0/G)$ for any discrete, torsion-free subgroup $G \subset SL_n(\mathbb{R})$. We note that if $n = 2$ the space X_0 can be identified with the upper half plane $\mathscr{H} \subset \mathbb{C}$, with $SL_2(\mathbb{R})$ acting by linear fractional transformations:

$$\begin{pmatrix} a & b \\ c & d \end{pmatrix} z = \frac{az+b}{cz+d}.$$

Indeed, one checks that this defines a transitive action of $SL_2(\mathbb{R})$ on $\mathscr{H}$ and that the isotropy group at $z = i$ is $SO_2(\mathbb{R})$; this leads easily to the desired homeomorphism

$$\mathscr{H} \approx SL_2(\mathbb{R})/SO_2(\mathbb{R}) \approx X_0.$$

8. Finally, we state the general facts which were illustrated in the previous examples. Let L be a Lie group with only finitely many connected components. Then L has a maximal compact subgroup K (unique up to conjugacy), and the homogeneous space $X = L/K$ is diffeomorphic to $\mathbb{R}^d$, $d = \dim L - \dim K$. (Proofs can be found, for example, in Hochschild [1965], ch. XV.) Consequently, if $G \subset L$ is a discrete, torsion-free subgroup, then the quotient manifold X/G is a $K(G, 1)$ and $H_* G \approx H_*(X/G)$.

Exercises

1. Let Y be a path-connected space. If Y has a contractible, regular covering space X with covering group G, show that $H_* Y \approx H_* G$. [Hint: The singular chain complex $C_*^{\text{sing}}(X)$ provides a free resolution of $\mathbb{Z}$ over $\mathbb{Z}G$ and $C_*^{\text{sing}}(X)_G \approx C_*^{\text{sing}}(Y)$.]

2. (This exercise requires some elementary homotopy theory.) Let Y be a $K(G, 1)$*-space*, i.e., a path-connected space with $\pi_1 Y = G$ and $\pi_i Y = 0$ for $i > 1$. Prove that $H_* Y \approx H_* G$. [Hint: This is clear if Y is a CW-complex; in the general case, replace Y by a CW-complex which is weakly homotopy equivalent to Y, cf. Spanier [1966], §7.8. Alternatively, one can directly construct a chain homotopy equivalence $C_*^{\text{sing}}(Y) \simeq C_* G$, cf. Eilenberg–MacLane [1945].] Note that exercise 1 is a special case of the present exercise.

3. Fix an integer $n \geq 1$. For any integer $N \geq 2$, let $\Gamma(N)$ be the kernel of the canonical map $GL_n(\mathbb{Z}) \to GL_n(\mathbb{Z}/N\mathbb{Z})$, i.e., $\Gamma(N) = \{g \in GL_n(\mathbb{Z}) : g \equiv 1 \bmod N\}$, where 1 denotes the identity matrix. The group $\Gamma(N)$ is called the *principal congruence subgroup of* $GL_n(\mathbb{Z})$ *of level* N. Note that $\Gamma(N)$ has finite index in $GL_n(\mathbb{Z})$, since $GL_n(\mathbb{Z}/N\mathbb{Z})$ is finite. The purpose of this exercise is to prove that $\Gamma(N)$ is torsion-free for $N \geq 3$.

   (a) Let p be a fixed prime and let A be an $n \times n$ matrix of integers such that $A \equiv 1 \bmod p$. If $A \neq 1$, then there is a unique positive integer $d = d(A)$ such that

$$A \equiv 1 \bmod p^d \quad \text{and} \quad A \not\equiv 1 \bmod p^{d+1}.$$

Show that $d(A^q) = d(A)$ for any prime $q \neq p$. If p is odd or $d(A) \geq 2$, show that $d(A^p) = d(A) + 1$. [Hint: Write $A = 1 + p^d B$ with $B \not\equiv 0 \bmod p$, and look at the binomial expansion of $(1 + p^d B)^l$, $l = p$ or q.]

(b) Deduce that $\Gamma(N)$ is torsion-free for $N \geq 3$ and that $\Gamma(2)$ has only 2-torsion. [Hint: Suppose $A \in \Gamma(N)$ has prime order, and apply (a) with p a prime divisor of N.]

5 Hopf's Theorems

The purpose of this section is to prove the results stated as 0.1 and 0.2 in the introduction. We will need to use the Hurewicz theorem (cf. Spanier [1966], ch. 7, §5), which says that if $\pi_i X = 0$ for $i < n$ (where $n \geq 2$), then $H_i X = 0$ for $0 < i < n$ and the Hurewicz map $h\colon \pi_n X \to H_n X$ is an isomorphism. (In fact, an examination of our proofs will show that we only need the surjectivity of h, which is considerably easier to prove; indeed, it follows directly from Spanier's Thm. 7.4.8.)

We begin by observing that for any group G and integer n one can compute $H_i G$ for $i \leq n + 1$ from a *partial* projective resolution of length n:

(5.1) Lemma. *Let $F_n \to \cdots \to F_0 \to \mathbb{Z} \to 0$ be an exact sequence of $\mathbb{Z}G$-modules where each F_i is projective. Then $H_i G \approx H_i(F_G)$ for $i < n$ and there is an exact sequence*

$$0 \to H_{n+1}(G) \to (H_n F)_G \to H_n(F_G) \to H_n(G) \to 0.$$

PROOF. Extend F to a full resolution F^+ by choosing a projective module F'_{n+1} mapping onto $H_n F$, etc.:

$$\cdots \to F'_{n+2} \to F'_{n+1} \to F_n \to F_{n-1} \to \cdots \to F_0 \to \mathbb{Z} \to 0.$$
$$\searrow \quad \nearrow$$
$$H_n F$$

It is easy to see (either by direct inspection or by considering the homology exact sequence associated to the exact sequence $0 \to F_G \to F_G^+ \to F_G^+/F_G \to 0$ of chain complexes) that $H_i(F_G^+) = H_i(F_G)$ for $i < n$ and that there is an exact sequence

$$0 \to H_{n+1}(F_G^+) \to A \to H_n(F_G) \to H_n(F_G^+) \to 0,$$

where $A = \operatorname{coker}\{(F'_{n+2})_G \to (F'_{n+1})_G\}$. By 2.2 this cokernel can be identified with $(H_n F)_G$, whence the lemma. □

(5.2) Theorem. *For any connected CW-complex Y there is a canonical map $\psi\colon H_* Y \to H_* \pi$ ($\pi = \pi_1 Y$). If $\pi_i Y = 0$ for $1 < i < n$ (for some $n \geq 2$) then ψ is an isomorphism $H_i Y \xrightarrow{\approx} H_i \pi$ for $i < n$, and the sequence*

$$\pi_n Y \xrightarrow{h} H_n Y \xrightarrow{\psi} H_n \pi \to 0$$

is exact.

(Note that the hypothesis always holds with $n = 2$, so 5.2 yields the result 0.1 stated in the introduction.)

Proof. Let X be the universal cover of Y and let F be a projective resolution of $\mathbb{Z}$ over $\mathbb{Z}\pi$. Since $C_*(X)$ is a complex of free $\mathbb{Z}\pi$-modules augmented over $\mathbb{Z}$, the fundamental lemma (I.7.4) gives us a chain map (over $\mathbb{Z}\pi$) $C_*(X) \to F$, well-defined up to homotopy. Taking co-invariants we obtain a map $C_*(Y) \to F_\pi$, which induces the desired map $\psi\colon H_* Y \to H_* \pi$. One knows that $\pi_i X \xrightarrow{\approx} \pi_i Y$ for $i > 1$ (cf. Spanier [1966], 7.3.7), so our hypothesis implies that $\pi_i X = 0$ for $i < n$. Hence $H_i X = 0$ for $0 < i < n$ and the Hurewicz map $h\colon \pi_n X \to H_n X$ is an isomorphism. We therefore have a partial free resolution

$$C_n(X) \to \cdots \to C_0(X) \to \mathbb{Z} \to 0,$$

whose n-th homology group is the group $Z_n X$ of n-cycles of X. Lemma 5.1 now implies that $H_i Y \xrightarrow{\approx} H_i \pi$ for $i < n$ and that there is an exact sequence

$$Z_n X \to Z_n Y \xrightarrow{\tilde{\psi}} H_n \pi \to 0.$$

The map $\tilde{\psi}$ which arises here is easily seen to be the composite $Z_n Y \to H_n Y \xrightarrow{\psi} H_n \pi$. Consequently the sequence

$$H_n X \to H_n Y \xrightarrow{\psi} H_n \pi \to 0$$

is exact, and the desired exact sequence

$$\pi_n Y \xrightarrow{h} H_n Y \xrightarrow{\psi} H_n \pi \to 0$$

now follows from the diagram

$$\begin{array}{ccc} \pi_n X & \xrightarrow[\approx]{h} & H_n X \\ {\scriptstyle\approx}\downarrow & & \downarrow \\ \pi_n Y & \xrightarrow{h} & H_n Y. \end{array}$$

□

We turn now to formula 0.2 of the introduction.

(5.3) Theorem. *If* $G = F/R$ *where* F *is free, then* $H_2 G \approx R \cap [F, F]/[F, R]$.

Proof. Let $F = F(S)$, let Y be a bouquet of circles indexed by S, and let $\tilde{Y}$ be the connected regular covering space of Y corresponding to the normal subgroup R of $F(S) = \pi_1 Y$. Choosing a basepoint $\tilde{v}$ in $\tilde{Y}$ lying over the vertex of Y, we identify $G = F/R$ with the group of covering transformations of $\tilde{Y}$ as in Chapter I, Appendix, Al. For any $f \in F$ we regard f as a combinatorial path in Y and we denote by $\tilde{f}$ the lifting of f to $\tilde{Y}$ starting at $\tilde{v}$. [By a *combinatorial path* in a CW-complex we mean a sequence $e_1, \ldots, e_n$ of oriented 1-cells such that the initial vertex of e_{i+1} is equal to the final vertex of e_i for $i = 1, \ldots, n - 1$.] This path $\tilde{f}$, then, ends at the vertex $\bar{f}\tilde{v}$, where $\bar{f}$ is the image of f in G.

The complex $C_* \tilde{Y}$ is a complex of free $\mathbb{Z}G$-modules, and it provides a partial resolution $C_1\tilde{Y} \to C_0\tilde{Y} \to \mathbb{Z} \to 0$. We can therefore apply 5.1 to obtain $H_2 G \approx \ker\{(H_1\tilde{Y})_G \to H_1 Y\}$. Now $H_1\tilde{Y} \approx (\pi_1\tilde{Y})_{ab} \approx R_{ab}$, and I claim that the composite isomorphism $H_1\tilde{Y} \approx R_{ab}$ is an isomorphism of G-modules, where the G-action on R_{ab} is induced by the conjugation action of F on R. [This makes sense because the conjugation action of R on itself induces the trivial action of R on R_{ab}.] To verify the claim, one checks the definitions and finds that the isomorphism $R_{ab} \xrightarrow{\approx} H_1\tilde{Y}$ is induced by a map $d: R \to H_1\tilde{Y}$ defined as follows: For any $r \in R$ the lifting $\tilde{r}$ is a closed path in $\tilde{Y}$; taking the sum of the oriented 1-cells which occur in $\tilde{r}$, we obtain a 1-cycle in $\tilde{Y}$, hence an element of $H_1\tilde{Y}$, and this element is by definition dr. Now if $f \in F$ and $r \in R$, the lifting of frf^{-1} is the path

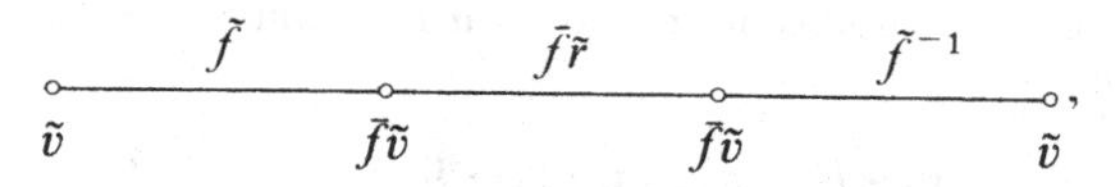

hence $d(frf^{-1}) = \bar{f}dr$. This shows that the isomorphism $R_{ab} \xrightarrow{\approx} H_1\tilde{Y}$ is an isomorphism of G-modules, as claimed.

We therefore have a diagram

$$\begin{array}{ccccc} (H_1\tilde{Y})_G & \approx & (R_{ab})_G & = & R/[F, R] \\ \downarrow & & \downarrow & & \downarrow \\ H_1 Y & \approx & F_{ab} & = & F/[F, F] \end{array}$$

where the second and third vertical arrows are induced by the inclusion $R \hookrightarrow F$. Hence $H_2 G \approx \ker\{R/[F, R] \to F/[F, F]\} = R \cap [F, F]/[F, R]$. □

(The G-module R_{ab} which occurred in the above proof is called the *relation module* associated to the presentation of G as F/R.)

Note that the chain complex $C_*\tilde{Y}$ can be described explicitly, exactly as in I.4.3. We therefore obtain, as a corollary of the above proof, the following result:

(5.4) Proposition. *If $G = F(S)/R$ then there is an exact sequence*

$$0 \to R_{ab} \xrightarrow{\theta} \mathbb{Z}G^{(S)} \xrightarrow{\partial} \mathbb{Z}G \xrightarrow{\varepsilon} \mathbb{Z} \to 0$$

of G-modules, where $\mathbb{Z}G^{(S)}$ is free with basis $(e_s)_{s \in S}$, and $\partial e_s = \bar{s} - 1$.

(Here $\bar{s}$, as usual, denotes the image of s in G.)

The map θ which occurs here has been described explicitly in the proof of 5.3 in terms of path lifting. We will see in exercise 3d below that θ can also be described purely algebraically, in terms of the "free differential calculus."

Remarks

1. It is possible to give an algebraic proof of 5.4 (see exercise 4d of §IV.2) and then deduce Hopf's formula 5.3 directly from 5.4.

2. Hopf's formula suggests that, roughly speaking, H_2G consists of commutator relations $\prod [a_i, b_i] = 1$ in G, modulo those relations that hold trivially. See C. Miller [1952] for a precise formulation and proof of this statement.

EXERCISES

1. In the situation of 5.2, suppose in addition that Y is n-dimensional; prove that there is an exact sequence

$$0 \to H_{n+1}\pi \to (\pi_n Y)_\pi \to H_n Y \to H_n \pi \to 0.$$

Here $(\pi_n Y)_\pi$ makes sense because π acts on $\pi_n Y$, cf. Spanier [1966], ch. 7, sec. 3; this action corresponds to the obvious action of π on $H_n X$ (X = universal cover of Y) under the isomorphism $\pi_n Y \overset{\approx}{\leftarrow} \pi_n X \overset{\approx}{\rightarrow} H_n X$. [Remark: Without any dimension restrictions on Y one can prove that there is an exact sequence

$$H_{n+1} Y \to H_{n+1}\pi \to (\pi_n Y)_\pi \to H_n Y \to H_n \pi \to 0.$$

This will be proved in exercise 6 of §VII.7 by a spectral sequence argument; the reader may want to try to give a direct proof now.]

2. Let $G = \langle S; r_1, r_2, \ldots \rangle$, i.e., $G = F(S)/R$ where R is the normal closure in $F(S)$ of $r_1, r_2, \ldots$.

(a) Show that the relation module R_{ab} is generated (as G-module) by the images of $r_1, r_2, \ldots$.

(b) Show that these elements *freely* generate R_{ab} as $\mathbb{Z}G$-module if and only if the 2-complex associated to the given presentation of G is a $K(G, 1)$. (By the 2-complex associated to the presentation we mean the complex $(\bigvee_{s\in S} S_s^1) \bigcup_{r_1} e_1^2 \bigcup_{r_2} e_2^2 \cup \cdots$, where the 2-cell e_i^2 is attached to $\bigvee S_s^1$ by the map $S^1 \to \bigvee S_s^1$ corresponding to $r_i \in \pi_1(\bigvee S_s^1) = F(S)$.) Thus, for example, Lyndon's theorem about one-relator groups $G = \langle S; r \rangle$ which we stated in example 3 of §4 can be interpreted as saying that R_{ab} is freely generated by the image of r, provided r is not a power. (It is in this form, in fact, that Lyndon stated and proved his theorem.)

*(c) Let $G = \langle S; r \rangle$ be an *arbitrary* one-relator group and write $r = u^n$ in $F(S)$, where $n \geq 1$ is maximal. One can show (cf. Lyndon–Schupp [1977], IV.5.2) that the image t of u in G has order exactly n, and we denote by C the cyclic group of order n generated by t. If $n > 1$ then R_{ab} is not freely generated by r mod $[R, R]$, since this generator is clearly fixed by C. But Lyndon [1950] proved that no other relations hold, i.e., that the obvious surjection $\mathbb{Z}[G/C] \to R_{ab}$ is an isomorphism. Show that this result can be interpreted topologically, as follows. Let Y, $\tilde{Y}$, and $\tilde{v}$ be as in the

proof of 5.3 and consider the lifting $\tilde{r}$ of the loop r in Y. Since $r = u^n$ and $\tilde{u}$ ends at $t\tilde{v}$, $\tilde{r}$ is the composite path pictured schematically as follows:

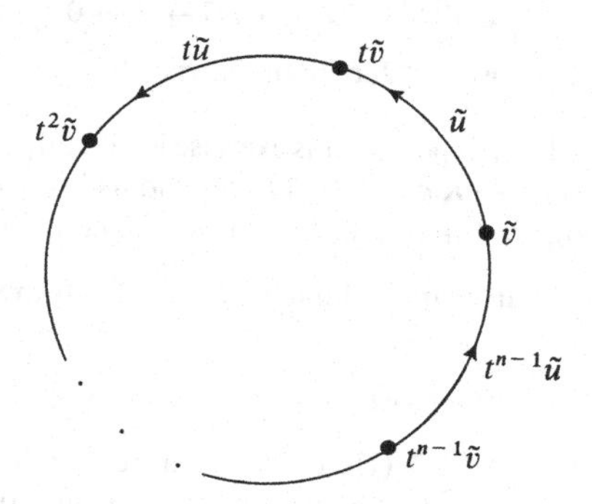

Thus the map $S^1 \to \tilde{Y}$ corresponding to $\tilde{r}$ is compatible with the action of the cyclic group C, where C acts on S^1 as a group of rotations as in §I.6. Deduce that we can form a 2-dimensional G-complex X by attaching 2-cells to $\tilde{Y}$ along the loops $g\tilde{r}$, where g ranges over a set of representatives for the cosets G/C; if σ is the 2-cell attached along $\tilde{r}$, then the isotropy group G_σ is equal to C, with C acting on σ as a group of rotations. Show that Lyndon's theorem about R_{ab} stated above is equivalent to the statement that X is contractible. [Note: In the terminology of Lyndon–Schupp [1977], ch. III, X is the *Cayley complex* associated to the presentation $\langle S; r\rangle$.]

3. In this exercise you will construct the Fox "free derivatives" by using the ideas introduced in the proof of 5.3. See the exercises in §IV.2 for a purely algebraic treatment of the same results. If G is a group and M a G-module, then a *derivation* (or *crossed homomorphism*) from G to M is a function $d\colon G \to M$ such that $d(gh) = dg + gdh$ for all $g, h \in G$.

   (a) Let the notation be as in 5.3 and its proof. Show that the definition of dr given in that proof can be used (almost verbatim) to define a function $d\colon F \to C_1\tilde{Y}$ which satisfies $d(f_1 f_2) = df_1 + \bar{f}_1 df_2$ for all $f_1, f_2 \in F$. Thus if we regard the G-module $C_1\tilde{Y}$ as an F-module via the canonical homomorphism $F \to G$, then $d\colon F \to C_1\tilde{Y}$ is a derivation.

   (b) For any free group $F = F(S)$, show that there is an F-module Ω which admits a derivation $d\colon F \to \Omega$ such that Ω is a free $\mathbb{Z}F$-module with basis $(ds)_{s\in S}$. [Hint: Apply (a) with $R = \{1\}$.] We call df for $f \in F$ the *total (free) derivative* of f. The coefficient of ds when df is expressed in terms of the basis (ds) is called the *partial derivative* of f with respect to s and is denoted $\partial f/\partial s$; thus $df = \sum_{s\in S} (\partial f/\partial s)ds$, where $\partial f/\partial s \in \mathbb{Z}F$. Show that $\partial/\partial s\colon F \to \mathbb{Z}F$ is a derivation and satisfies $\partial t/\partial s = \delta_{s,t}$ $(t \in S)$. These properties completely characterize $\partial/\partial s$. (For example, if $S = \{s, t\}$, you should be able to use these properties to compute $\partial(ts^{-1}ts^2)/\partial s$.)

   (c) With F as in (b), show that *any* derivation $d\colon F \to M$, where M is an F-module, satisfies $df = \sum_{s\in S} (\partial f/\partial s)ds$.

   (d) Show that the map $\theta\colon R_{ab} \to \mathbb{Z}G^{(S)}$ in 5.4 is induced by the map $R \to \mathbb{Z}G^{(S)}$ given by $r \mapsto \sum_{s\in S} (\overline{\partial r/\partial s})e_s$ where $\overline{\partial r/\partial s}$ is the image of $\partial r/\partial s$ under the canonical map

$\mathbb{Z}F \to \mathbb{Z}G$. If R is the normal closure of a subset $T \subseteq F$, deduce that there is a partial free resolution

$$\mathbb{Z}G^{(T)} \xrightarrow{\partial_2} \mathbb{Z}G^{(S)} \xrightarrow{\partial_1} \mathbb{Z}G \xrightarrow{\varepsilon} \mathbb{Z} \to 0$$

such that the matrix of ∂_2 is the "Jacobian matrix" $(\overline{\partial t/\partial s})_{t \in T, s \in S}$.

*4. Let $G = F/R$ as in 5.3. The purpose of this exercise is to establish explicit formulas for the isomorphism $\varphi: H_2 G \xrightarrow{\approx} R \cap [F, F]/[F, R]$ and its inverse, where we view $H_2 G$ as the second homology group of the standard chain complex $C_*(G)$.

(a) Exhibit a specific chain map in dimensions ≤ 2 from the bar resolution to the partial resolution

$$(*) \qquad \mathbb{Z}G^{(R)} \to C_1\tilde{Y} \to C_0\tilde{Y} \to \mathbb{Z} \to 0,$$

where $\tilde{Y}$ is as in the proof of 5.3. (Here $\mathbb{Z}G^{(R)}$ is a free module with basis $(e_r)_{r \in R}$; it maps onto $R_{ab} = H_1\tilde{Y} \subset C_1\tilde{Y}$ in the obvious way.) Deduce that φ can be computed as follows: Choose for each $g \in G$ an element $f(g) \in F$ such that $\overline{f(g)} = g$. Given $g, h \in G$, write $f(g)f(h) = f(gh)r(g, h)$ where $r(g, h) \in R$. Then there is an abelian group homomorphism $C_2(G) \to R_{ab}$ given by $[g|h] \mapsto r(g, h) \bmod [R, R]$, and this induces the isomorphism $\varphi: H_2 G \to R \cap [F, F]/[F, R]$ by passage to subquotients.

(b) Exhibit a chain map from $(*)$ to the bar resolution, and deduce that φ^{-1} can be computed as follows: Let $D: F \to C_2(G)$ be the unique derivation such that $Ds = [1|\bar{s}]$ for each free generator $s \in S$, where the group $C_2(G)$ is regarded as an F-module by $f \cdot [g|h] = [\bar{f}g|h]$. (Explicitly, $Df = \sum_{s \in S} [\overline{\partial f/\partial s}|\bar{s}]$, where the symbol $[\cdot|\cdot]$ is understood to be $\mathbb{Z}$-bilinear.) Then $(D|R): R \to C_2(G)$ is a homomorphism which induces φ^{-1} by passage to subquotients.

(c) Let $a_1, \ldots, a_g, b_1, \ldots, b_g$ be elements of F such that the element $r = \prod_{i=1}^{g} [a_i, b_i]$ is in R. Then $\varphi^{-1}(r \bmod [F, R])$ is represented by the cycle

$$\sum_{i=1}^{g} \{[I_{i-1}|\bar{a}_i] + [I_{i-1}\bar{a}_i|\bar{b}_i] - [I_{i-1}\bar{a}_i\bar{b}_i\bar{a}_i^{-1}|\bar{a}_i] - [I_i|\bar{b}_i]\},$$

where $I_i = [\bar{a}_1, \bar{b}_1] \cdots [\bar{a}_i, \bar{b}_i]$. [Hint: It suffices to prove this in the universal example where F is the free group on $a_1, \ldots, a_g, b_1, \ldots, b_g$ and R is the normal closure of r. In this case, $\varphi^{-1}(r \bmod [F, R])$ is easily computed by the formula of (b).]

5. (a) Let G be a group which admits a presentation with n generators and m relations. Let $r = \mathrm{rk}_{\mathbb{Z}}(G_{ab}) = \dim_{\mathbb{Q}}(\mathbb{Q} \otimes G_{ab})$. Prove that the abelian group $H_2 G$ can be generated by $m - n + r$ elements. [Hint: Let Y be the 2-complex associated to the presentation. Computing the Euler characteristic $\chi(Y)$ in two different ways, one finds $1 - n + m = 1 - \mathrm{rk}(H_1 Y) + \mathrm{rk}(H_2 Y) = 1 - r + \mathrm{rk}(H_2 Y)$, whence $\mathrm{rk}(H_2 Y) = m - n + r$. Now $H_2 Y$ is a free abelian group, being a subgroup of the group of cellular 2-chains of Y, and we have a surjection $H_2 Y \twoheadrightarrow H_2 G$.]

Parts (b) and (c) below illustrate typical applications of (a).

(b) Let G be a perfect group (i.e., a group such that $G_{ab} = 0$) which admits a finite presentation with the same number of generators as relations. Prove that $H_2 G = 0$.

(c) Let G be a perfect group such that $H_2 G = \mathbb{Z}_2 \oplus \mathbb{Z}_2$. Show that any n-generator presentation of G must involve at least $n + 2$ relations.

6. (a) By a *group extension* we mean a short exact sequence $1 \to N \to G \to Q \to 1$ of groups. Deduce from Hopf's formula that such an extension gives rise to a 5-term exact sequence

$$H_2G \xrightarrow{\alpha} H_2Q \xrightarrow{\beta} (H_1N)_Q \xrightarrow{\gamma} H_1G \xrightarrow{\delta} H_1Q \to 0,$$

where the Q-action on $H_1N = N_{ab}$ is induced by the conjugation action of G on N, and γ and δ are induced by the maps $N \hookrightarrow G \twoheadrightarrow Q$. [Hint: Write $G = F/R$ and $Q = F/S$, where $R \subseteq S \subseteq F$. The desired sequence is then

$$\frac{R \cap [F, F]}{[F, R]} \to \frac{S \cap [F, F]}{[F, S]} \to \frac{N}{[G, N]} \to \frac{G}{[G, G]} \to \frac{G}{N \cdot [G, G]} \to 0,$$

which is easily proved to be exact.] Remark: We will give another derivation of this 5-term exact sequence, and a generalization of it, in §VII.6. It has interesting applications to the study of the lower central series, due independently to Stallings [1965a] and Stammbach [1966]. See also Stammbach [1973], Chapter IV, for further developments along these lines.

(b) Conversely, show that Hopf's formula can be deduced from the 5-term exact sequence. [Apply (a) to $1 \to R \to F \to G \to 1$ and recall that $H_2F = 0$.]

*7. (a) Recall that the 3-sphere S^3 has a group structure. [It is the multiplicative group of quaternions of norm 1.] If G is a finite subgroup of S^3, deduce from the results of this section that $H_2G = 0$. [Hint: S^3/G is a closed, orientable 3-manifold with finite fundamental group, hence $H_2(S^3/G) = 0$.] Remark: A list of the groups G to which this applies is given in Wolf [1974] and recalled briefly in example 2 of VI.9.2 below. We will also be able to give there a more elementary solution of the present exercise, cf. exercise 3 of §VI.9.

(b) One of the most interesting groups to which (a) applies is the "binary icosahedral group." This is a group G of order 120 which maps onto A_5, the alternating group on 5 letters, with central kernel of order 2. Moreover, G is perfect. Using these facts, the result of (a), and the 5-term exact sequence, deduce that $H_2(A_5) = \mathbb{Z}_2$.

(c) (R. Strebel) It is known that A_5 admits the presentation $\langle x, y, z; x^2 = y^3 = z^5 = xyz = 1\rangle$. Consider now the abstract group $G = \langle x, y, z; x^2 = y^3 = z^5 = xyz\rangle$, and let C be the cyclic subgroup of G generated by the central element $x^2 = y^3 = z^5 = xyz$ of G. Thus $G/C = A_5$. Determine the order of C and hence that of G. [Hint: From the given presentation of G you can show that G is perfect, hence $H_2G = 0$ by exercise 5b above. The 5-term exact sequence now yields $H_2(A_5) \xrightarrow{\approx} H_1(C) = C$, so C has order 2 and hence G has order 120. In fact, with a little more work you can show that G is the binary icosahedral group.] Similar methods can be used to analyze other abstract groups defined by presentations closely related to presentations of known groups. You might look, for instance, at some of the examples in Coxeter–Moser [1980] in connection with Miller's generalization of the polyhedral groups; the treatment there can be simplified by the use of the method of the present exercise.

Remark. It is clear from this exercise that H_2 is closely related to the theory of central extensions. It would be possible to develop this connection systematically on the basis of Hopf's formula and the 5-term exact sequence, but we will instead deduce it from the general theory of group extensions, to be discussed in Chapter IV. See exercise 7 in §IV.3.

6 Functoriality

The standard chain complex $C_*(G)$ is clearly functorial in G, hence $H_* G$ is a (covariant) functor of G. This functoriality can also be described in terms of arbitrary resolutions, as follows: Given a homomorphism $\alpha: G \to G'$ and projective resolutions F and F' of $\mathbb{Z}$ over $\mathbb{Z}G$ and $\mathbb{Z}G'$, respectively, we can regard F' as a complex of G-modules via α. Then F' is acyclic (although not projective, in general, over $\mathbb{Z}G$), so the fundamental lemma (I.7.4) gives us an augmentation-preserving G-chain map $\tau: F \to F'$, well-defined up to homotopy. The condition that τ be a G-map is expressed by the formula

$$\tau(gx) = \alpha(g)\tau(x) \tag{6.1}$$

for $g \in G$, $x \in F$. Clearly τ induces a map $F_G \to F'_{G'}$, well-defined up to homotopy, hence we obtain a well-defined map $\alpha_*: H_* G \to H_* G'$.

(6.2) Proposition. *Fix $g_0 \in G$ and let $\alpha: G \to G$ be given by $\alpha(g) = g_0 g g_0^{-1}$. Then $\alpha_*: H_* G \to H_* G$ is the identity.*

PROOF. Let F be a projective resolution of $\mathbb{Z}$ over $\mathbb{Z}G$ and define $\tau: F \to F$ by $\tau(x) = g_0 x$. Then τ commutes with the boundary operator and satisfies 6.1, so τ can be used to compute α_*. But clearly τ induces the identity map on F_G, whence the proposition. □

(6.3) Corollary. *If G is a group and N is a normal subgroup, then the conjugation action of G on N induces an action of G/N on $H_* N$.*

EXERCISES

1. Given $N \lhd G$ as in 6.3, let F be a projective resolution of $\mathbb{Z}$ over $\mathbb{Z}G$ and consider the complex F_N. Then F_N is a complex of G/N-modules (§2, exercise 3a), hence $H_*(F_N)$ inherits an action of G/N. Show that $H_*(F_N) = H_* N$ and that the resulting G/N-action on $H_* N$ agrees with that defined in 6.3. [Hint: Given $g \in G$, the action of g on $H_* N$ can be computed via the map $\tau: F \to F$ given by $\tau(x) = gx$.]

2. For any finite set A let $\Sigma(A)$ be the group of permutations of A. If $|A| \leq |B|$ (where $|\cdot|$ denotes cardinality), choose an injection $i: A \hookrightarrow B$ and consider the injection $\Sigma(A) \hookrightarrow \Sigma(B)$ obtained by extending a permutation of A to be the identity on $B - iA$. Show that the induced map $H_* \Sigma(A) \to H_* \Sigma(B)$ is independent of the choice of i. In particular, if $|A| = |B|$, then $H_* \Sigma(A)$ is canonically isomorphic to $H_* \Sigma(B)$.

3. (a) Under the isomorphism $H_1(\ ) \approx (\ )_{ab}$ of §3, show that $H_1(\alpha): H_1(G) \to H_1(G')$ corresponds to the map $G_{ab} \to G'_{ab}$ obtained from α by passage to the quotient; in other words, the isomorphism $H_1(\ ) \approx (\ )_{ab}$ is natural. In particular, the action of G/N on $H_1(N)$ in 6.3 above agrees with that defined in exercise 6 of §5. [Hint: Use the bar resolution to compute $H_1(\alpha)$.]

(b) Prove the following naturality property of Hopf's isomorphism 5.3: Suppose $G = F/R$ and $G' = F'/R'$ with F and F' free, and suppose $\alpha: G \to G'$ lifts to $\tilde{\alpha}: F \to F'$. Then the diagram

$$\begin{array}{ccc} H_2(G) & \approx & R \cap [F, F]/[F, R] \\ {\scriptstyle H_2(\alpha)}\downarrow & & \downarrow \\ H_2(G') & \approx & R' \cap [F', F']/[F', R'] \end{array}$$

commutes, where the right-hand vertical arrow is induced by $\tilde{\alpha}$. [Hint: Let Y and $\tilde{Y}$ be associated to the presentation $G = F/R$ as in the proof of 5.3, and similarly let Y' and $\tilde{Y}'$ correspond to $G' = F'/R'$. Then $\tilde{\alpha}$ yields a map $Y \to Y'$, which yields a map $C_*(\tilde{Y}) \to C_*(\tilde{Y}')$, which can be extended to a map τ of resolutions, which can be used to compute $H_2(\alpha)$.]

Remark. This exercise allows one to interpret in terms of the functoriality of H_1 and H_2 three of the four maps which occur in the 5-term exact sequence of exercise 6 of §5.

7 The Homology of Amalgamated Free Products

As an illustration of the topological interpretation of group homology, we will derive in this section a Mayer–Vietoris sequence for computing the homology of an amalgamated free product. We begin by reviewing the necessary group theory.

Suppose we are given groups G_1, G_2, and A and homomorphisms $\alpha_1: A \to G_1$ and $\alpha_2: A \to G_2$. Eventually we will assume further that α_1 and α_2 are injective, so that A can be viewed as a common subgroup of G_1 and G_2, but for the moment we do not make this assumption. By the *amalgamated free product* (or *amalgamated sum*, or *amalgam*) of G_1 and G_2 along A we mean a group G which fits into a commutative square

(7.1)
$$\begin{array}{ccc} A & \xrightarrow{\alpha_2} & G_2 \\ {\scriptstyle \alpha_1}\downarrow & & \downarrow{\scriptstyle \beta_2} \\ G_1 & \xrightarrow[\beta_1]{} & G \end{array}$$

with the following *universal mapping property*: Given a group H and homomorphisms $\gamma_i: G_i \to H$ $(i = 1, 2)$ with $\gamma_1\alpha_1 = \gamma_2\alpha_2$, there is a unique map $\varphi: G \to H$ such that $\varphi\beta_i = \gamma_i$. We write $G = G_1 *_A G_2$, and we say that the square 7.1 is an *amalgamation* diagram.

The universal property above shows that amalgamation is the group-theoretic analogue of pasting two topological spaces together along a common subspace. The Seifert–van Kampen theorem, which the reader has probably seen in some form, makes the analogy precise via the π_1-functor. We will need the following simple version of that theorem:

(7.2) Theorem. *Let X be a CW-complex which is the union of two connected subcomplexes X_1 and X_2 whose intersection Y is connected and non-empty. Then the square*

$$\begin{array}{ccc} \pi_1 Y & \longrightarrow & \pi_1 X_2 \\ \downarrow & & \downarrow \\ \pi_1 X_1 & \longrightarrow & \pi_1 X \end{array}$$

is an amalgamation diagram, where all fundamental groups are computed at a fixed vertex $y \in Y$ and all maps are induced by inclusions. Thus

$$\pi_1 X = \pi_1 X_1 *_{\pi_1 Y} \pi_1 X_2.$$

This is an easy consequence of the usual combinatorial description of the fundamental group of a CW-complex. Details can be found in Schubert [1968], III.5.8, or Cohen [1978], II.2.3. See also exercise 2 below for a proof based on covering space theory.

We can express this theorem more concisely by saying that the functor

$$\pi_1: (\text{connected, pointed complexes}) \to (\text{groups})$$

preserves amalgamations. In order to study the homology of amalgamations of groups, we would like to have a result going in the other direction, saying that the "functor" $K(-, 1)$: (groups) $\to$ (complexes) preserves amalgamations. This turns out to be true as long as the maps α_1 and α_2 of 7.1 are injective:

(7.3) Theorem (Whitehead). *Any amalgamation diagram* 7.1 *with α_1 and α_2 injective can be realized by a diagram*

$$\begin{array}{ccc} Y & \hookrightarrow & X_2 \\ \downarrow & & \downarrow \\ X_1 & \hookrightarrow & X \end{array}$$

of $K(\pi, 1)$-complexes such that $X = X_1 \cup X_2$ and $Y = X_1 \cap X_2$.

The proof will require three elementary lemmas:

(7.4) Lemma. *If α_1 and α_2 are injective then so are β_1 and β_2. Thus G_1, G_2, and A can be regarded as subgroups of G.*

This is part of the "normal form theorem" for amalgamations. See, for instance, Serre [1977a], Lyndon–Schupp [1977], or Cohen [1978].

(7.5) Lemma. *Let $X' \hookrightarrow X$ be an inclusion of connected CW-complexes such that the induced map $\pi' \to \pi$ of fundamental groups is injective. Let $p: \tilde{X} \to X$ be the universal cover of X. Then each connected component of $p^{-1}X'$ is simply* *connected (hence it is a copy of the universal cover of X'). Moreover, these components are permuted transitively by the action of π on $\tilde{X}$, and π' is the isotropy group of one of them; in other words, $\pi_0(p^{-1}X') \approx \pi/\pi'$.*

PROOF. For any basepoint in $p^{-1}X'$ we have a diagram

$$\begin{array}{ccc} \pi_1(p^{-1}X') & \longrightarrow & \pi_1\tilde{X} \\ \downarrow & & \downarrow \\ \pi' & \hookrightarrow & \pi, \end{array}$$

where the vertical maps are induced by p and the horizontal maps by inclusions. Since $\pi_1\tilde{X} = \{1\}$, the first assertion follows at once. The second assertion, which we will not make serious use of, is left as an exercise for the interested reader. □

(7.6) Lemma. *Any diagram $G_1 \leftarrow A \rightarrow G_2$ of groups can be realized by a diagram $X_1 \hookleftarrow Y \hookrightarrow X_2$ of $K(\pi, 1)$-complexes.*

PROOF. According to exercise 4 of §I.4 or exercise 3 of §I.5, $K(\pi, 1)$-complexes can be constructed functorially. We can therefore realize the group homomorphisms by cellular maps $X_1 \leftarrow Y \rightarrow X_2$ of $K(\pi, 1)$'s. Taking mapping cylinders if necessary (cf. Spanier [1966], 1.4), we can make these maps inclusions. □

PROOF OF 7.3. Start with $X_1 \hookleftarrow Y \hookrightarrow X_2$ as in 7.6 and form the adjunction space $X = X_1 \cup_Y X_2$, i.e., X is obtained from the disjoint union $X_1 \coprod X_2$ by identifying the two copies of Y. Then $\pi_1 X = G_1 *_A G_2 = G$ by 7.2, so we need only show that the universal cover $\tilde{X}$ satisfies $H_i\tilde{X} = 0$ for $i > 1$. Let $\tilde{X}_1, \tilde{X}_2$, and $\tilde{Y}$ be the inverse images of X_1, X_2, and Y in $\tilde{X}$. Since X_1, X_2, and Y have acyclic universal covers, it follows from 7.4 and 7.5 that $\tilde{X}_1, \tilde{X}_2$, and $\tilde{Y}$ have trivial homology in positive dimensions. The Mayer–Vietoris sequence associated to the square

$$\begin{array}{ccc} \tilde{Y} & \hookrightarrow & \tilde{X}_2 \\ \downarrow & & \downarrow \\ \tilde{X}_1 & \hookrightarrow & \tilde{X} \end{array}$$

therefore shows that $H_i\tilde{X} = 0$ for $i > 1$. □

(7.7) Corollary. *Given $G = G_1 *_A G_2$ where $A \hookrightarrow G_1$ and $A \hookrightarrow G_2$, there is a "Mayer–Vietoris" sequence*

$$\cdots \rightarrow H_n A \rightarrow H_n G_1 \oplus H_n G_2 \rightarrow H_n G \rightarrow H_{n-1} A \rightarrow \cdots.$$

This is immediate from 7.3.

Remark. As a bi-product of the proof of 7.3 we obtain an exact sequence of permutation modules

$$(7.8) \qquad 0 \to \mathbb{Z}[G/A] \to \mathbb{Z}[G/G_1] \oplus \mathbb{Z}[G/G_2] \to \mathbb{Z} \to 0.$$

Indeed, this is just the low-dimensional part of the Mayer–Vietoris sequence which we used in the proof of 7.3. (The second part of 7.5 allows one to identify $H_0(\tilde{Y})$, $H_0(\tilde{X}_1)$, and $H_0(\tilde{X}_2)$ with permutation modules.) We will give another derivation of 7.8, again based on topological ideas, in an appendix to this chapter. It is also possible to prove 7.8 algebraically, cf. Bieri [1976], Prop. 2.8, or Swan [1969], Lemma 2.1. Moreover, it is possible to give a purely algebraic proof of 7.7, using 7.8 as the starting point. We will explain this in the exercise of §III.6 below.

EXERCISES

1. Let $G = G_1 *_A G_2$ with $A \to G_1$ and $A \to G_2$ not necessarily injective. Let $\bar{G}_1$, $\bar{G}_2$ and $\bar{A}$ be the images of G_1, G_2, and A in G. Show that $G = \bar{G}_1 *_{\bar{A}} \bar{G}_2$. Thus any amalgam as defined at the beginning of this section is isomorphic to one in which the maps $A \to G_1$ and $A \to G_2$ are injective.

*2. Give a proof of 7.2 based on the classification of regular covering spaces, as stated in the appendix to Chapter I, A2. [Hint: For any group H, a pointed regular H-covering $\tilde{X}$ of X is specified by giving pointed regular H-coverings $\tilde{X}_1$ of X_1 and $\tilde{X}_2$ of X_2 such that the induced coverings of Y are isomorphic (as pointed regular H-coverings).]

3. It is a classical fact that $SL_2(\mathbb{Z}) \approx \mathbb{Z}_4 *_{\mathbb{Z}_2} \mathbb{Z}_6$. (See Serre [1977a] for an indication of an easy proof of this; see also example 3 of §VIII.9 below.) Use the Mayer–Vietoris sequence to calculate $H_*(SL_2(\mathbb{Z}))$. [Suggestion: You can save a lot of work by considering separately the 2-torsion and the 3-torsion in the Mayer–Vietoris sequence. As far as 2-torsion is concerned, $SL_2(\mathbb{Z})$ behaves like $\mathbb{Z}_4 *_{\mathbb{Z}_2} \mathbb{Z}_2 = \mathbb{Z}_4$, whereas it behaves like $\{1\} *_{\{1\}} \mathbb{Z}_3 = \mathbb{Z}_3$ with respect to 3-torsion.]

Appendix. Trees and Amalgamations

The results of §7 were based on a topological interpretation of amalgamated free products in terms of "amalgamations" of topological spaces. The purpose of this appendix is to describe a different topological interpretation of amalgamated free products, due to Serre [1977a]. In particular, we will obtain another proof of 7.8.

Recall that a *graph* is a 1-dimensional CW-complex and that a *path* in a graph is a sequence $e_1, \ldots, e_n$ of oriented edges such that the final vertex of e_i equals the initial vertex of e_{i+1} for $1 \leq i < n$. The path is a *loop* if the final vertex of e_n equals the initial vertex of e_1. We allow the case $n = 0$, in which case the path is called *trivial*. Finally, the path is called *reduced* if $e_{i+1} \neq \bar{e}_i$ for $0 \leq i < n$, where $\bar{e}_i$ is the same geometric edge as e_i but with the opposite orientation.

The graph X is called a *tree* if it satisfies the following conditions, which are easily seen to be equivalent:

(i) X is contractible.
(ii) X is simply connected.
(iii) X is acyclic.
(iv) X is connected and contains no non-trivial reduced loops.

Suppose a group G acts as a group of automorphisms of a tree X, and let e be an edge of X with vertices v and w. We will say that e is a *fundamental domain* for the G-action if every edge of X is equivalent to e mod G and every vertex of X is equivalent to either v or w but not both. In other words, we require that the subgraph

$$\overset{v}{\circ}\underset{e}{\rule{10em}{0.4pt}}\overset{w}{\circ}$$

map isomorphically onto the orbit graph X/G. Note, in this case, that the isotropy groups G_e, G_v, and G_w satisfy

$$G_e = G_v \cap G_w.$$

Indeed, we have $G_e \supseteq G_v \cap G_w$ since e is the only edge with vertices v and w [X is a tree]; and the opposite inclusion holds since no element of G can interchange v and w [they are inequivalent mod G].

The theorem of Serre that we wish to state says that actions of G of this type (i.e., with an edge as fundamental domain) are essentially the same as decompositions of G as an amalgamated free product:

(A1) Theorem. *Let a group G act on a tree X. Suppose e is an edge with vertices v and w such that e is a fundamental domain for the action. Then $G = G_v *_{G_e} G_w$. Conversely, given an amalgamation $G = G_1 *_A G_2$ (where $A \hookrightarrow G_1$ and $A \hookrightarrow G_2$), there exists a tree on which G acts as above, with G_1, G_2, and A as the isotropy groups G_v, G_w, and G_e.*

Sketch of proof. Note that two edges g_1e and g_2e of X ($g_i \in G$) have a vertex in common if and only if $g_1^{-1}g_2 \in G_v$ or G_w. This allows one to relate reduced paths in X to reduced words ("normal forms") in $G_v *_{G_e} G_w$. Consider, for example, a reduced path of length 4 starting with e, oriented so that v is its first vertex. Such a path must have the form

$$\overset{v}{\circ}\underset{e}{\rule{6em}{0.4pt}}\overset{w}{\bullet}\underset{ge}{\rule{6em}{0.4pt}}\overset{gv}{\circ}\underset{ghe}{\rule{6em}{0.4pt}}\overset{ghw}{\bullet}\underset{ghke}{\rule{6em}{0.4pt}}\overset{ghkv}{\circ}$$

where $g, k \in G_w$ and $h \in G_v$. Moreover, $g, h, k \notin G_e$. Thus the path gives rise to the reduced word ghk in $G_v *_{G_e} G_w$. Since X has no loops, we must have

$ghk \neq 1$ in G. Generalizing this argument, one sees that the canonical map $\varphi: G_v *_{G_e} G_w \to G$ is injective. Similarly, it follows from the connectivity of X that φ is surjective. [Given $g \in G$, consider a path ending at gv, with e as its first edge.] This proves the first part of the theorem.

Conversely, given $G = G_1 *_A G_2$, we have no choice as to how to construct X. Namely, we must take $G/G_1 \coprod G/G_2$ as the set of vertices of X and G/A as the set of edges. There are canonical maps $\alpha: G/A \to G/G_1$ and $\beta: G/A \to G/G_2$, by which we attach the edges to the vertices, i.e., an edge $e \in G/A$ joins the vertices $\alpha(e)$ and $\beta(e)$. In this way we obtain a graph X on which G acts with an edge as fundamental domain and with G_1, G_2, and A as the isotropy groups. Since $G = G_1 *_A G_2$, the first part of the proof shows that X is connected and has no non-trivial reduced loops, so that it is a tree. □

It is apparent from this sketch of the proof that the theorem is not particularly deep; indeed, the existence of the tree X associated to $G_1 *_A G_2$ is little more than a reformulation of the normal form theorem for amalgamated free products. Nevertheless, the tree is a very convenient tool for keeping track of the combinatorics of normal forms. It is often considerably easier to prove things about G by using X than it is to work directly with normal forms.

For example, we obtain the promised proof of 7.8 by noting that the sequence in 7.8 is simply the augmented chain complex of X. Its exactness therefore follows from the acyclicity of X.

Here are two other applications of A1:

(A2) Corollary. *Let $F \subseteq G_1 *_A G_2$ be a subgroup which intersects every conjugate of G_1 and G_2 trivially. Then F is a free group.*

Proof. The hypothesis implies that F acts freely on X. The latter being simply connected, it follows that X is the universal cover of the orbit graph X/F and $F \approx \pi_1(X/F)$. But it is well-known that the fundamental group of a graph is free. □

(A3) Corollary. *Let $H \subseteq G_1 *_A G_2$ be a finite subgroup. Then H is conjugate to a subgroup of G_1 or G_2.*

Proof. We must show that H fixes some vertex of X. But this follows from the elementary fact that every finite group of automorphisms of a tree has a fixed point. See Serre [1977a], I.4.3, for more details. □

CHAPTER III

Homology and Cohomology with Coefficients

0 Preliminaries on $\otimes_G$ and Hom_G

Recall that the tensor product $M \otimes_R N$ is defined whenever M is a right R-module and N is a left R-module. It is the quotient of $M \otimes_{\mathbb{Z}} N$ (which we denote $M \otimes N$) obtained by introducing the relations $mr \otimes n = m \otimes rn$ $(m \in M, r \in R, n \in N)$.

In case R is a group ring $\mathbb{Z}G$, we can avoid having to consider both left and right modules by using the anti-automorphism $g \mapsto g^{-1}$ of G. Thus we can regard any left G-module M as a right G-module by setting $mg = g^{-1}m$ $(m \in M, g \in G)$, and in this way we can make sense out of the tensor product $M \otimes_{\mathbb{Z}G} N$ (also denoted $M \otimes_G N$) of two left G-modules.

Note that $M \otimes_G N$ is obtained from $M \otimes N$ by introducing the relations $g^{-1}m \otimes n = m \otimes gn$. If we replace m by gm, these relations take the form $m \otimes n = gm \otimes gn$, and we see that

$$M \otimes_G N = (M \otimes N)_G, \tag{0.1}$$

where G acts "diagonally" on $M \otimes N$: $g \cdot (m \otimes n) = gm \otimes gn$. In particular, this shows that $- \otimes_G -$ is commutative:

$$M \otimes_G N \approx N \otimes_G M.$$

Warning. The passage between left and right modules as above is convenient, but it can sometimes be confusing, for instance if M naturally admits both a left and a right G-action (e.g., $M = \mathbb{Z}G'$, where $G \subseteq G'$). In such cases we will revert to the standard notation $M \otimes_{\mathbb{Z}G} N$ if we want to indicate that the tensor product is to be formed with respect to the given right action of G on M rather than the right action obtained from the left action.

The diagonal G-action used above is quite general and can be used whenever a functor of one or several abelian groups is applied to G-modules. Consider, for example, the functor $\mathrm{Hom}(\ ,\ ) = \mathrm{Hom}_{\mathbb{Z}}(\ ,\ )$. If M and N are G-modules, then the action of G on M and N induces by functoriality a "diagonal" action of G on $\mathrm{Hom}(M, N)$, given by

$$(gu)(m) = g \cdot u(g^{-1}m)$$

for $g \in G$, $u \in \mathrm{Hom}(M, N)$, $m \in M$. [The use of g^{-1} here is needed because of the contravariance of Hom in the first variable. In effect, we compensate for the contravariance by converting M to a right module.]

Note that $gu = u$ if and only if u commutes with the action of g. Thus

$$\mathrm{Hom}_G(M, N) = \mathrm{Hom}(M, N)^G. \tag{0.2}$$

This observation has already occurred implicitly in exercise 5 of §I.8, where we used an averaging procedure to convert an arbitrary element of $\mathrm{Hom}(M, N)$ to an element of $\mathrm{Hom}(M, N)^G = \mathrm{Hom}_G(M, N)$.

EXERCISES

1. Let F be a flat $\mathbb{Z}G$-module and M a G-module which is $\mathbb{Z}$-torsion-free (i.e., $\mathbb{Z}$-flat). Show that $F \otimes M$ (with diagonal G-action) is $\mathbb{Z}G$-flat. [Hint: $(F \otimes M) \otimes_G - = (F \otimes M \otimes -)_G = F \otimes_G (M \otimes -)$, which is an exact functor.]

2. Let F be a projective $\mathbb{Z}G$-module and M a $\mathbb{Z}$-free G-module. Show that $F \otimes M$ (with diagonal G-action) is projective. [Hint: $\mathrm{Hom}_G(F \otimes M, -) = \mathrm{Hom}(F \otimes M, -)^G = \mathrm{Hom}(F, \mathrm{Hom}(M, -))^G = \mathrm{Hom}_G(F, \mathrm{Hom}(M, -))$, which is an exact functor. See exercise 3 of §5 below for an alternative proof.]

1 Definition of $H_*(G, M)$ and $H^*(G, M)$

Let F be a projective resolution of $\mathbb{Z}$ over $\mathbb{Z}G$ and let M be a G-module. We define the *homology of G with coefficients in M* by

$$H_*(G, M) = H_*(F \otimes_G M). \tag{1.1}$$

Here $F \otimes_G M$ can be thought of as the complex obtained from F by applying the functor $- \otimes_G M$. Thus 1.1 is a natural generalization of the definition of $H_* G$ in Chapter II; indeed, we recover the latter by taking $M = \mathbb{Z}$:

$$H_*(G, \mathbb{Z}) = H_* G. \tag{1.2}$$

As in Chapter II, $H_*(G, M)$ is well-defined up to canonical isomorphism.

The complex $F \otimes_G M$ can also be thought of as the tensor product of chain complexes (cf. §I.0), where M is regarded as a chain complex concentrated in dimension 0. From this point of view there is a certain asymmetry in 1.1 which did not appear in the context of Chapter II. A more symmetric looking definition is obtained by choosing projective resolutions of *both* $\mathbb{Z}$ and M, say $\varepsilon: F \to \mathbb{Z}$ and $\eta: P \to M$, and setting

(1.3) $$H_*(G, M) = H_*(F \otimes_G P).$$

Fortunately, 1.3 is consistent with 1.1; for η induces a weak equivalence $F \otimes \eta: F \otimes_G P \to F \otimes_G M$ by I.8.6.

Note that I.8.6 also gives us a weak equivalence $\varepsilon \otimes P: F \otimes_G P \to \mathbb{Z} \otimes_G P$. Thus

(1.4) $$H_*(G, M) = H_*(P_G).$$

Clearly, then, we have considerable flexibility in the choice of a chain complex from which to compute $H_*(G, M)$. For the moment we content ourselves with a trivial example of such a computation:

(1.5) $$H_0(G, M) \approx M_G.$$

This follows from the right exactness of the tensor product. [Apply $- \otimes_G M$ to $F_1 \to F_0 \to \mathbb{Z} \to 0$ and use 1.1, or apply $(\ )_G$ to $P_1 \to P_0 \to M \to 0$ and use 1.4.]

We turn now to *cohomology with coefficients*, which is defined via $\mathcal{H}om$ rather than $\otimes$. Choose a projective resolution $F \to \mathbb{Z}$ as above and consider the complex $\mathcal{H}om_G(F, M)$, where M is again regarded as a chain complex concentrated in dimension 0. Checking the definition of $\mathcal{H}om$ in §I.0, we see that $\mathcal{H}om_G(F, M)_n = \operatorname{Hom}_G(F_{-n}, M)$. It is therefore reasonable to regard $\mathcal{H}om_G(F, M)$ as a *co*chain complex by the usual indexing conventions, i.e., by setting

$$\mathcal{H}om_G(F, M)^n = \mathcal{H}om_G(F, M)_{-n} = \operatorname{Hom}_G(F_n, M).$$

It is then a non-negative cochain complex with coboundary operator δ given by

$$(\delta u)(x) = (-1)^{n+1}u(\partial x)$$

for $u \in \mathcal{H}om_G(F, M)^n$, $x \in F_{n+1}$. This formula is most easily remembered in the form

(1.6) $$\langle \delta u, x\rangle + (-1)^{\deg u}\langle u, \partial x\rangle = 0,$$

cf. exercise 3 of §I.0.

We now define

(1.7) $$H^*(G, M) = H^*(\mathcal{H}om_G(F, M)).$$

Remark. Because of the sign in the definition of δ above, $\mathcal{H}om_G(F, M)$ is *not* the same as the complex $\operatorname{Hom}_G(F, M)$ obtained from F by applying the contravariant functor $\operatorname{Hom}_G(-, M)$ dimension-wise. This somewhat annoying fact of nature is not serious, for changing the sign of a coboundary operator does not change cocycles, coboundaries, or cohomology. Our convention of using $\mathscr{H}om$ instead of Hom is dictated by our desire to be able to use the properties of $\mathscr{H}om$ developed systematically in the exercises of §I.0. The reader is warned, however, that $H^*(G, M)$ is often defined in the literature by means of Hom instead of $\mathscr{H}om$.

Note that the exact sequence $F_1 \to F_0 \to \mathbb{Z} \to 0$ yields an exact sequence $0 \to \mathrm{Hom}_G(\mathbb{Z}, M) \to \mathrm{Hom}_G(F_0, M) \to \mathrm{Hom}_G(F_1, M)$. Since $\mathrm{Hom}_G(\mathbb{Z}, M) = M^G$, this gives

$$H^0(G, M) = M^G. \tag{1.8}$$

There exist analogues for $H^*(G, —)$ of 1.3 and 1.4, but these involve the notion of *injective resolution*, which we have only mentioned briefly (cf. exercise 4 of §I.7). The interested reader can find these analogues in exercise 4b below, which will not be needed elsewhere in this book.

EXAMPLES

1. Suppose G is infinite cyclic with generator t. Then we have a resolution (I.4.5)

$$0 \to \mathbb{Z}G \xrightarrow{t-1} \mathbb{Z}G \to \mathbb{Z} \to 0,$$

hence $H_*(G, M)$ is the homology of

$$\cdots \to 0 \to M \xrightarrow{t-1} M$$

and $H^*(G, M)$ is the cohomology of

$$M \xrightarrow{t-1} M \to 0 \to \cdots.$$

Thus $H_0(G, M) = H^1(G, M) = M_G$, $H_1(G, M) = H^0(G, M) = M^G$, and $H_i(G, M) = H^i(G, M) = 0$ for $i > 1$.

2. Suppose G is cyclic of finite order n with generator t. Then we have a resolution (I.6.3)

$$\cdots \xrightarrow{N} \mathbb{Z}G \xrightarrow{t-1} \mathbb{Z}G \xrightarrow{N} \mathbb{Z}G \xrightarrow{t-1} \mathbb{Z}G \to \mathbb{Z} \to 0,$$

where $N = \sum_{i=0}^{n-1} t^i$. Hence $H_*(G, M)$ is the homology of

$$\cdots \xrightarrow{N} M \xrightarrow{t-1} M \xrightarrow{N} M \xrightarrow{t-1} M$$

and $H^*(G, M)$ is the cohomology of

$$M \xrightarrow{t-1} M \xrightarrow{N} M \xrightarrow{t-1} M \xrightarrow{N} \cdots.$$

Note that $N: M \to M$ satisfies $Ngm = Nm$ ($g \in G$, $m \in M$) and that $NM \subseteq M^G$; thus N induces a map $\overline{N}: M_G \to M^G$, called the *norm map*. [This is, in fact, true for any finite group G, where $N = \sum_{g \in G} g$.] We can now read off from the above that

$$H_i(G, M) = H^{i+1}(G, M) = M^G/NM = \operatorname{coker} \overline{N}$$

for i odd ($i \geq 1$) and that

$$H_i(G, M) = H^{i-1}(G, M) = \ker \bar{N}$$

for i even ($i \geq 2$).

3. For any group G we can always take our resolution F to be the bar resolution. In this case we write $C_*(G, M)$ for $F \otimes_G M$ [$=M \otimes_G F$] and $C^*(G, M)$ for $\mathcal{H}om_G(F, M)$. Thus an element of $C_n(G, M)$ can be uniquely expressed as a finite sum of elements of the form $m \otimes [g_1|\ldots|g_n]$, i.e., as a formal linear combination with coefficients in M of the symbols $[g_1|\ldots|g_n]$. The boundary operator $\partial: C_n(G, M) \to C_{n-1}(G, M)$ is given by

$$\begin{aligned}\partial(m \otimes [g_1|\cdots|g_n]) = mg_1 \otimes [g_2|\cdots|g_n] \\ - m \otimes [g_1g_2|\cdots|g_n] + \cdots + (-1)^n m \otimes [g_1|\cdots|g_{n-1}].\end{aligned}$$

Similarly, an element of $C^n(G, M)$ can be regarded as a function $f: G^n \to M$, i.e., as a function of n variables from G to M. The coboundary operator $\delta: C^{n-1}(G, M) \to C^n(G, M)$ is given, up to sign, by

$$\begin{aligned}(\delta f)(g_1, \ldots, g_n) = g_1 f(g_2, \ldots, g_n) \\ - f(g_1g_2, \ldots, g_n) + \cdots + (-1)^n f(g_1, \ldots, g_{n-1}).\end{aligned}$$

(Note: If $n = 0$ then G^n is, by convention, a one-element set, so that $C^0(G, M) \approx M$.) We could also use the normalized bar resolution here; the resulting normalized cochain complex $C_N^*(G, M) \subseteq C^*(G, M)$ is the subcomplex consisting of those cochains f such that $f(g_1, \ldots, g_n) = 0$ whenever some $g_i = 1$.

Finally, we remark that $H_*(G, M)$ and $H^*(G, M)$ have a topological interpretation analogous to that of $H_* G$:

$$H_*(G, M) = H_*(K(G, 1); \mathcal{M})$$

$$H^*(G, M) = H^*(K(G, 1); \mathcal{M}),$$

where $\mathcal{M}$ is the local coefficient system on $K(G, 1)$ associated to the G-module M. (See Eilenberg [1947], Chapter V, for the relevant facts about local coefficient systems.) We will not make use of these isomorphisms, so we leave the details to the interested reader. [Hint: Let F be the chain complex of the universal cover of a $K(G, 1)$ Y. Then $C_*(Y; \mathcal{M}) = F \otimes_G M$ and $C^*(Y; \mathcal{M}) = \mathcal{H}om_G(F, M)$.]

Exercises

1. Let G be a finite group, M a G-module, and $\bar{N}: M_G \to M^G$ the norm map defined in Example 2 above.

   (a) Show that $\ker \bar{N}$ and $\operatorname{coker} \bar{N}$ are annihilated by $|G|$. [Hint: Consider the obvious map $M^G \to M_G$.]

(b) Suppose M is a module of the form $M = \mathbb{Z}G \otimes A$, where A is an abelian group and G acts by $g \cdot (r \otimes a) = gr \otimes a$. [Such a module M is said to be *induced*.] Show that $\overline{N}: M_G \to M^G$ is an isomorphism.

(c) Show that $\overline{N}$ is an isomorphism if M is a projective $\mathbb{Z}G$-module.

2. Use the standard cochain complex $C^*(G, M)$ to show that $H^1(G, M)$ is isomorphic to the group of derivations from G to M (§II.5, exercise 3) modulo the subgroup of principal derivations. (A principal derivation is one of the form $dg = gm - m$, where m is a fixed element of M.) In particular, if G acts trivially on M then $H^1(G, M) \approx \mathrm{Hom}(G, M) = \mathrm{Hom}(G_{ab}, M) = \mathrm{Hom}(H_1 G, M)$.

3. Let A be an abelian group with trivial G-action. Show that $F \otimes_G A \approx F_G \otimes A$ and that $\mathcal{H}om_G(F, A) \approx \mathcal{H}om(F_G, A)$. Deduce universal coefficient sequences

$$0 \to H_n(G) \otimes A \to H_n(G, A) \to \mathrm{Tor}_1^{\mathbb{Z}}(H_{n-1}(G), A) \to 0$$

and

$$0 \to \mathrm{Ext}_{\mathbb{Z}}^1(H_{n-1}(G), A) \to H^n(G, A) \to \mathrm{Hom}(H_n(G), A) \to 0.$$

*4. (a) State and prove an analogue of I.8.5 for maps into a non-negative cochain complex of injectives.

(b) Let $\varepsilon: F \to \mathbb{Z}$ be a projective resolution and let $\eta: M \to Q$ be an injective resolution. Prove that ε and η induce weak equivalences

$$\mathcal{H}om_G(F, M) \to \mathcal{H}om_G(F, Q) \leftarrow \mathcal{H}om_G(\mathbb{Z}, Q),$$

so any of these three complexes can be used to compute $H^*(G, M)$. In particular,

$$H^*(G, M) \approx H^*(Q^G).$$

2 Tor and Ext

There are obvious generalizations of $H_*(G, —)$ and $H^*(G, —)$, called $\mathrm{Tor}_*^G(—, —)$ and $\mathrm{Ext}_G^*(—, —)$, obtained by removing the restriction in 1.1 and 1.7 that F be a resolution of the particular module $\mathbb{Z}$. For Tor, we take projective resolutions $F \to M$ and $P \to N$ of two arbitrary G-modules M and N and set

$$\mathrm{Tor}_*^G(M, N) = H_*(F \otimes_G N) = H_*(F \otimes_G P) = H_*(M \otimes_G P).$$

We recover $H_*(G, —)$ as $\mathrm{Tor}_*^G(\mathbb{Z}, —)$.

Similarly, if $F \to M$ is again a projective resolution, then we set

$$\mathrm{Ext}_G^*(M, N) = H^*(\mathcal{H}om_G(F, N)).$$

We recover $H^*(G, —)$ as $\operatorname{Ext}^*_G(\mathbb{Z}, —)$. The reader who has done exercise 4 of §1 will note that we could also take an injective resolution $N \to Q$ and write

$$\operatorname{Ext}^*_G(M, N) = H^*(\mathcal{H}om_G(F, Q)) = H^*(\mathcal{H}om_G(M, Q)).$$

We do not intend to systematically develop the properties of Tor and Ext, but it will be useful to record a few results about them for future reference. The first shows that one has considerably more flexibility in the choice of resolutions for computing Tor than is apparent from the definition.

(2.1) Proposition. *Let $\varepsilon: F \to M$ and $\eta: P \to N$ be resolutions, not necessarily projective. If either F or P is a complex of flat modules, then* $\operatorname{Tor}^G_*(M, N) \approx H_*(F \otimes_G P)$.

PROOF. Suppose, for instance, that F is flat, and let $\tilde{\eta}: \tilde{P} \to N$ be a projective resolution. By the fundamental lemma I.7.4, there is an augmentation-preserving chain map $f: \tilde{P} \to P$. Two applications of I.8.6 now yield

$$H_*(F \otimes_G P) \xleftarrow{\approx} H_*(F \otimes_G \tilde{P}) \xrightarrow{\approx} H_*(M \otimes_G \tilde{P}) = \operatorname{Tor}^G_*(M, N),$$

where the first map is induced by f and the second by ε. □

One way to construct a resolution of a module M is to start with a resolution $\varepsilon: F \to \mathbb{Z}$ and tensor it with M to obtain $\varepsilon \otimes M: F \otimes M \to \mathbb{Z} \otimes M = M$. This will be a resolution under mild hypotheses on F and/or M. Suppose, for instance, that F is $\mathbb{Z}$-free; then ε is a homotopy equivalence if we ignore the G-action (cf. I.7.6), so the same is true of $\varepsilon \otimes M$. [Alternatively, use the universal coefficient theorem to compute that

$$H_*(F \otimes M) = H_*(F) \otimes M = M.]$$

This method will be used now to show that Tor^G_* and Ext^*_G can often be computed in terms of $H_*(G, —)$ and $H^*(G, —)$.

(2.2) Proposition. *Let M and N be G-modules. If M is $\mathbb{Z}$-torsion-free then*

$$\operatorname{Tor}^G_*(M, N) \approx H_*(G, M \otimes N),$$

where G acts diagonally on $M \otimes N$. If M is $\mathbb{Z}$-free, then

$$\operatorname{Ext}^*_G(M, N) \approx H^*(G, \operatorname{Hom}(M, N)),$$

where G acts diagonally on $\operatorname{Hom}(M, N)$.

PROOF. Let $\varepsilon: F \to \mathbb{Z}$ be a projective resolution, and consider the resolution $\varepsilon \otimes M: F \otimes M \to M$. This is a flat resolution if M is $\mathbb{Z}$-torsion-free and a projective resolution if M is $\mathbb{Z}$-free, by exercises 1 and 2 of §0. So if M is $\mathbb{Z}$-torsion-free, then we have

$$\begin{aligned}\operatorname{Tor}_*^G(M, N) &\approx H_*((F \otimes M) \otimes_G N) \qquad \text{by 2.1}\\ &= H_*((F \otimes M \otimes N)_G)\\ &= H_*(F \otimes_G (M \otimes N))\\ &= H_*(G, M \otimes N).\end{aligned}$$

And if M is $\mathbb{Z}$-free, then

$$\begin{aligned}\operatorname{Ext}_G^*(M, N) &= H^*(\mathcal{H}om_G(F \otimes M, N)) \qquad \text{by definition of Ext}\\ &= H^*(\mathcal{H}om(F \otimes M, N)^G)\\ &= H^*(\mathcal{H}om(F, \operatorname{Hom}(M, N))^G)\\ &= H^*(\mathcal{H}om_G(F, \operatorname{Hom}(M, N))\\ &= H^*(G, \operatorname{Hom}(M, N)).\end{aligned}$$ □

We will outline alternative proofs of this proposition in exercise 2 of §7 below.

Finally, we remark that Tor and Ext can be defined over an arbitrary ring R, using the same definitions as we gave for G-modules. One need only be careful about which modules are left modules and which are right ones. Thus $\operatorname{Tor}_*^R(M, N)$ is defined when M is a right module and N is a left module, whereas $\operatorname{Ext}_R^*(M, N)$ is defined when M and N are either both left modules or both right modules.

EXERCISE

Show that $\operatorname{Tor}_*^G(M, N) \approx \operatorname{Tor}_*^G(N, M)$. [Hint: See exercise 5 of §I.0.]

3 Extension and Co-extension of Scalars

Before proceeding further with the study of $H_*(G, M)$ and $H^*(G, M)$, we digress to discuss, in this and the next two sections, some module-theoretic constructions which play a fundamental role in the homology and cohomology theory of groups. Let $\alpha: R \to S$ be a ring homomorphism. Then any S-module can be regarded as an R-module via α, and we obtain in this way a functor from S-modules to R-modules, called *restriction of scalars*. The purpose of this section is to study two constructions which go in the opposite direction, from R-modules to S-modules.

For any (left) R-module M, consider the tensor product $S \otimes_R M$, where S is regarded as a right R-module by $s \cdot r = s\alpha(r)$. Since the natural left action of S on itself commutes with this right action of R on S, we can make $S \otimes_R M$ a (left) S-module by setting $s \cdot (s' \otimes m) = ss' \otimes m$. This S-module is said to be obtained from M by *extension of scalars* from R to S.

Note that there is a natural map $i: M \to S \otimes_R M$ given by $i(m) = 1 \otimes m$. Since $1 \otimes rm = \alpha(r) \otimes m = \alpha(r) \cdot (1 \otimes m)$ for $r \in R$, we have

(3.1) $$i(rm) = \alpha(r)i(m);$$

in other words, i is an R-module map, where the S-module $S \otimes_R M$ is regarded as an R-module by restriction of scalars. Moreover, the following *universal mapping property* holds:

(3.2) Given an S-module N and an R-module map $f: M \to N$, there is a unique S-module map $g: S \otimes_R M \to N$ such that $gi = f$:

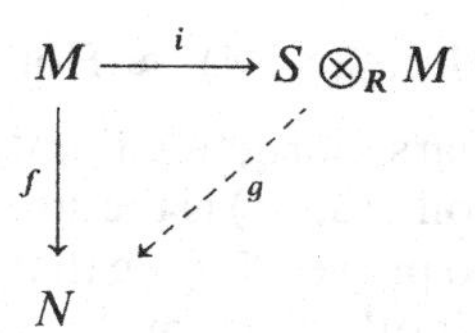

[Thus we have

(3.3) $$\mathrm{Hom}_S(S \otimes_R M, N) \approx \mathrm{Hom}_R(M, N),$$

showing that the extension of scalars functor (R-modules) $\to$ (S-modules) is left adjoint to the restriction of scalars functor (S-modules) $\to$ (R-modules).] Heuristically, 3.2 says that $S \otimes_R M$ is the smallest S-module which receives an R-module map from M. To prove 3.2 we need only note that g, if it exists, must satisfy $g(s \otimes m) = sg(1 \otimes m) = sg(i(m)) = sf(m)$. This proves uniqueness and tells us how to define g in order to prove existence; the remaining details are left to the reader.

Note that 3.2 can be applied, in particular, with $M = N$ (regarded as an R-module by restriction of scalars) and $f = \mathrm{id}_N$. We obtain, then, for any S-module N, a canonical S-module map

(3.4) $$S \otimes_R N \to N,$$

given by $s \otimes n \mapsto sn$. This map is *surjective*; moreover, as an R-module map it is a *split surjection.*

We now consider a dual construction, which uses Hom instead of $\otimes$. Given a (left) R-module M, consider the abelian group $\mathrm{Hom}_R(S, M)$, where S is regarded as a left R-module by $r \cdot s = \alpha(r)s$. Since the natural right action of S on itself commutes with this left action of R on S, we can make $\mathrm{Hom}_R(S, M)$ a left S-module by setting $(sf)(s') = f(s's)$ for $f \in \mathrm{Hom}_R(S, M)$. [Note: It is because of the contravariance of Hom in the first variable that the *right* action of S on itself induces a *left* action of S on $\mathrm{Hom}_R(S, M)$.] This S-module is said to be obtained from M by *co-extension of scalars* from R to S.

There is a natural map $\pi: \mathrm{Hom}_R(S, M) \to M$, given by $\pi(f) = f(1)$. Note that $\pi(\alpha(r)f) = (\alpha(r)f)(1) = f(\alpha(r)) = rf(1) = r\pi(f)$, so π is an

R-module map if the S-module $\mathrm{Hom}_R(S, M)$ is regarded as an R-module by restriction of scalars. Moreover, we have:

(3.5) Given an S-module N and an R-module map $f: N \to M$, there is a unique S-module map $g: N \to \mathrm{Hom}_R(S, M)$ such that $\pi g = f$:

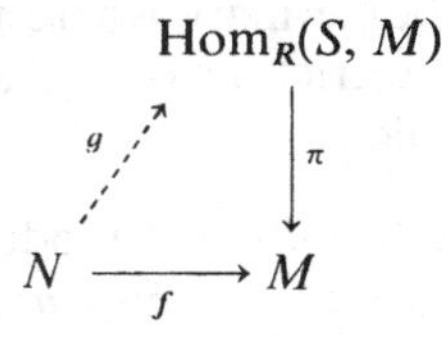

[Thus

$$\mathrm{Hom}_S(N, \mathrm{Hom}_R(S, M)) \approx \mathrm{Hom}_R(N, M), \tag{3.6}$$

so that co-extension of scalars is right adjoint to restriction of scalars.] Heuristically, 3.5 says that $\mathrm{Hom}_R(S, M)$ is the smallest S-module which maps to M by an R-module map. To prove 3.5, note that g must satisfy $sg(n) = g(sn)$ for $s \in S$, $n \in N$; evaluating both sides at 1, we find $g(n)(s) = g(sn)(1) = \pi(g(sn)) = f(sn)$; existence and uniqueness of g follow easily.

Taking $M = N$ (regarded as an R-module) and $f = \mathrm{id}_N$, we obtain from 3.5 a canonical S-module map

$$N \to \mathrm{Hom}_R(S, N), \tag{3.7}$$

given by $n \mapsto (s \mapsto sn)$. This map is *injective*; moreover, as an R-module map it is a *split injection.*

Examples of extension and co-extension of scalars have already occurred in these notes. Namely, if α is the augmentation map $\varepsilon: \mathbb{Z}G \to \mathbb{Z}$, then the "extension of scalars" functor ($\mathbb{Z}G$-modules) $\to$ ($\mathbb{Z}$-modules) is simply $M \mapsto M_G$, and the "co-extension of scalars" functor is $M \mapsto M^G$. More generally, exercise 3 of §II.2 treated the extension of scalars relative to a surjection $\mathbb{Z}G \to \mathbb{Z}[G/H]$, where $H \lhd G$. What we are primarily interested in, however, is the case where α is an inclusion of the form $\mathbb{Z}H \hookrightarrow \mathbb{Z}G$, where $H \subseteq G$. This case will be discussed in detail in §5 below.

Exercises

1. Show that extension of scalars takes projective R-modules to projective S-modules. [Hint: If P is a projective R-module, 3.3 shows that $\mathrm{Hom}_S(S \otimes_R P, -)$ is an exact functor.]

2. Recall from exercise 4 of §I.7 that an R-module Q is called *injective* if the functor $\mathrm{Hom}_R(-, Q)$ is exact. Show that co-extension of scalars takes injective R-modules to injective S-modules. [Hint: If Q is an injective R-module, 3.6 shows that $\mathrm{Hom}_S(-, \mathrm{Hom}_R(S, Q))$ is an exact functor.]

3. If S is flat as a right R-module (i.e., if $S \otimes_R -$ is an exact functor), show that restriction of scalars takes injective S-modules to injective R-modules. [Similar hint.]

4. If S is projective as a left R-module, show that restriction of scalars takes projective S-modules to projective R-modules. [Similar hint.]

The reader familiar with adjoint functors may wish to formulate a general statement of which 1–4 are special cases. [Note: Exercises 1 and 4 can also be done in a more concrete way, using the fact that a projective module is a direct summand of a free module. A special case of exercise 4 was already done in this way in exercise 2 of §I.8.]

4 Injective Modules

Recall from exercise 4 of §I.7 that an R-module Q is called *injective* if it satisfies the following equivalent conditions, dual to those by which we defined "projective":

(i) $\mathrm{Hom}_R(—, Q)$ is exact.
(ii) Every mapping problem

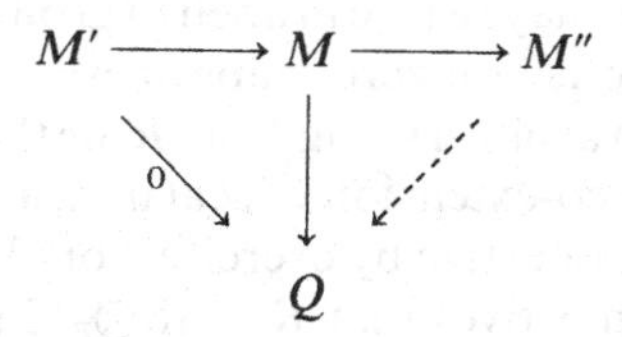

with exact row can be solved.
(iii) For any inclusion $M' \hookrightarrow M$, every map $M' \to Q$ can be extended to a map $M \to Q$:

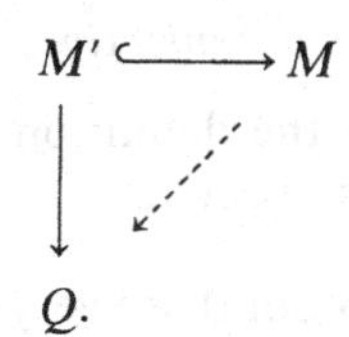

The purpose of this section is to prove that every module can be embedded in an injective module; this is analogous to the (obvious) fact that every module is a quotient of a projective module.

(4.1) Proposition. *An R-module Q is injective if and only if every map $I \to Q$, where I is a left ideal of R, extends to a map $R \to Q$.*

PROOF. The "only if" part is trivial. To prove the "if" part, suppose that every map $I \to Q$ extends to R for all ideals $I \subset R$. It follows at once that if $C' \subset C$, where C is a cyclic R-module, then every map $C' \to Q$ extends to a map $C \to Q$. Now let $M' \subset M$ be an arbitrary inclusion. Given $f: M' \to Q$, there exists by Zorn's lemma a maximal extension $F: M'' \to Q$ of f, where $M' \subseteq M'' \subseteq M$. If $M'' \neq M$ then there is a cyclic module $C \subseteq M$ such that $C \nsubseteq M''$.

But we know that $F|C \cap M''$ extends to C, so F extends to $M'' + C$, contradicting the maximality of F. Thus $M'' = M$ and Q is injective. □

(4.2) Corollary. *If R is a principal ideal domain, then an R-module Q is injective if and only if it is divisible, i.e., if and only if $rQ = Q$ for every $r \neq 0$ in R.*

The proof is immediate.

As an example of the corollary, the $\mathbb{Z}$-module $\mathbb{Q}$ is divisible and hence injective. Similarly, $\mathbb{Q}/\mathbb{Z}$ is injective. The latter module is particularly interesting, because of the following result:

(4.3) Proposition. *If A is a $\mathbb{Z}$-module and $0 \neq a \in A$, then there is a map $f: A \to \mathbb{Q}/\mathbb{Z}$ such that $f(a) \neq 0$.*

Proof. Since $\mathbb{Q}/\mathbb{Z}$ contains $(1/n)\mathbb{Z}/\mathbb{Z} \approx \mathbb{Z}/n\mathbb{Z}$ for every $n \neq 0$, it is clear that there is a map $f_0: \mathbb{Z}a \to \mathbb{Q}/\mathbb{Z}$ such that $f_0(a) \neq 0$. But $\mathbb{Q}/\mathbb{Z}$ is injective, so f_0 extends to a map $f: A \to \mathbb{Q}/\mathbb{Z}$. □

This shows that $\mathbb{Q}/\mathbb{Z}$ plays a role in abelian group theory dual to that played by $\mathbb{Z}$; for the latter admits non-zero maps *to* every non-zero abelian group.

Returning now to an arbitrary ring R, let R' be the R-module $\mathrm{Hom}_{\mathbb{Z}}(R, \mathbb{Q}/\mathbb{Z})$ obtained from $\mathbb{Q}/\mathbb{Z}$ by co-extension of scalars with respect to the ring homomorphism $\mathbb{Z} \to R$. It is injective by exercise 2 of §3. As we will see, it plays the role in the theory of injectives that $R = R \otimes_{\mathbb{Z}} \mathbb{Z}$ plays in the theory of projectives. By a *co-free* module we mean an arbitrary direct product of copies of R'. The following result is the analogue of I.7.2:

(4.4) Lemma. *Co-free modules are injective.*

Proof. It is immediate from the definition of "injective" that an arbitrary direct product of injectives is injective. □

Recall now the standard proof that every module M is a quotient of a free module F; one takes $F = \bigoplus R$, where there is one summand for each element of M, i.e., for each R-module map $R \to M$. Dually, we can consider $Q = \prod_{f \in \mathscr{F}} R'$, where $\mathscr{F} = \mathrm{Hom}_R(M, R')$. There is an obvious map $i: M \to Q$, whose f-component for $f \in \mathscr{F}$ is the map f.

(4.5) Theorem. *The map i is a monomorphism. Consequently, any R-module can be embedded in an injective module.*

Proof. If $0 \neq m \in M$, there is by 4.3 a $\mathbb{Z}$-module map $f_0: M \to \mathbb{Q}/\mathbb{Z}$ with $f_0(m) \neq 0$. By 3.5, f_0 factors through an R-module map $f: M \to R'$. It follows that $f(m) \neq 0$, and hence that $i(m) \neq 0$. □

See Cartan–Eilenberg [1956], I.3.3, for a different proof of the second assertion of 4.5, based directly on 4.1.

EXERCISE

Let $R = \mathbb{Z}/n\mathbb{Z}$. Show that R is an injective R-module. Deduce:

(a) If A is an abelian group such that $nA = 0$, and $C \subseteq A$ is a cyclic subgroup of order n, then C is a direct summand of A.

(b) If A is as in (a), then A is a direct sum of cyclic groups.

5 Induced and Co-induced Modules

In this section we apply the constructions of §3 to ring homomorphisms of the form $\mathbb{Z}H \hookrightarrow \mathbb{Z}G$, where $H \subset G$. In this case extension of scalars (resp. co-extension of scalars) is called *induction* (resp. *co-induction*) from H to G. We will often write

$$\mathbb{Z}G \otimes_{\mathbb{Z}H} M = \operatorname{Ind}_H^G M$$

and

$$\operatorname{Hom}_{\mathbb{Z}H}(\mathbb{Z}G, M) = \operatorname{Coind}_H^G M$$

for an H-module M. Note that a module of the form $\operatorname{Ind}_{\{1\}}^G M$ is precisely what we called an *induced module* in exercise 1b of §1 above. Similarly, a module of the form $\operatorname{Coind}_{\{1\}}^G M$ will be called *co-induced.*

Since the right translation action of H on G is free, $\mathbb{Z}G$ is a free right $\mathbb{Z}H$-module; as basis we can take any set E of representatives for the left cosets gH. It follows that $\mathbb{Z}G \otimes_{\mathbb{Z}H} M$, as abelian group, admits a decomposition

$$\mathbb{Z}G \otimes_{\mathbb{Z}H} M = \bigoplus_{g \in E} g \otimes M,$$

where $g \otimes M = \{g \otimes m : m \in M\}$ and $g \otimes M \approx M$ via $g \otimes m \leftrightarrow m$. In particular, since we can take 1 as the representative of its coset, it follows that the canonical H-map $i: M \to \mathbb{Z}G \otimes_{\mathbb{Z}H} M$ defined in §3 maps M isomorphically onto its image $1 \otimes M$. We can therefore use i to regard M as an H-submodule of $\operatorname{Ind}_H^G M$. Moreover, the summand $g \otimes M$ which occurs above is simply the transform of this submodule under the action of g, since $g \cdot (1 \otimes m) = g \otimes m$. We have therefore established:

(5.1) Proposition. *The G-module $\operatorname{Ind}_H^G M$ contains M as an H-submodule and is the direct sum of the transforms gM, where g ranges over any set of representatives for the left cosets of H in G.*

More briefly, the second assertion of the proposition says:

$$\operatorname{Ind}_H^G M = \bigoplus_{g \in G/H} gM. \tag{5.2}$$

This makes sense because M is mapped onto itself by the action of H, so that the subgroup gM of $\operatorname{Ind}_H^G M$ depends only on the class of g in G/H.

The description 5.2 completely characterizes G-modules of the form $\operatorname{Ind}_H^G M$. More precisely, suppose N is a G-module whose underlying abelian group is a direct sum $\bigoplus_{i \in I} M_i$. Assume that the G-action transitively permutes the summands, in the sense that there is a transitive action of G on I such that $gM_i = M_{gi}$ for all $g \in G$ and $i \in I$. Then we have:

(5.3) Proposition. *Let N be a G-module as above, let M be one of the summands M_i, and let $H \subseteq G$ be the isotropy group of i. Then M is an H-module and $N \approx \operatorname{Ind}_H^G M$.*

PROOF. It is obvious that M is an H-submodule of N, and 3.2 implies that the inclusion $M \hookrightarrow N$ extends to a G-map $\operatorname{Ind}_H^G M \to N$. Clearly φ maps the summand gM of $\operatorname{Ind}_H^G M$ isomorphically onto the corresponding summand M_{gi} of N, so φ is an isomorphism. □

(5.4) Corollary. *Let N be a G-module whose underlying abelian group is of the form $\bigoplus_{i \in I} M_i$. Assume that the G-action permutes the summands according to some action of G on I. Let G_i be the isotropy group of i and let E be a set of representatives for I mod G. Then M_i is a G_i-module and there is a G-isomorphism $N \approx \bigoplus_{i \in E} \operatorname{Ind}_{G_i}^G M_i$.*

PROOF. We have $I = \coprod_{i \in E} Gi$, so $N = \bigoplus_{i \in E} \bigoplus_{j \in Gi} M_j$; now apply 5.3 to the inner sum. □

(5.5) EXAMPLES. (a) The permutation module $\mathbb{Z}[G/H]$ is isomorphic to $\operatorname{Ind}_H^G \mathbb{Z}$, with H acting trivially on $\mathbb{Z}$. This can be seen directly from the definition of $\operatorname{Ind}_H^G \mathbb{Z}$, or, alternatively, by writing $\mathbb{Z}[G/H]$ as a direct sum of copies of $\mathbb{Z}$ and applying 5.3.

(b) Let X be a G-CW-complex and consider the G-module $C_n(X)$. This is a direct sum of copies of $\mathbb{Z}$, one for each n-cell of X, and the summands are permuted by the G-action. Hence 5.4 gives

$$C_n(X) \approx \bigoplus_{\sigma \in \Sigma_n} \operatorname{Ind}_{G_\sigma}^G \mathbb{Z}_\sigma,$$

where Σ_n is a set of representatives for the G-orbits of n-cells,

$$G_\sigma = \{g \in G : g\sigma = \sigma\},$$

and $\mathbb{Z}_\sigma$ is the "orientation module" associated to σ, i.e., $\mathbb{Z}_\sigma$ is an infinite cyclic group whose two generators correspond to the two orientations of σ. (Thus $g \in G_\sigma$ acts on $\mathbb{Z}_\sigma$ as $+1$ if g preserves the orientation of σ and -1 otherwise.) Note that if G acts freely on X, then the isomorphism above simply reduces to our observation in §I.4 that $C_n(X)$ is a free $\mathbb{Z}G$-module with one basis element for each $\sigma \in \Sigma_n$.

Before stating our next result we remark that the summand gM of $\mathrm{Ind}_H^G M$ is closed under the action of gHg^{-1}, hence gM is a gHg^{-1}-module. Note that the action of g^{-1} gives a bijection $f: gM \xrightarrow{\approx} M$ such that $f(kn) = g^{-1}kg \cdot f(n)$ for $k \in gHg^{-1}$, $n \in gM$. Hence gM can be identified with the gHg^{-1}-module obtained from the H-module M by "restriction of scalars" via the conjugation isomorphism $gHg^{-1} \xrightarrow{\approx} H$.

For any G-module N we denote by $\mathrm{Res}_H^G N$ the H-module obtained by restriction of scalars from G to H.

(5.6) Proposition. (*a*) *Let N be a G-module. Then*

$$\mathrm{Ind}_H^G \mathrm{Res}_H^G N \approx \mathbb{Z}[G/H] \otimes N,$$

where G acts diagonally on the tensor product.

(*b*) *Let H and K be subgroups of G and let E be a set of representatives for the double cosets KgH. For any H-module M, there is a K-isomorphism*

$$\mathrm{Res}_K^G \mathrm{Ind}_H^G M \approx \bigoplus_{g \in E} \mathrm{Ind}_{K \cap gHg^{-1}}^K \mathrm{Res}_{K \cap gHg^{-1}}^{gHg^{-1}} gM.$$

In particular, if $H \lhd G$, then there is an H-isomorphism

$$\mathrm{Res}_H^G \mathrm{Ind}_H^G M \approx \bigoplus_{g \in G/H} gM.$$

Remark. In view of the observations preceding the statement of the proposition, we can identify the module $\mathrm{Res}_{K \cap gHg^{-1}}^{gHg^{-1}} gM$ which occurs above with the module $\mathrm{Res}_{K \cap gHg^{-1}}^{H} M$, where the latter restriction is with respect to the conjugation map $K \cap gHg^{-1} \to H$, $k \mapsto g^{-1}kg$.

PROOF OF 5.6. (a) will be left to the reader; it is the special case $M = \mathbb{Z}$ of exercise 2a below. For (b) we note that the summands of $\mathrm{Ind}_H^G M = \bigoplus_{g \in G/H} gM$ are permuted by the action of K according to the natural action of K on G/H. Since the K-orbits in G/H correspond to double cosets KgH and the isotropy group in K of the coset gH is $K \cap gHg^{-1}$, (b) follows from 5.4. □

We will have several occasions to use 5.6a in the special case $H = \{1\}$. The content of the result in this case is:

(5.7) Corollary. *Let M be a G-module and let M_0 be its underlying abelian group. Then $\mathbb{Z}G \otimes M$ (with diagonal G-action) is canonically isomorphic to the induced module $\mathbb{Z}G \otimes M_0$. In particular, $\mathbb{Z}G \otimes M$ is a free $\mathbb{Z}G$-module if M is free as a $\mathbb{Z}$-module.*

The interested reader can verify that the proof of 5.6a yields the specific isomorphism $\mathbb{Z}G \otimes M_0 \to \mathbb{Z}G \otimes M$ given by $g \otimes m \mapsto g \otimes gm$, with inverse $g \otimes m \mapsto g \otimes g^{-1}m$.

Co-induction has properties analogous to the properties of induction given above, but with $\bigoplus$ replaced by $\prod$. The precise statements, however, are somewhat more awkward and require some notation: If $\pi_1: N \twoheadrightarrow M_1$ and $\pi_2: N \twoheadrightarrow M_2$ are surjections of abelian groups, then we write $\pi_1 \sim \pi_2$ if there is an isomorphism $h: M_1 \xrightarrow{\approx} M_2$ such that $h\pi_1 = \pi_2$, or, equivalently, if $\ker \pi_1 = \ker \pi_2$. If $\pi: N \twoheadrightarrow M$ is a surjection and N has a G-module structure, then we denote by πg for $g \in G$ the surjection $N \twoheadrightarrow M$ defined by $(\pi g)(x) = \pi(gx)$. Finally, by a *direct product decomposition* of N we mean a family of surjections $(\pi_i: N \twoheadrightarrow M_i)_{i \in I}$ such that the corresponding map $N \to \prod_{i \in I} M_i$ is an isomorphism. It is now easy to state and prove analogues for co-induction of 5.1, 5.3, 5.4, and 5.6. For example, the analogue of 5.1 is that the underlying abelian group of $\operatorname{Coind}_H^G M$ admits a direct product decomposition $(\pi g)_{g \in H \backslash G}$, where $\pi: \operatorname{Coind}_H^G M \to M$ is the canonical H-module map defined in §3. (Note that $\pi h \sim \pi$ for $h \in H$, so the equivalence class of the surjection πg depends only on the class of g in $H \backslash G$.) Similarly, the analogue of 5.3 is:

(5.8) Proposition. *Let N be a G-module which, as an abelian group, admits a direct product decomposition $(\pi_i: N \twoheadrightarrow M_i)_{i \in I}$. Assume that there is a transitive right action of G on I such that $\pi_i g \sim \pi_{ig}$ for all $i \in I$ and $g \in G$. Let $\pi: N \to M$ be one of the π_i and let $H \subseteq G$ be the isotropy group of i. Then M inherits an H-module structure from N, and $N \approx \operatorname{Coind}_H^G M$.*

The reader can similarly formulate the analogues of 5.4 and 5.6.

Finally, since a direct product indexed by a finite set can be identified with the corresponding direct sum, the following result is not surprising:

(5.9) Proposition. *If $(G: H) < \infty$ then $\operatorname{Ind}_H^G M \approx \operatorname{Coind}_H^G M$.*

Proof. There is an H-map $\varphi_0: M \to \operatorname{Hom}_H(\mathbb{Z}G, M)$ given by

$$\varphi_0(m)(g) = \begin{cases} gm & g \in H \\ 0 & g \notin H. \end{cases}$$

This extends to a G-map $\varphi: \mathbb{Z}G \otimes_{\mathbb{Z}H} M \to \operatorname{Hom}_H(\mathbb{Z}G, M)$ by 3.2. Choosing coset representatives and expressing $\mathbb{Z}G \otimes_{\mathbb{Z}H} M$ (resp. $\operatorname{Hom}_H(\mathbb{Z}G, M)$) as a direct sum (resp. direct product) of copies of M, one sees that φ can be identified with the canonical inclusion of the sum into the product. Consequently, φ is an isomorphism if $(G: H) < \infty$. Alternatively, one can set $\psi(f) = \sum_{g \in G/H} g \otimes f(g^{-1})$ and verify that $\psi = \varphi^{-1}$. □

Exercises

1. Prove that induction is "invariant under conjugation," in the sense that $\operatorname{Ind}_H^G M \approx \operatorname{Ind}_{gHg^{-1}}^G gM$ for any H-module M. Deduce that the K-module

$$\operatorname{Ind}_{K \cap gHg^{-1}}^K \operatorname{Res}_{K \cap gHg^{-1}}^{gHg^{-1}} gM$$

which occurs in 5.6b depends (up to isomorphism) only on the class of g in $K \backslash G/H$.

2. (a) For any H-module M and G-module N prove that

$$N \otimes \mathrm{Ind}_H^G M \approx \mathrm{Ind}_H^G(\mathrm{Res}_H^G N \otimes M),$$

where the tensor product on the left has the diagonal G-action and that on the right has the diagonal H-action. [Hint: Write the left side as $\bigoplus (N \otimes gM)$ and apply 5.3.]

(b) State and prove similar results for $\mathrm{Hom}(\mathrm{Ind}_H^G M, N)$ and $\mathrm{Hom}(N, \mathrm{Coind}_H^G M)$.

3. Use 5.7 to give a new proof of the result of exercise 2 of §0.

4. (a) If $(G:H) = \infty$, prove that $(\mathrm{Ind}_H^G M)^G = 0$ for any H-module M.

*(b) If G is finitely generated and $(G:H) = \infty$, prove that $(\mathrm{Coind}_H^G M)_G = 0$ for any H-module M. [See Strebel [1977] for an algebraic proof of this. A topological proof can be given as follows. We may assume that G is a free group. In this case let Y be a finite, connected 1-complex with $\pi_1 Y = G$ and let $\tilde{Y}$ be the covering complex corresponding to H. Then M (resp. $\mathrm{Coind}_H^G M$) can be regarded as a local coefficient system on $\tilde{Y}$ (resp. Y), and it is easy to see that $(\mathrm{Coind}_H^G M)_G = H_0(Y, \mathrm{Coind}_H^G M) = H_0^{(\infty)}(\tilde{Y}, M)$, where $H_*^{(\infty)}$ denotes homology based on infinite chains. The result now follows from the following elementary fact: If X is an infinite, locally finite, connected complex, then $H_0^{(\infty)}(X, M) = 0$ for any local coefficient system M.]

*(c) Given $H \subseteq G$, prove that the following conditions are equivalent:
   (i) There is a finitely generated subgroup $G' \subseteq G$ such that $(G': G' \cap gHg^{-1}) = \infty$ for all $g \in G$. [If $H \lhd G$, for example, this just says that the group G/H contains a finitely generated infinite subgroup, i.e., G/H is not locally finite.]
   (ii) $(\mathrm{Coind}_H^G M)_G = 0$ for all H-modules M.
   (iii) The element of $(\mathrm{Coind}_H^G \mathbb{Z})_G$ represented by the augmentation map $\varepsilon \in \mathrm{Hom}_H(\mathbb{Z}G, \mathbb{Z}) = \mathrm{Coind}_H^G \mathbb{Z}$ is zero. [Hint: Use (b) and the double-coset formula.]

*5. If G is finite and k is a field, show that a kG-module is projective iff it is injective. [Hint: Start by showing that the free module kG is injective; this follows from the analogue over k of 5.9, which implies that $\mathrm{Hom}_{kG}(-, kG) \approx \mathrm{Hom}_k(-, k)$.]

6 H_* and H^* as Functors of the Coefficient Module

Since $F \otimes_G -$ and $\mathcal{H}om_G(F, -)$ are covariant functors, it is clear that $H_*(G, -)$ and $H^*(G, -)$ are covariant functors of the coefficient module. The following proposition gives the basic properties of these functors.

(6.1) Proposition. (i) *There is a natural isomorphism* $H_0(G, M) \approx M_G$.

(i′) *There is a natural isomorphism* $H^0(G, M) \approx M^G$.

(ii) *For any exact sequence* $0 \to M' \xrightarrow{i} M \xrightarrow{j} M'' \to 0$ *of G-modules and any integer n there is a natural map* $\partial: H_n(G, M'') \to H_{n-1}(G, M')$ *such that the sequence*

$$\cdots \to H_1(G, M) \to H_1(G, M'') \xrightarrow{\partial} H_0(G, M') \to H_0(G, M) \to H_0(G, M'') \to 0$$

is exact. (The unlabelled arrows here represent the maps induced by i and j.)

(ii′) *For any exact sequence as in* (ii) *and any integer n there is a natural map* $\delta\colon H^n(G, M'') \to H^{n+1}(G, M')$ *such that the sequence*

$$0 \to H^0(G, M') \to H^0(G, M) \to H^0(G, M'') \xrightarrow{\delta} H^1(G, M') \to H^1(G, M) \to \cdots$$

is exact.

(iii) *If P is a projective* $\mathbb{Z}G$*-module then* $H_n(G, P) = 0$ *for* $n > 0$.

(iii′) *If Q is an injective* $\mathbb{Z}G$*-module then* $H^n(G, Q) = 0$ *for* $n > 0$.

Note: The naturality assertion in (ii) means that for any commutative diagram

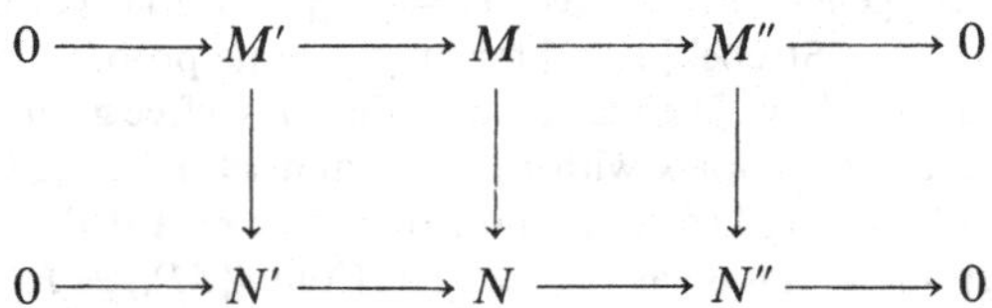

with exact rows, the square

$$\begin{array}{ccc} H_n(G, M'') & \xrightarrow{\partial} & H_{n-1}(G, M') \\ \downarrow & & \downarrow \\ H_n(G, N'') & \xrightarrow{\partial} & H_{n-1}(G, N') \end{array}$$

is commutative.

Proof of 6.1. (i) and (i′) were proved in §1. Given an exact sequence as in (ii) and a projective resolution of $\mathbb{Z}$ over $\mathbb{Z}G$, we have an exact sequence of chain complexes

$$0 \to M' \otimes_G F \to M \otimes_G F \to M'' \otimes_G F \to 0$$

since projectives are flat; the corresponding long exact homology sequence yields (ii). Similarly, (ii′) follows from the sequence of cochain complexes

$$0 \to \mathscr{H}om_G(F, M') \to \mathscr{H}om_G(F, M) \to \mathscr{H}om_G(F, M'') \to 0,$$

which is exact by the definition of "projective." Finally, (iii) and (iii′) are immediate consequences of the definitions of $H_*(G, —)$ and $H^*(G, —)$ and the exactness of $- \otimes_G P$ and $\mathrm{Hom}_G(—, Q)$. □

Properties (iii) and (iii′) are usually expressed by saying that projective modules are H_**-acyclic* and injective modules are H^**-acyclic*.

A functor T (say from R-modules to abelian groups) is said to be *effaceable* if every module M is a quotient of a module $\overline{M}$ such that $T(\overline{M}) = 0$. Clearly T is effaceable if $T(P) = 0$ for every projective P. Conversely, if T is effaceable then $T(P) = 0$ for every projective P. For let $\pi\colon \overline{P} \to P$ be a surjection with $T(\overline{P}) = 0$; since P is projective, π must split, hence $T(\pi)\colon T(\overline{P}) \to T(P)$ is a split surjection and $T(P) = 0$. Thus the content of (iii) above is that $H_n(G, —)$ is effaceable for $n > 0$. Similarly, (iii′) says that $H^n(G, —)$ is *co-effaceable* for $n > 0$, i.e., that every module M can be embedded in a module $\overline{\overline{M}}$ such that $H^n(G, \overline{\overline{M}}) = 0$.

(6.2) Proposition ("Shapiro's lemma"). *If $H \subseteq G$ and M is an H-module, then*

$$H_*(H, M) \approx H_*(G, \operatorname{Ind}_H^G M)$$

and

$$H^*(H, M) \approx H^*(G, \operatorname{Coind}_H^G M).$$

[Note: After we have discussed the functoriality of H_* and H^* with respect to group homomorphisms, we will be able to make 6.2 more precise by showing that the isomorphisms are induced by the inclusion $H \hookrightarrow G$ and the canonical H-maps $M \to \operatorname{Ind}_H^G M$ and $\operatorname{Coind}_H^G M \to M$. See exercise 2 of §8.]

PROOF. Let F be a projective resolution of $\mathbb{Z}$ over $\mathbb{Z}G$. Then F can also be regarded as a projective resolution of $\mathbb{Z}$ over $\mathbb{Z}H$, so

$$H_*(H, M) \approx H_*(F \otimes_{\mathbb{Z}H} M).$$

But $F \otimes_{\mathbb{Z}H} M \approx F \otimes_{\mathbb{Z}G} (\mathbb{Z}G \otimes_{\mathbb{Z}H} M) \approx F \otimes_G (\operatorname{Ind}_H^G M)$, whence the first isomorphism. The second isomorphism follows from the universal property of co-induction, which implies $\mathscr{H}om_H(F, M) \approx \mathscr{H}om_G(F, \operatorname{Coind}_H^G M)$, cf. 3.6. □

Taking $M = \mathbb{Z}$, for example, we conclude from 6.2 that

$$H_*(H) \approx H_*(G, \mathbb{Z}[G/H]). \tag{6.3}$$

(This shows that homology with coefficients is of interest even if one is primarily interested in ordinary integral homology). In case $(G\colon H) < \infty$, $\operatorname{Coind}_H^G M \approx \operatorname{Ind}_H^G M$ by 5.9, so we also have

$$H^*(H, \mathbb{Z}) \approx H^*(G, \mathbb{Z}[G/H]). \tag{6.4}$$

Similarly, still assuming $(G\colon H) < \infty$, we find

$$H^*(H, \mathbb{Z}H) \approx H^*(G, \mathbb{Z}G), \tag{6.5}$$

since $\mathbb{Z}G \otimes_{\mathbb{Z}H} \mathbb{Z}H \approx \mathbb{Z}G$. (It is also true, of course, that $H_*(H, \mathbb{Z}H) \approx H_*(G, \mathbb{Z}G)$, but this is of no interest in view of 6.1(iii).) Finally, we remark that Shapiro's lemma can be applied with $H = \{1\}$ to yield:

(6.6) Corollary. *Induced modules $\mathbb{Z}G \otimes A$ are H_*-acyclic. Co-induced modules* $\operatorname{Hom}(\mathbb{Z}G, A)$ *are H^*-acyclic.*

Note that 6.6 yields another proof that $H_n(G, —)$ is effaceable and $H^n(G, —)$ is co-effaceable for $n > 0$; for we have seen (3.4 and 3.7) that every module is a quotient of an induced module and that every module can be embedded in a co-induced module.

EXERCISE

(a) Show that the Mayer–Vietoris sequence II.7.7 can be deduced from the short exact sequence II.7.8 of permutation modules. [Hint: Use 6.1(ii) and 6.3.]

(b) More generally, derive in this way a Mayer–Vietoris sequence for homology and cohomology with coefficients. [Apply $- \otimes M$ and $\mathrm{Hom}(-, M)$ to II.7.8; use 5.6a and 6.2.]

7 Dimension Shifting

We saw in §6 that $H_n(G, —)$ is effaceable and $H^n(G, —)$ is co-effaceable for $n > 0$. The significance of this is that it allows us to use the following *dimension-shifting* technique: Given a module M, choose an H_*-acyclic module $\overline{M}$ which maps onto M (e.g., take $\overline{M} = \mathbb{Z}G \otimes M$), and let

$$K = \ker\{\overline{M} \to M\}.$$

Then the long exact sequence of 6.1(ii) yields

$$(7.1) \qquad H_n(G, M) \approx \begin{cases} H_{n-1}(G, K) & n > 1 \\ \ker\{H_0(G, K) \to H_0(G, \overline{M})\} & n = 1. \end{cases}$$

Thus a question about H_n can, in principle, be reduced to a question about H_{n-1}, provided we are willing to change the coefficient module. Iterating this procedure, we are ultimately reduced to H_0. Similarly, embedding M in an H^*-acyclic module $\overline{\overline{M}}$ (e.g., $\overline{\overline{M}} = \mathrm{Hom}(\mathbb{Z}G, M)$) and letting

$$C = \mathrm{coker}\{M \to \overline{\overline{M}}\},$$

we find

$$(7.2) \qquad H^n(G, M) \approx \begin{cases} H^{n-1}(G, C) & n > 1 \\ \mathrm{coker}\{H^0(G, \overline{\overline{M}}) \to H^0(G, C)\} & n = 1. \end{cases}$$

This argument shows, at least heuristically, that a "homology theory" H_* having properties analogous to (ii) and (iii) of 6.1 is completely determined by H_0. Similarly, a "cohomology theory" H^* satisfying (ii′) and (iii′) is determined by H^0. The rest of this section will be devoted to a precise formulation and proof of these assertions.

Let R be an arbitrary ring and $T = (T_n)_{n \in \mathbb{Z}}$ a family of covariant functors from R-modules to abelian groups. Assume that we are given "connecting homomorphisms" $\partial\colon T_n(M'') \to T_{n-1}(M')$ for every short exact sequence $0 \to M' \to M \to M'' \to 0$ of R-modules. We require that ∂ be *natural* (in the sense explained after the statement of Proposition 5.1) and that all composites be zero in the sequence

$$(\dagger)\qquad \cdots \to T_{n+1}(M'') \xrightarrow{\partial} T_n(M') \to T_n(M) \to T_n(M'') \xrightarrow{\partial} T_{n-1}(M') \to \cdots.$$

We will then say that T (or, more precisely, (T, ∂)) is a *∂-functor*. If, in addition, $T_n = 0$ for $n < 0$ and (†) is exact for every short exact sequence $0 \to M' \to M \to M'' \to 0$, then we will say that T is a *homological functor*. Thus, for example, the content of 6.1(ii) is that $H_*(G, —)$ is a homological functor on the category of G-modules.

If S and T are ∂-functors, then a *map* from S to T is a family φ of natural transformations $\varphi_n: S_n \to T_n$ such that the square

$$\begin{array}{ccc} S_n(M'') & \xrightarrow{\partial} & S_{n-1}(M') \\ \downarrow{\scriptstyle \varphi_n(M'')} & & \downarrow{\scriptstyle \varphi_{n-1}(M')} \\ T_n(M'') & \xrightarrow{\partial} & T_{n-1}(M') \end{array}$$

commutes for every short exact sequence $0 \to M' \to M \to M'' \to 0$ and every n.

(7.3) Theorem. *Let H be a homological functor such that H_n is effaceable for $n > 0$. If T is an arbitrary ∂-functor and φ_0: $T_0 \to H_0$ is a natural transformation, then φ_0 extends uniquely to a map φ: $T \to H$ of ∂-functors. This map φ is an isomorphism if and only if the following three conditions hold:*

(i) *φ_0 is an isomorphism.*
(ii) *T is homological.*
(iii) *T_n is effaceable for $n > 0$.*

Thus a homological functor H which is effaceable in positive dimensions is uniquely determined (up to canonical isomorphism of ∂-functors) by H_0. In particular, it follows that $H_*(G, —)$ is uniquely determined by the properties (i)–(iii) of 6.1.

PROOF OF 7.3. We wish to construct φ_n: $T_n \to H_n$ so that

$$(*)\qquad \begin{array}{ccc} T_n(M'') & \xrightarrow{\partial} & T_{n-1}(M') \\ \downarrow{\scriptstyle \varphi_n(M'')} & & \downarrow{\scriptstyle \varphi_{n-1}(M')} \\ H_n(M'') & \xrightarrow{\partial} & H_{n-1}(M') \end{array}$$

commutes for every short exact sequence $0 \to M' \to M \to M'' \to 0$. For $n \leq 0$ there is no choice and all diagrams (∗) commute automatically. Assume, then, that $n > 0$ and that φ_{n-1} has been defined. For any module M choose a short exact sequence $0 \to K \to P \to M \to 0$ with P projective, and consider the diagram

$$(**)\qquad \begin{array}{ccccccc} & & T_n(M) & \xrightarrow{\partial} & T_{n-1}(K) & \longrightarrow & T_{n-1}(P) \\ & & \Big\downarrow{\scriptstyle \varphi_n(M)} & & \Big\downarrow{\scriptstyle \varphi_{n-1}(K)} & & \Big\downarrow{\scriptstyle \varphi_{n-1}(P)} \\ 0 & \longrightarrow & H_n(M) & \xrightarrow{\partial} & H_{n-1}(K) & \longrightarrow & H_{n-1}(P) \end{array}$$

where the dotted arrow represents the map we want to define. Note that the composite on the top row is zero, that the bottom row is exact (because $H_n(P) = 0$), and that the right-hand square commutes (by naturality of φ_{n-1}). Consequently, there is a unique map $\varphi_n(M)$ which makes $(**)$ commute. Since this definition of φ_n was forced on us by $(*)$, the uniqueness assertion of the theorem is clear. It is also clear from $(**)$ that if conditions (i)–(iii) hold then we can prove inductively that φ_n is an isomorphism. It remains, therefore, to complete the proof of the existence of φ_n by proving that φ_n is well-defined and natural and that all squares of the form $(*)$ commute. We will need the following lemma.

(7.4) Lemma. *Let $\varphi_n(M)$ be defined as above by means of a short exact sequence $0 \to K \to P \to M \to 0$. For any diagram*

$$\begin{array}{ccccccccc} & & & & & & M & & \\ & & & & & & \Big\downarrow{\scriptstyle f} & & \\ 0 & \longrightarrow & N' & \longrightarrow & N & \longrightarrow & N'' & \longrightarrow & 0 \end{array}$$

with exact row, the diagram

$$\begin{array}{ccccc} T_n(M) & \xrightarrow{T_n(f)} & T_n(N'') & \xrightarrow{\partial} & T_{n-1}(N') \\ \Big\downarrow{\scriptstyle \varphi_n(M)} & & & & \Big\downarrow{\scriptstyle \varphi_{n-1}(N')} \\ H_n(M) & \xrightarrow{H_n(f)} & H_n(N'') & \xrightarrow{\partial} & H_{n-1}(N') \end{array}$$

commutes.

Proof. Since P is projective, the given diagram can be completed to a commutative diagram

$$\begin{array}{ccccccccc} 0 & \longrightarrow & K & \longrightarrow & P & \longrightarrow & M & \longrightarrow & 0 \\ & & \Big\downarrow{\scriptstyle g} & & \Big\downarrow & & \Big\downarrow{\scriptstyle f} & & \\ 0 & \longrightarrow & N' & \longrightarrow & N & \longrightarrow & N'' & \longrightarrow & 0. \end{array}$$

Consider now the diagram

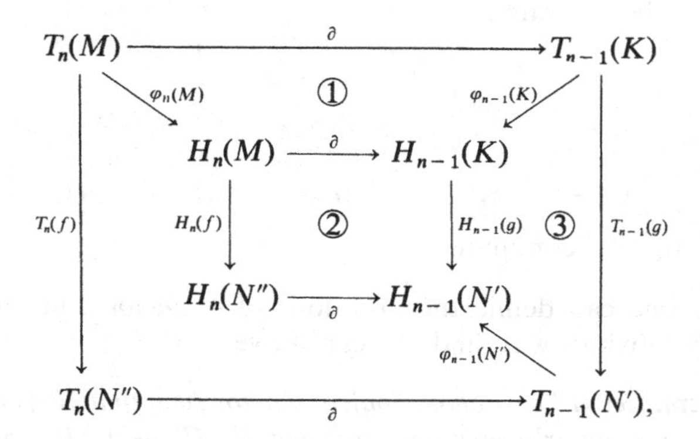

where the maps labelled ∂ in ① are the "∂-maps" associated to $0 \to K \to P \to M \to 0$. Then ① commutes by $(**)$, ② commutes because H is a ∂-functor, and ③ commutes by naturality of φ_{n-1}. Moreover, the outer square commutes because T is a ∂-functor. The lemma follows at once. □

Returning now to the proof of the theorem, let $f: M_1 \to M_2$ be an arbitrary map of R-modules. Choose exact sequences $0 \to K_i \to P_i \to M_i \to 0$ with P_i projective $(i = 1, 2)$, and use them to define $\varphi_n(M_i): T_n(M_i) \to H_n(M_i)$ as in $(**)$. Applying the lemma to the diagram

$$\begin{array}{ccccccccc} & & & & & & M_1 & & \\ & & & & & & \downarrow{\scriptstyle f} & & \\ 0 & \longrightarrow & K_2 & \longrightarrow & P_2 & \longrightarrow & M_2 & \longrightarrow & 0, \end{array}$$

we conclude that the outer rectangle is commutative in the diagram

$$\begin{array}{ccccc} T_n(M_1) & \xrightarrow{T_n(f)} & T_n(M_2) & \xrightarrow{\partial} & T_{n-1}(K_2) \\ \downarrow{\scriptstyle \varphi_n(M_1)} & & \downarrow{\scriptstyle \varphi_n(M_2)} & & \downarrow{\scriptstyle \varphi_{n-1}(K_2)} \\ H_n(M_1) & \xrightarrow{H_n(f)} & H_n(M_2) & \xrightarrow{\partial} & H_{n-1}(K_2). \end{array}$$

Now the right-hand square is commutative by $(**)$, and

$$\partial: H_n(M_2) \to H_{n-1}(K_2)$$

is injective; hence the left-hand square commutes. This shows that φ_n is natural. It also shows that $\varphi_n(M)$ is well-defined, independent of the choice of exact sequence $0 \to K \to P \to M \to 0$. [Take $f = \mathrm{id}_M$.] Finally, if $0 \to M' \to M \to M'' \to 0$ is an arbitrary exact sequence, then we can apply the lemma to the diagram

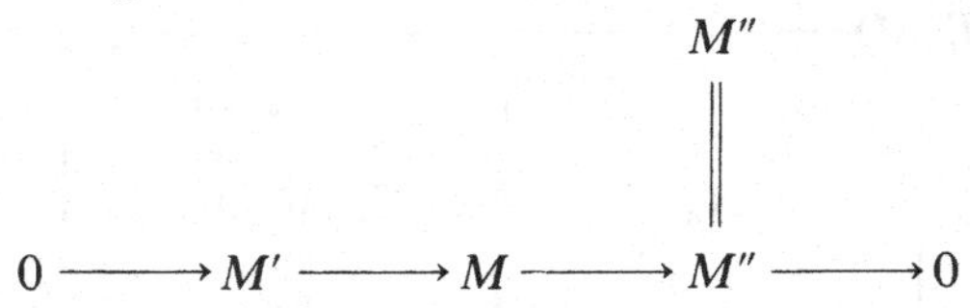

to conclude that $(*)$ commutes. □

Similarly one can define the notations of *δ-functor* and *cohomological functor* in the obvious way, and one can prove:

(7.5) Theorem. *Let H be a cohomological functor such that H^n is co-effaceable for $n > 0$. If T is an arbitrary δ-functor and $\varphi^0: H^0 \to T^0$ is a natural transformation, then φ^0 extends uniquely to a map $\varphi: H \to T$ of δ-functors. This map φ is an isomorphism if and only if the following three conditions hold:*

(i) *φ^0 is an isomorphism.*
(ii) *T is cohomological.*
(iii) *T^n is co-effaceable for $n > 0$.*

EXERCISES

1. (a) Prove the following generalization of 1.4: Let $\cdots \to C_1 \xrightarrow{\partial} C_0 \xrightarrow{\varepsilon} M \to 0$ be exact, where each C_i is H_*-acyclic; then $H_*(G, M) \approx H_*(C_G)$. [Hint: Apply the dimension-shifting argument using the exact sequences $0 \to \ker \varepsilon \to C_0 \to M \to 0$, $0 \to \ker \partial \to C_1 \to \ker \varepsilon \to 0$, etc.]

   (b) Similarly, let $0 \to M \to C^0 \to C^1 \to \cdots$ be exact, where each C^i is H^*-acyclic. Prove that $H^*(G, M) \approx H^*(C^G)$.

2. Use the results of this section to give a new proof of Proposition 2.2. [Method 1: If M is $\mathbb{Z}$-torsion-free, then $\mathrm{Tor}_*^G(M, -)$ and $H_*(G, M \otimes -)$ are homological functors which agree in dimension 0 and are effaceable in positive dimensions. A similar argument applies to $\mathrm{Ext}_G^*(M, -)$ and $H^*(G, \mathrm{Hom}(M, -))$. Method 2: Compute $\mathrm{Tor}_*^G(M, N)$ in terms of a projective resolution of M. Express this computation in the form $\mathrm{Tor}_*^G(M, N) = H_*(C_G)$ and apply exercise 1a. Similarly, deduce the result on Ext from 1b.]

3. In the discussion at the beginning of this section, show that the surjection $\overline{M} \twoheadrightarrow M$ and the injection $M \hookrightarrow \overline{\overline{M}}$ can always be taken to be $\mathbb{Z}$-split [Hint: Use 3.4 and 3.7.]

8 H_* and H^* as Functors of Two Variables

Let $\mathscr{C}$ be the following category: An object of $\mathscr{C}$ is a pair (G, M), where G is a group and M is a G-module; a map in $\mathscr{C}$ from (G, M) to (G', M') is a pair (α, f), where $\alpha: G \to G'$ is a map of groups and $f: M \to M'$ is a map of abelian groups such that $f(gm) = \alpha(g)f(m)$ for $g \in G$, $m \in M$. (In other words, f is a G-module homomorphism if M' is regarded as a G-module via α.) Given (α, f), let F (resp. F') be a projective resolution of $\mathbb{Z}$ over $\mathbb{Z}G$ (resp. $\mathbb{Z}G'$), and let $\tau: F \to F'$ be a chain map compatible with α as in §II.6. Then there is a chain map $\tau \otimes f: F \otimes_G M \to F' \otimes_{G'} M'$, and $\tau \otimes f$ induces a well-defined map $(\alpha, f)_*: H_*(G, M) \to H_*(G', M')$. In this way H_* becomes a covariant functor on $\mathscr{C}$. In case $M = M'$ and $f = \mathrm{id}_{M'}$, we will simply write α_* for $(\alpha, f)_*: H_*(G, M') \to H_*(G', M')$. Note that the general map $(\alpha, f)_*$ can always be written as the composite

$$H_*(G, M) \xrightarrow{H_*(G, f)} H_*(G, M') \overset{\alpha_*}{\to} H_*(G', M'),$$

where $H_*(G, f)$ makes sense because f is a map of G-modules.

As an example of this functoriality we consider the maps induced by conjugation. Given $H \subseteq G$, a G-module M, and an element $g \in G$, we denote by $c(g): (H, M) \to (gHg^{-1}, M)$ the isomorphism in $\mathscr{C}$ given by

$$(h \mapsto ghg^{-1}, m \mapsto gm).$$

To compute $c(g)_*: H_*(H, M) \to H_*(gHg^{-1}, M)$ on the chain level, we choose a projective resolution F of $\mathbb{Z}$ over $\mathbb{Z}G$ and use F to compute the homology of H and gHg^{-1}. We can take $\tau: F \to F$ to be multiplication by g as in the proof of II.6.2, and we then find that $c(g)_*$ is induced by the chain map $F \otimes_H M \to F \otimes_{gHg^{-1}} M$ given by $x \otimes m \mapsto gx \otimes gm$. Note that this is just the map obtained from the usual diagonal action of g on $F \otimes M$ by passage to quotients.

For $z \in H_*(H, M)$ we set $gz = c(g)_* z \in H_*(gHg^{-1}, M)$. In view of the above description of $c(g)_*$ on the chain level, we have:

(8.1) Proposition. *If $h \in H$ then $hz = z$ for all $z \in H_*(H, M)$.*

(8.2) Corollary. *If $H \lhd G$ and M is a G-module, then the conjugation action of G on (H, M) induces an action of G/H on $H_*(H, M)$.*

The situation for cohomology is similar. Let $\mathscr{D}$ be the category with the same objects as $\mathscr{C}$, but where a map $(G, M) \to (G', M')$ is now a pair

$$(\alpha: G \to G', f: M' \to M);$$

as before we require that f be a G-module map, i.e., that $f(\alpha(g)m') = gf(m')$ for $g \in G$, $m' \in M'$. If F and F' are resolutions for G and G' and $\tau: F \to F'$ is a chain map compatible with α, then there is a chain map

$$\mathscr{H}om(\tau, f): \mathscr{H}om_{G'}(F', M') \to \mathscr{H}om_G(F, M),$$

which induces $(\alpha, f)^*: H^*(G', M') \to H^*(G, M)$. Thus H^* is a contravariant functor on $\mathscr{D}$. In case $M = M'$ and $f = \mathrm{id}_{M'}$, we will write α^* for $(\alpha, f)^*$; note that the general map $(\alpha, f)^*$ is the composite

$$H^*(G', M') \overset{\alpha^*}{\to} H^*(G, M') \xrightarrow{H^*(G, f)} H^*(G, M).$$

Let $\mathscr{C}_0$ be the subcategory of $\mathscr{C}$ consisting of all the objects (G, M) and of those maps (α, f) such that f is bijective. Then $\mathscr{C}_0$ is isomorphic to a subcategory of $\mathscr{D}$ by $(\alpha, f) \mapsto (\alpha, f^{-1})$, so H^* can be regarded as a contravariant functor on $\mathscr{C}_0$. In particular, the conjugation maps $c(g)$ discussed above are in $\mathscr{C}_0$, so we have an isomorphism $c(g)^*: H^*(gHg^{-1}, M) \to H^*(H, M)$ for any $g \in G$, $H \subseteq G$, and G-module M. If $z \in H^*(H, M)$ then we set

$$gz = (c(g)^*)^{-1}(z) \in H^*(gHg^{-1}, M).$$

In terms of a projective resolution F of $\mathbb{Z}$ over $\mathbb{Z}G$, the map $z \mapsto gz$ is induced by the cochain map $\mathscr{H}om_H(F, M) \to \mathscr{H}om_{gHg^{-1}}(F, M)$ given by

$$f \mapsto [x \mapsto gf(g^{-1}x)].$$

Note that this is just the usual diagonal action of g on $\mathscr{H}om(F, M)$, restricted to the subcomplex $\mathscr{H}om_H(F, M)$. In case $g \in H$ we clearly obtain the identity map, hence:

(8.3) Proposition. *If $h \in H$ then $hz = z$ for all $z \in H^*(H, M)$.*

(8.4) Corollary. *If $H \lhd G$ and M is a G-module, then the conjugation action of G on (H, M) induces an action of G/H on $H^*(H, M)$.*

EXERCISES

1. If H is central in G and M is an abelian group with trivial G-action, show that G/H acts trivially on $H_*(H, M)$ and $H^*(H, M)$.

2. Let $\alpha: H \hookrightarrow G$ be an inclusion, let M be an H-module, and let $i: M \to \operatorname{Ind}_H^G M$ and $\pi: \operatorname{Coind}_H^G M \to M$ be the canonical H-maps. Show that the isomorphisms of Shapiro's lemma (6.2) are given by $(\alpha, i)_*: H_*(H, M) \to H_*(G, \operatorname{Ind}_H^G M)$ and $(\alpha, \pi)^*: H^*(G, \operatorname{Coind}_H^G M) \to H^*(H, M)$. [Hint: You can take τ to be the identity map.]

9 The Transfer Map

Given an inclusion $\alpha: H \hookrightarrow G$ and a G-module M, we have seen that there are maps $\alpha^*: H^*(G, M) \to H^*(H, M)$ and $\alpha_*: H_*(H, M) \to H_*(G, M)$. We will often write $\alpha^* = \operatorname{res}_H^G$ and call it a *restriction map*. Similarly, we write $\alpha_* = \operatorname{cor}_H^G$ and call it a *corestriction map*. The purpose of this section is to show that if $(G: H) < \infty$ then there are maps going in the other direction, called *transfer maps*. This extremely useful observation is due to Eckmann [1953] and Artin–Tate [unpublished]. The existence of the transfer maps is somewhat more subtle than that of α_* and α^*; in particular, they are *not* induced by maps in the categories $\mathscr{C}$ and $\mathscr{D}$ of §8. [The name "transfer map" comes from the fact that on H_1 it coincides with the transfer map of classical group theory, due to Schur [1902]; see exercise 2 below.]

We will explain the transfer maps from five different points of view, all of which are useful:

(A) For any G-module M and any $H \subseteq G$ we have a canonical surjection of G-modules

$$\mathbb{Z}G \otimes_{\mathbb{Z}H} M \to M \tag{9.1}$$

and a canonical injection

$$M \to \operatorname{Hom}_{\mathbb{Z}H}(\mathbb{Z}G, M), \tag{9.2}$$

cf. 3.4 and 3.7. Applying $H_*(G, —)$ to 9.1 and using Shapiro's lemma, we obtain a map $H_*(H, M) \to H_*(G, M)$; this is easily seen to coincide with α_*. [Use exercise 2 of §8.] Similarly, applying $H^*(G, —)$ to 9.2 and using Shapiro's lemma we obtain $\alpha^*: H^*(G, M) \to H^*(H, M)$. If we assume now that $(G:H) < \infty$, then $\mathbb{Z}G \otimes_{\mathbb{Z}H} M \approx \operatorname{Hom}_{\mathbb{Z}H}(\mathbb{Z}G, M)$ by 5.9. Consequently, we can apply $H^*(G, —)$ to 9.1 and $H_*(G, —)$ to 9.2 to obtain the transfer maps going in the other direction. These maps are often denoted

$$\operatorname{cor}_H^G: H^*(H, M) \to H^*(G, M)$$

and

$$\operatorname{res}_H^G: H_*(G, M) \to H_*(H, M)$$

or simply tr_H^G if it is clear from the context whether we are talking about H_* or H^*.

(B) For any $H \subseteq G$ we can regard both $H_*(G, —)$ and $H_*(H, —)$ as homological functors on the category of G-modules, and both are easily seen to be effaceable in positive dimensions. Assuming now that $(G: H) < \infty$, we define $\operatorname{tr}: M_G \to M_H$ by $\operatorname{tr}(\overline{m}) = \sum_{g \in H\backslash G} \overline{\overline{gm}}$, where $\overline{m}$ (resp. $\overline{\overline{m}}$) denotes the image of m in M_G (resp. M_H). (The sum makes sense because $\overline{\overline{gm}}$ depends only on the class of g in $H\backslash G$.) We now define res: $H_*(G, —) \to H_*(H, —)$ to be the unique extension of tr to a map of homological functors, cf. Theorem 7.3. To see that this agrees with the map defined in (A), one need only verify that the latter is compatible with ∂ and equals tr in dimension zero. Details are left to the reader. Similarly, cor: $H^*(H, —) \to H^*(G, —)$ can be defined as the unique map of cohomological functors which in dimension zero is the map $\operatorname{tr}: M^H \to M^G$ defined by $\operatorname{tr}(m) = \sum_{g \in G/H} gm$. [To verify co-effaceability of $H^n(H, —)$ on the category of G-modules, use exercise 3 of §3; alternatively, note that a co-induced G-module is also co-induced as an H-module.]

(C) Let F be a projective resolution of $\mathbb{Z}$ over $\mathbb{Z}G$. Then $F \otimes_H M = (F \otimes M)_H$ and $F \otimes_G M = (F \otimes M)_G$. We can now define res: $H_*(G, M) \to H_*(H, M)$ to be the map induced by the chain map $\operatorname{tr}: (F \otimes M)_G \to (F \otimes M)_H$, where tr is as in (B). This defines a map of homological functors which equals tr on H_0, hence it agrees with definition (B). Similarly, $\mathcal{H}om_G(F, M) = \mathcal{H}om(F, M)^G$ and $\mathcal{H}om_H(F, M) = \mathcal{H}om(F, M)^H$, where G acts diagonally on $\mathcal{H}om$ (cf. §0). Hence there is a cochain map tr: $\mathcal{H}om(F, M)^H \to \mathcal{H}om(F, M)^G$ which induces cor: $H^*(H, M) \to H^*(G, M)$.

(D) If we wish to compute the transfer maps of (C) using one resolution $F(G)$ for G and a different one $F(H)$ for H, then we need an (augmentation-preserving) H-chain map $\tau: F(G) \to F(H)$ in order to realize the canonical isomorphism $H_*(F(G) \otimes_H M) \xrightarrow{\approx} H_*(F(H) \otimes_H M)$. The transfer maps will then be induced by the composites

(9.3) $$F(G) \otimes_G M \xrightarrow{\text{tr}} F(G) \otimes_H M \xrightarrow{\tau \otimes M} F(H) \otimes_H M$$

and

(9.4) $$\mathcal{H}om_H(F(H), M) \xrightarrow{\mathcal{H}om(\tau, M)} \mathcal{H}om_H(F(G), M) \xrightarrow{\text{tr}} \mathcal{H}om_G(F(G), M).$$

Taking $F(G)$ and $F(H)$ to be the standard resolutions, for example, we can easily write down such a map τ as follows. Choose a set of representatives for the right cosets Hg. Then there is a unique map $\rho: G \to H$ of left H-sets which sends every coset representative to 1; explicitly, if $\bar{g}$ denotes the representative of Hg, then $\rho g = g\bar{g}^{-1}$. Using the "homogeneous" description of the standard resolutions, we can now define $\tau: F(G) \to F(H)$ by $\tau(g_0, \ldots, g_n) = (\rho g_0, \ldots, \rho g_n)$. The interested reader can translate this into the bar notation and can write out the corresponding transfer maps $C_*(G, M) \to C_*(H, M)$ and $C^*(H, M) \to C^*(G, M)$ given by 9.3 and 9.4.

(E) If $\tilde{Y} \to Y$ is a covering map of finite degree, then there are transfer maps $H_*(Y, M) \to H_*(\tilde{Y}, M)$ and $H^*(\tilde{Y}, M) \to H^*(Y, M)$ for any coefficient system M on Y. In terms of cellular chain complexes, the homology transfer is induced by $\sigma \otimes m \mapsto \sum_{\tilde{\sigma}} \tilde{\sigma} \otimes m$, where σ is an oriented cell of Y and $\tilde{\sigma}$ ranges over the oriented cells of $\tilde{Y}$ lying over σ. Similarly, the cohomology transfer is induced by the cochain map $f \mapsto [\sigma \mapsto \sum_{\tilde{\sigma}} f(\tilde{\sigma})]$. If Y is a $K(G, 1)$ and $\tilde{Y}$ is the covering space corresponding to $H \subseteq G$, then $\tilde{Y}$ is a $K(H, 1)$ and these transfer maps agree with the transfer maps $H_*(G, M) \to H_*(H, M)$ and $H^*(H, M) \to H^*(G, M)$. To see this, one need only apply definition (C) with F equal to the chain complex of the universal cover of Y.

We now give some properties of the transfer maps. Let $H(—, —)$ denote either H_* or H^*.

(9.5) Proposition. (i) *Given* $K \subseteq H \subseteq G$ *with* $(G: K) < \infty$,

$$\text{cor}_K^G = \text{cor}_H^G \circ \text{cor}_K^H \quad \text{and} \quad \text{res}_K^G = \text{res}_K^H \circ \text{res}_H^G.$$

(ii) *Given* $(G: H) < \infty$ *and* $z \in H(G, M)$, $\text{cor}_H^G \, \text{res}_H^G \, z = (G: H)z$.

(iii) *Given* $H, K \subseteq G$ *with* $(G: H) < \infty$ *if* $H(—, —) = H^*$ *and* $(G: K) < \infty$ *if* $H(—, —) = H_*$, *we have*

$$\text{res}_K^G \, \text{cor}_H^G \, z = \sum_{g \in E} \text{cor}_{K \cap gHg^{-1}}^K \, \text{res}_{K \cap gHg^{-1}}^{gHg^{-1}} \, gz$$

for any $z \in H(H, M)$, *where* E *is a set of representatives for the double cosets* KgH. *In particular, if* $H \lhd G$ *and* $(G:H) < \infty$, *then* $\operatorname{res}_H^G \operatorname{cor}_H^G z = \sum_{g \in G/H} gz = Nz$, *where* N *is the norm element of* $\mathbb{Z}[G/H]$.

(See §8 for the definition of gz. Note that in the cohomology case we can write $\operatorname{res}^{gHg^{-1}}_{K \cap gHg^{-1}} gz = \operatorname{res}^{H}_{K \cap gHg^{-1}} z$, where the "restriction map" on the right is with respect to the conjugation map $(k \mapsto g^{-1}kg, m \mapsto g^{-1}m)$, regarded as a map $(K \cap gHg^{-1}, M) \to (H, M)$ in $\mathscr{C}_0$.)

Proof of 9.5: (i) and (ii) follow easily from the definitions and will be left to the reader. (iii) can be deduced from the double coset formula 5.6b and definition (A) of the transfer, but it is easier to work directly with definition (C). We will give the proof in the case $H(\text{—}, \text{—}) = H^*$; the homology case is similar. Let $u \in \mathscr{H}om(F, M)^H$ represent z. Then $\operatorname{res}_K^G \operatorname{cor}_H^G z$ is represented by $\sum_{g \in G/H} gu \in \mathscr{H}om(F, M)^G \subseteq \mathscr{H}om(F, M)^K$. Grouping together those terms gu corresponding to a given K-orbit in G/H, we find

$$\sum_{g \in G/H} gu = \sum_{g \in E} \sum_{k \in K/K_g} kgu,$$

where $K_g = \{k \in K: kgH = gH\} = K \cap gHg^{-1}$. Since the inner sum on the right clearly represents $\operatorname{cor}_{K_g}^K \operatorname{res}_{K_g}^{gHg^{-1}} gz$, this proves (iii). □

Exercises

1. Show that $\operatorname{res}_H^G: H(G, M) \to H(H, M)$ and $\operatorname{cor}_H^G: H(H, M) \to H(G, M)$ are "invariant under conjugation," in the sense that $g \cdot \operatorname{res}_H^G w = \operatorname{res}_{gHg^{-1}}^G w$ for $w \in H(G, M)$ and $\operatorname{cor}_{gHg^{-1}}^G gz = \operatorname{cor}_H^G z$ for $z \in H(H, M)$.

2. The transfer map $H_1(G) \to H_1(H)$ can be regarded as a map $G_{ab} \to H_{ab}$. Using definition (D), show that this map is given by

$$g \bmod [G, G] \mapsto \prod_{g' \in E} g'g\overline{g'g}^{-1} \bmod [H, H],$$

where E is a set of representatives for the right cosets Hg and $\bar{x}$ is the representative of Hx. This formula is usually taken as the definition of the transfer map in group theory texts, cf. Hall [1959], formula (14.2.4).

10 Applications of the Transfer

To simplify the notation we will state the results of this section only for cohomology; similar results hold for homology.

(10.1) Proposition. *Let* M *be a* G*-module and* $H \subseteq G$ *a subgroup of finite index such that* $H^n(H, M) = 0$ *for some* n. *Then* $H^n(G, M)$ *is annihilated by* $(G:H)$. *In particular, if* $(G:H)$ *is invertible in* M *(i.e., if multiplication by* $(G:H)$ *is an isomorphism), then* $H^n(G, M) = 0$.

PROOF. The first assertion follows immediately from 9.5(ii). The second follows from the first, since $(G: H)$ is invertible in $H^n(G, M)$ if it is invertible in M. □

Note that 10.1 applies to any $n > 0$ if G is finite and $H = \{1\}$. Hence:

(10.2) Corollary. *If G is finite, then $H^n(G, M)$ is annihilated by $|G|$ for all $n > 0$. If $|G|$ is invertible in M (e.g., if M is a $\mathbb{Q}G$-module), then $H^n(G, M) = 0$ for all $n > 0$.*

It follows that $H^n(G, M)$ admits a primary decomposition

$$H^n(G, M) = \bigoplus_p H^n(G, M)_{(p)},$$

where p ranges over the primes dividing $|G|$ and $H^n(G, M)_{(p)}$ is the p-primary component of $H^n(G, M)$. Now fix a prime p and let H be a p-Sylow subgroup of G. Since $(G: H)$ is relatively prime to p, 9.5(ii) implies that $\operatorname{cor}_H^G \operatorname{res}_H^G$ is an isomorphism on $H^n(G, M)_{(p)}$. In particular, the restriction map induces a monomorphism $H^n(G, M)_{(p)} \hookrightarrow H^n(H, M)$. In order to describe the image of this monomorphism, we need some terminology.

If G is an arbitrary group, $H \subseteq G$ is a subgroup, and M is a G-module, we will say that an element $z \in H^*(H, M)$ is *G-invariant* if $\operatorname{res}^H_{H \cap gHg^{-1}} z = \operatorname{res}^{gHg^{-1}}_{H \cap gHg^{-1}} gz$ for all $g \in G$. In case $H \lhd G$, this just says that $z \in H^*(H, M)^{G/H}$. Note that if $z = \operatorname{res}_H^G w$ for some $w \in H^*(G, M)$ then z is G-invariant; for $gz = \operatorname{res}^G_{gHg^{-1}} w$ by exercise 1 of §9, hence $\operatorname{res}^{gHg^{-1}}_{H \cap gHg^{-1}} gz = \operatorname{res}^G_{H \cap gHg^{-1}} w = \operatorname{res}^H_{H \cap gHg^{-1}} z$.

We can now state:

(10.3) Theorem. *Let G be a finite group and H a p-Sylow subgroup. For any G-module M and any $n > 0$, res_H^G maps $H^n(G, M)_{(p)}$ isomorphically onto the set of G-invariant elements of $H^n(H, M)$. In particular, if $H \lhd G$ then*

$$H^n(G, M)_{(p)} \approx H^n(H, M)^{G/H}.$$

PROOF. We have already shown that res_H^G maps $H^n(G, M)_{(p)}$ monomorphically into the G-invariants, so it remains to show that any invariant z is in the image. Consider the element $w = \operatorname{cor}_H^G z \in H^n(G, M)$. Since $H^n(H, M)$ is annihilated by $|H|$, $w \in H^n(G, M)_{(p)}$. We now compute $\operatorname{res}_H^G w$ by the double coset formula 9.5(iii); since z is invariant, we obtain

$$\begin{aligned}
\operatorname{res}_H^G w &= \sum_{g \in H\backslash G/H} \operatorname{cor}^H_{H \cap gHg^{-1}} \operatorname{res}^{gHg^{-1}}_{H \cap gHg^{-1}} gz \\
&= \sum_{g \in H\backslash G/H} \operatorname{cor}^H_{H \cap gHg^{-1}} \operatorname{res}^H_{H \cap gHg^{-1}} z \\
&= \sum_{g \in H\backslash G/H} (H: H \cap gHg^{-1})z \\
&= (G: H)z.
\end{aligned}$$

(The last equality is obtained by decomposing the set G/H into H-orbits and noting that $(H: H \cap gHg^{-1})$ is the cardinality of the H-orbit of gH.) Since $(G: H)$ is prime to p, it follows that $z = \operatorname{res}_H^G w'$, where

$$w' = w/(G: H) \in H^n(G, M)_{(p)}. \qquad \square$$

This proof, which should be thought of as an averaging argument, can be used in other situations where division by $(G: H)$ is possible. For example, one proves in the same way:

(10.4) Proposition. *Let G be arbitrary, M a G-module, and $H \subseteq G$ a subgroup of finite index. If $(G:H)$ is invertible in M then res_H^G maps $H^*(G, M)$ isomorphically onto the set of G-invariants in $H^*(H, M)$. In particular, if $H \lhd G$ then $H^*(G, M) \approx H^*(H, M)^{G/H}$.*

Exercises

1. Use 10.3 to compute $H^*(G, \mathbb{Z})$, where G is the symmetric group on three letters. [This will be easier after we have cup products available, cf. exercise 5 of §V.3, but it is doable now.]

2. Let $H \subseteq G$ be a subgroup of finite index. For any G-module M and any H-H double coset C, define a G-endomorphism $f(C)$ of $\mathbb{Z}G \otimes_{\mathbb{Z}H} M$ by $1 \otimes m \mapsto \sum_{g \in C/H} g \otimes g^{-1}m$. Using Shapiro's lemma, we can view the map $H^*(G, f(C))$ as an endomorphism $T(C)$ of $H^*(H, M)$.

   (a) If $C = HgH$, show that $T(C)z = \operatorname{cor}_{H \cap gHg^{-1}}^H \operatorname{res}_{H \cap gHg^{-1}}^{gHg^{-1}} gz$. [Hint: It suffices to check this in dimension 0.]

   (b) If $z \in H^*(H, M)$ is G-invariant, show that $T(C)z = a(C)z$, where $a(C) = |C/H| = (H: H \cap gHg^{-1})$.

   (c) In the situation of 10.3 and 10.4, show that the image of res_H^G is the set of z such that $T(C)z = a(C)z$ for all double cosets C.

CHAPTER IV
Low-Dimensional Cohomology and Group Extensions

1 Introduction

An *extension* of a group G by a group N is a short exact sequence of groups

$$(*) \qquad 1 \to N \to E \to G \to 1.$$

[Warning: Some people call this an extension of N by G.] A second extension $1 \to N \to E' \to G \to 1$ of G by N is said to be *equivalent* to $(*)$ if there is a map $E \to E'$ making the diagram

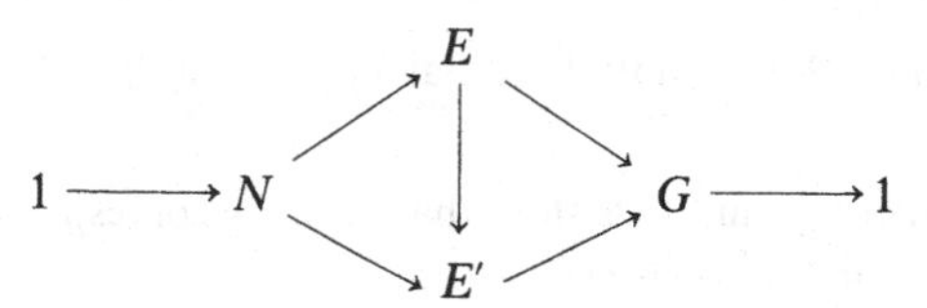

commute. Note that such a map is necessarily an isomorphism.

The main problem in the subject of group extensions is to classify the extensions of G by N up to equivalence. Roughly speaking, then, we are trying to understand all possible ways of building a group E with N as a normal subgroup and G as the quotient. This problem turns out to involve the cohomology functors $H^i(G, -)$ for $i = 1, 2, 3$.

For a while we will consider only the case where the kernel N is an abelian group A (written additively). A special feature of this case is that an extension

$$0 \to A \xrightarrow{i} E \xrightarrow{\pi} G \to 1$$

gives rise to an *action* of G on A, making A a G-module. For E acts on A by conjugation since A is embedded as a normal subgroup of E; and the conjugation action of A on itself is trivial, so there is an induced action of $E/A = G$

on A. Explicitly, given $g \in G$ we choose $\tilde{g} \in E$ such that $\pi(\tilde{g}) = g$, and the action of g on A is then characterized by

$$(1.1) \qquad i(ga) = \tilde{g}i(a)\tilde{g}^{-1} \qquad (a \in A).$$

It is often convenient to rewrite 1.1 as a *commutation rule*

$$(1.2) \qquad \tilde{g}i(a) = i(ga)\tilde{g}.$$

In particular, this shows that $i(A)$ is central in E if and only if the G-action is trivial. In this case the extension is called a *central extension*.

In view of the G-module structure on A, we can refine the classification problem by fixing a G-module A and trying to classify the extensions of G by A which give rise to the given action of G on A. As we will see in §3, this problem has a very simple solution: The equivalence classes of such extensions are in 1-1 correspondence with the elements of $H^2(G, A)$.

We begin by reviewing the simplest class of extensions, namely, the *split extensions.*

2 Split Extensions

Fix a G-module A and let

$$(*) \qquad 0 \to A \xrightarrow{i} E \xrightarrow{\pi} G \to 1$$

be an extension which gives rise to the given action of G on A. We say that $(*)$ *splits* if there is a homomorphism $s: G \to E$ such that $\pi s = \mathrm{id}_G$. The following characterization of split extensions is probably well-known to the reader, but we will write out the proof in detail since it motivates the proof of the main result of the next section.

(2.1) Proposition. *The following conditions on the extension* $(*)$ *are equivalent:*

(i) $(*)$ *splits.*

(ii) *E has a subgroup $\tilde{G}$ which is mapped by π isomorphically onto G, i.e., which satisfies $E = i(A) \cdot \tilde{G}$ and $i(A) \cap \tilde{G} = \{1\}$.*

(iii) *E has a subgroup $\tilde{G}$ such that every element $e \in E$ is uniquely expressible in the form $e = i(a)\tilde{g}$ $(a \in A, \tilde{g} \in \tilde{G})$.*

(iv) $(*)$ *is equivalent to the extension*

$$0 \to A \xrightarrow{i'} A \rtimes G \xrightarrow{\pi'} G \to 1,$$

where $A \rtimes G$ is the semi-direct product of G and A relative to the given action, and i' and π' are the canonical inclusion and projection maps.

(The definition of $A \rtimes G$ will be recalled in the course of the proof of 2.1.)

PROOF. Clearly (i) $\Leftrightarrow$ (ii) $\Leftrightarrow$ (iii). To prove that these conditions imply (iv), let $s: G \to E$ be a splitting and note that we have a set-theoretic bijection $A \times G \approx E$, given by $(a, g) \mapsto i(a)s(g)$. There is therefore a unique group law on the set $A \times G$ such that this bijection is an isomorphism. To calculate this group law, we must express a typical product $[i(a)s(g)] \cdot [i(b)s(h)]$ in E in the form $i(-)s(-)$. Using the commutation rule 1.2, we find

$$\begin{aligned} i(a)s(g)i(b)s(h) &= i(a)i(gb)s(g)s(h) \\ &= i(a + gb)s(gh). \end{aligned}$$

Thus the group law on $A \times G$ is given by

$$(a, g) \cdot (b, h) = (a + gb, gh).$$

The set $A \times G$ with this group law is, by definition, the semi-direct product $A \rtimes G$, and (iv) follows at once. Finally, it is trivial that (iv) $\Rightarrow$ (i). □

Proposition 2.1 says that there is only one split extension of G by A (up to equivalence) associated to the given action of G on A. Nevertheless, there is an interesting "classification" problem involving split extensions: Given that an extension (*) splits, classify all possible splittings.

In case G acts trivially on A, for example, so that the group E is isomorphic to the direct product $A \times G$, then the splittings are obviously in 1-1 correspondence with homomorphisms $G \to A$. In the general case, I claim that splittings correspond to *derivations* (also called *crossed homomorphisms*). These are functions $d: G \to A$ satisfying

(2.2) $$d(gh) = dg + g \cdot dh$$

for all $g, h \in G$. To see this we may assume that (*) is the canonical split extension

$$0 \to A \to A \rtimes G \to G \to 1.$$

A function $s: G \to A \rtimes G$ with $\pi s = \mathrm{id}$ then has the form $s(g) = (dg, g)$, where d is a function $G \to A$. We have

$$s(g)s(h) = (dg + g \cdot dh, gh),$$

so s will be a homomorphism if and only if d is a derivation. This proves the claim.

Remark. The term "derivation" seems more reasonable if we think of G as acting on A on the right by the trivial action, in addition to the given left action of G on A. The equation 2.2 then takes the familiar form

$$d(gh) = dg \cdot h + g \cdot dh,$$

as in the product rule for derivatives.

Two splittings s_1, s_2 will be said to be *A-conjugate* if there is an element $a \in A$ such that $s_1(g) = i(a)s_2(g)i(a)^{-1}$ for all $g \in G$. Since

$$(a, 1)(b, g)(a, 1)^{-1} = (a + b - ga, g)$$

in $A \rtimes G$, this conjugacy relation becomes

$$d_1 g = a + d_2 g - ga$$

in terms of the derivations d_1, d_2 corresponding to s_1, s_2. Thus d_1 and d_2 correspond to A-conjugate splittings if and only if their difference $d_2 - d_1$ is a function $G \to A$ of the form $g \mapsto ga - a$ for some fixed $a \in A$. Such a function is said to be a *principal derivation.*

We can summarize the preceding paragraph by saying that the A-conjugacy classes of splittings of a split extension of G by A correspond to the elements of the quotient group $\mathrm{Der}(G, A)/P(G, A)$, where $\mathrm{Der}(G, A)$ is the abelian group of derivations $G \to A$ and $P(G, A)$ is the group of principal derivations. On the other hand, a glance at the standard cochain complex $C^*(G, A)$ shows that $\mathrm{Der}(G, A)$ is the group of 1-cocycles and $P(G, A)$ is the group of 1-coboundaries (cf. exercise 2 of §III.1). We have therefore established:

(2.3) Proposition. *For any G-module A, the A-conjugacy classes of splittings of the split extension*

$$0 \to A \to A \rtimes G \to G \to 1$$

are in 1-1 *correspondence with the elements of* $H^1(G, A)$.

Exercises

The purpose of these exercises is to develop the theory of derivations and to give algebraic proofs of some results that were proved topologically in Chapters I and II.

1. If d is a derivation, show that $d(1) = 0$.

2. Let I be the augmentation ideal of $\mathbb{Z}G$ and let $D: G \to I$ be the derivation defined by $Dg = g - 1$ for $g \in G$. [Regarded as a derivation $G \to \mathbb{Z}G$, this is simply the principal derivation corresponding to $1 \in \mathbb{Z}G$.] Show that D is the *universal derivation* on G, in the following sense. Given any G-module A and any derivation $d: G \to A$, there is a unique G-module map $f: I \to A$ such that $d = fD$:

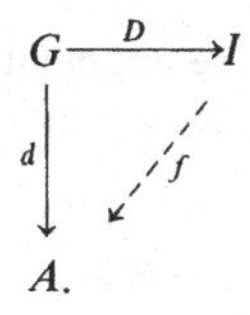

Thus $\operatorname{Der}(G, A) \approx \operatorname{Hom}_{\mathbb{Z}G}(I, A)$. [Hint: A derivation $d: G \to A$ extends to an additive map $\bar{d}: \mathbb{Z}G \to A$ such that

$$\bar{d}(rs) = \bar{d}r \cdot \varepsilon(s) + r \cdot \bar{d}s$$

for all $r, s \in \mathbb{Z}G$, where $\varepsilon: \mathbb{Z}G \to \mathbb{Z}$ is the augmentation. The restriction of $\bar{d}$ to I is a module homomorphism such that $\bar{d}(g - 1) = dg$ for all $g \in G$.]

3. Let $F = F(S)$ be the free group generated by a set S.

(a) If A is an F-module and $(a_s)_{s \in S}$ is a family of elements of A, show that there is a unique derivation $d: F \to A$ such that $ds = a_s$ for all $s \in S$. [Hint: Derivations $F \to A$ correspond to splittings of $0 \to A \to A \rtimes F \to F \to 1$; now use the universal mapping property of F.]

(b) Deduce from (a) and exercise 2 that the augmentation ideal of $\mathbb{Z}F$ is a free $\mathbb{Z}F$-module with basis $(Ds)_{s \in S}$. Thus we have reproved I.4.4 as well as exercise 3b of §II.5. As in the latter, we can now define the partial derivatives $\partial/\partial s: F \to \mathbb{Z}F$ by $Df = \sum_{s \in S} (\partial f/\partial s) Ds$ for $f \in F$.

(c) Reprove, from the present point of view, exercise 3c of §II.5. [Hint: The formula to be proved is true for the universal derivation, hence for any derivation.]

4. Let $G = F/R$, where $F = F(S)$ and R is the normal closure of a subset $T \subseteq F$.

(a) For any G-module A, show that derivations $G \to A$ correspond to derivations $d: F \to A$ such that $d(T) = 0$. [Hint: A splitting of $0 \to A \to A \rtimes G \to G \to 1$ can be defined by constructing a suitable map $F \to A \rtimes G$.]

(b) Using exercise 3, restate (a) as follows: Derivations $G \to A$ correspond to families $(a_s)_{s \in S}$ of elements of A such that $\sum_{s \in S} (\overline{\partial t/\partial s}) a_s = 0$ for all $t \in T$, where $\overline{\partial t/\partial s}$ is the image of $\partial t/\partial s$ under the quotient map $\mathbb{Z}F \to \mathbb{Z}G$.

(c) Using exercise 2, restate (b) as follows: Let I be the augmentation ideal of $\mathbb{Z}G$. Then there is an exact sequence

$$\mathbb{Z}G^{(T)} \xrightarrow{\partial_2} \mathbb{Z}G^{(S)} \xrightarrow{\partial_1} I \to 0,$$

where $\partial_1 e_s = \bar{s} - 1$ and $\partial_2 e_t = \sum_{s \in S} (\overline{\partial t/\partial s}) e_s$. (This reproves the second part of exercise 3d of §II.5.)

(d) Make (c) more precise by proving that there is an exact sequence

$$(*) \qquad 0 \to R_{ab} \xrightarrow{\theta} \mathbb{Z}G^{(S)} \xrightarrow{\partial} \mathbb{Z}G \xrightarrow{\varepsilon} \mathbb{Z} \to 0,$$

where $\partial e_s = \bar{s} - 1$ and $\theta(r \bmod [R, R]) = \sum_{s \in S} (\overline{\partial r/\partial s}) e_s$. (This reproves II.5.4 and the first part of exercise 3d of §II.5.) [Hint: We have a free resolution

$$(**) \qquad 0 \to I \to \mathbb{Z}F \xrightarrow{\varepsilon} \mathbb{Z} \to 0$$

of $\mathbb{Z}$ over $\mathbb{Z}F$, where I now denotes the augmentation ideal of $\mathbb{Z}F$. Taking R-coinvariants, we obtain a complex whose homology is $H_* R$. Since $I \approx \mathbb{Z}F^{(S)}$, this yields an exact sequence

$$0 \to H_1 R \xrightarrow{\varphi} \mathbb{Z}G^{(S)} \to \mathbb{Z}G \to \mathbb{Z} \to 0.$$

On the other hand, we can calculate $H_* R$ from the standard (bar) resolution of $\mathbb{Z}$ over $\mathbb{Z}R$, and this is what we used to show $H_1 R \approx R_{ab}$. Now it is easy to map the standard resolution to (**), by $[\ \ ] \mapsto 1$ in dimension 0, $[r] \mapsto r - 1 = Dr$ in dimension 1, and $[r_1|\cdots|r_n] \mapsto 0$ for $n > 1$. Deduce that there is a commutative diagram

$$\begin{array}{ccc} R & \xrightarrow{D|R} & I \\ \downarrow & & \downarrow \\ H_1 R & \xrightarrow{\varphi} & \mathbb{Z}G^{(S)} \end{array}$$

where the vertical arrows are quotient maps. Since the composite $F \xrightarrow{D} I \to \mathbb{Z}G^{(S)}$ is a derivation such that $s \mapsto e_s$, the exactness of (*) follows easily.]

3 The Classification of Extensions with Abelian Kernel

Let A be a fixed G-module. All extensions of G by A to be considered in this section will be assumed to give rise to the given action of G on A. To analyze an extension

(3.1) $$0 \to A \xrightarrow{i} E \xrightarrow{\pi} G \to 1,$$

we choose a set-theoretic cross-section of π, i.e., a function $s: G \to E$ such that $\pi s = \mathrm{id}_G$. For simplicity, we will assume that s satisfies the *normalization condition*

(3.2) $$s(1) = 1.$$

If s is a homomorphism, then the extension splits and we know its structure by 2.1. In the general case, however, there is a function $f: G \times G \to A$ which measures the failure of s to be a homomorphism. Indeed, for any $g, h \in G$, the elements $s(gh)$ and $s(g)s(h)$ of E both map to gh in G, so they differ by an element of $i(A)$. Thus we can define f by the equation

(3.3) $$s(g)s(h) = i(f(g, h))s(gh).$$

Note that 3.2 implies that f is *normalized*, in the sense that

(3.4) $$f(g, 1) = 0 = f(1, g)$$

for all $g \in G$. The function f is often called the *factor set* associated to 3.1 and s.

I claim now that the extension 3.1 can be completely recovered from the G-module structure on A and the function f. Indeed, since $s(G)$ is a set of coset representatives for $i(A)$ in E, we have a bijection $A \times G \to E$ given by $(a, g) \mapsto i(a)s(g)$. To compute the group law on $A \times G$ which makes this bijection an isomorphism of groups, consider two elements (a, g), (b, h) in $A \times G$. Using the commutation rule 1.2, we find

$$\begin{aligned} i(a)s(g)i(b)s(h) &= i(a)i(gb)s(g)s(h) \\ &= i(a + gb)i(f(g, h))s(gh) \\ &= i(a + gb + f(g, h))s(gh). \end{aligned}$$

Thus the group law on $A \times G$ is given by

$$(a, g)(b, h) = (a + gb + f(g, h), gh). \tag{3.5}$$

We will denote by E_f the set $A \times G$ with the product 3.5. Note that 3.5 looks like the product in $A \rtimes G$, "perturbed" by f.

Since $i(a) = i(a)s(1)$ for any $a \in A$, the composite $A \xrightarrow{i} E \approx E_f$ is the canonical inclusion

$$a \mapsto (a, 1). \tag{3.6}$$

And the composite $E_f \approx E \xrightarrow{\pi} G$ is obviously the canonical projection

$$(a, g) \mapsto g. \tag{3.7}$$

Thus the original extension 3.1 is equivalent to the extension

$$0 \to A \to E_f \to G \to 1 \tag{3.8}$$

defined by 3.5, 3.6, and 3.7 entirely in terms of G, A, and f.

It is natural to ask at this point whether we could start with an arbitrary function $f: G \times G \to A$ satisfying 3.4 and define a group E_f by means of 3.5. The answer is no. Indeed, if we define a product by 3.5 and compute the triple products $[(a, g)(b, h)](c, k)$ and $(a, g)[(b, h)(c, k)]$, we find that 3.5 is associative if and only if f satisfies the identity

$$f(g, h) + f(gh, k) = gf(h, k) + f(g, hk) \tag{3.9}$$

for all $g, h, k \in G$. If 3.9 holds, however, then we do in fact have a group. For (3.4) implies that $(0, 1)$ is a 2-sided identity, and it is easy to prove the existence of inverses. [Given $(a, g) \in E_f$, solve the equations $(a, g)(b, g^{-1}) = (0, 1)$ and $(b', g^{-1})(a, g) = (0, 1)$ for b and b'. One finds that (a, g) has left inverse $(-g^{-1}a - f(g^{-1}, g), g^{-1})$ and right inverse $(-g^{-1}a - g^{-1}f(g, g^{-1}), g^{-1})$. These two inverses are necessarily equal because of associativity.] Moreover, one checks easily that 3.6 and 3.7 define homomorphisms making the resulting sequence 3.8 exact, that 3.8 gives rise to the given action of G on A, and that the factor set associated to 3.8 (with the canonical cross-section $g \mapsto (0, g)$) is the original function f.

What we have established, then, is essentially a 1-1 correspondence

$$\begin{pmatrix} \text{extensions 3.1 with} \\ \text{a normalized section} \end{pmatrix} \leftrightarrow \begin{pmatrix} \text{functions } G \times G \to A \\ \text{satisfying 3.4 and 3.9} \end{pmatrix}.$$

Now the identity 3.9 can be rewritten in a form which should look familiar:

$$(3.10) \qquad gf(h, k) - f(gh, k) + f(g, hk) - f(g, h) = 0.$$

Indeed, f can be regarded as a 2-cochain of the standard complex $C^*(G, A)$ for computing $H^*(G, A)$ (cf. §III.1), and 3.10 says precisely that f is a cocycle. Moreover, 3.4 simply says that f is in the *normalized* cochain complex $C_N^*(G, A) \subseteq C^*(G, A)$. Thus

$$\begin{pmatrix}\text{extensions 3.1 with}\\ \text{a normalized section}\end{pmatrix} \leftrightarrow \begin{pmatrix}\text{normalized 2-cocycles of}\\ G \text{ with coefficients in } A\end{pmatrix}$$

Finally, I claim that changing the choice of the section in 3.1 corresponds precisely to modifying the cocycle f in $C_N^*(G, A)$ by a coboundary. In fact, an arbitrary normalized section of 3.1 is given by

$$(3.11) \qquad g \mapsto i(c(g))s(g),$$

where $c: G \to A$ is a function such that $c(1) = 0$, i.e., c is an arbitrary element of $C_N^1(G, A)$. And we can easily compute the factor set associated to 3.11; for we have

$$\begin{aligned} i(c(g))s(g)i(c(h))s(h) &= i(c(g))i(gc(h))s(g)s(h) \\ &= i(c(g) + gc(h))i(f(g, h))s(gh) \\ &= i(c(g) + gc(h) + f(g, h) - c(gh))i(c(gh))s(gh). \end{aligned}$$

Thus the new factor set is

$$(g, h) \mapsto c(g) + gc(h) + f(g, h) - c(gh),$$

which is equal to $f + \delta c$, as claimed.

We have therefore proved:

(3.12) Theorem. *Let A be a G-module and let $\mathscr{E}(G, A)$ be the set of equivalence classes of extensions of G by A giving rise to the given action of G on A. Then there is a bijection*

$$\mathscr{E}(G, A) \approx H^2(G, A).$$

As a simple application of 3.12 and 2.3 we will prove the following theorem of group theory:

(3.13) Corollary. *Let E be a finite group of order mn, where m and n are relatively prime, and suppose that E contains an abelian normal subgroup A of order m. Then E contains subgroups of order n, and any two such subgroups are conjugate.*

PROOF. Let $G = E/A$ and consider the extension

$$0 \to A \to E \to G \to 1.$$

Since $|A|$ and $|G|$ are relatively prime, $H^2(G, A) = 0$ (cf. III.10.2). Thus $\mathscr{E}(G, A)$ contains a single element and hence the extension splits. This proves the existence part of the corollary. The uniqueness (up to conjugacy) follows similarly from 2.3, once one notes that any $\tilde{G} \subseteq E$ of order n must map isomorphically onto G and hence split the extension. $\square$

Remark. The corollary can be generalized to the case where A is non-abelian, cf. Zassenhaus [1958], IV.7. The existence part of this generalization, as well as the uniqueness if either A or E/A is assumed to be solvable, is proved by a straightforward inductive argument based on 3.13 and the Sylow theorems. One then completes the uniqueness proof by appealing to the Feit–Thompson theorem, which says that every group of odd order is solvable and hence that either A or E/A is solvable.

Further group-theoretic applications of 3.12 will be given in exercises 3–7 below and in §4.

Exercises

1. (Functorial properties of $\mathscr{E}(G, A)$)

   (a) Given an extension $0 \to A \to E \to G \to 1$ and a group homomorphism $\alpha\colon G' \to G$, show that there is an extension $0 \to A \to E' \to G' \to 1$, characterized up to equivalence by the fact that it fits into a commutative diagram

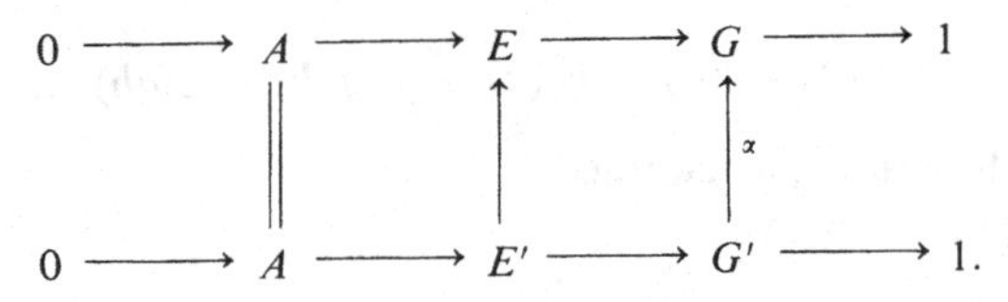

   Deduce that α induces a map $\mathscr{E}(G, A) \to \mathscr{E}(G', A)$, which corresponds under the bijection of 3.12 to $H^2(\alpha, A)\colon H^2(G, A) \to H^2(G', A)$. [Hint: Take E' to be the fiber-product (or "pull-back") $E \times_G G'$, which by definition is the set of pairs $(e, g') \in E \times G'$ such that e and g' have the same image in G.]

   (b) Given an extension $0 \to A \to E \to G \to 1$ and a homomorphism $f\colon A \to A'$ of G-modules, show that there is an extension $0 \to A' \to E' \to G \to 1$, characterized up to equivalence by the fact that it fits into a commutative diagram

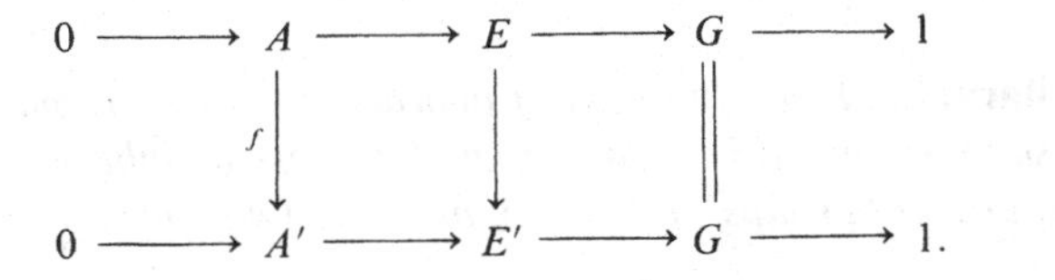

   Deduce that f induces a map $\mathscr{E}(G, A) \to \mathscr{E}(G, A')$, which corresponds under the bijection of 3.12 to $H^2(G, f)\colon H^2(G, A) \to H^2(G, A')$. [Hint: Take E' to be the largest quotient of $A' \rtimes E$ such that the left-hand square above commutes.]

2. (Interpretation of $\delta\colon H^1 \to H^2$)

(a) Let $0 \to A' \to A \to A'' \to 0$ be a short exact sequence of G-modules. Given a derivation $d\colon G \to A''$, show that there is an extension $0 \to A' \to E \to G \to 1$, characterized by the fact that it fits into a commutative diagram

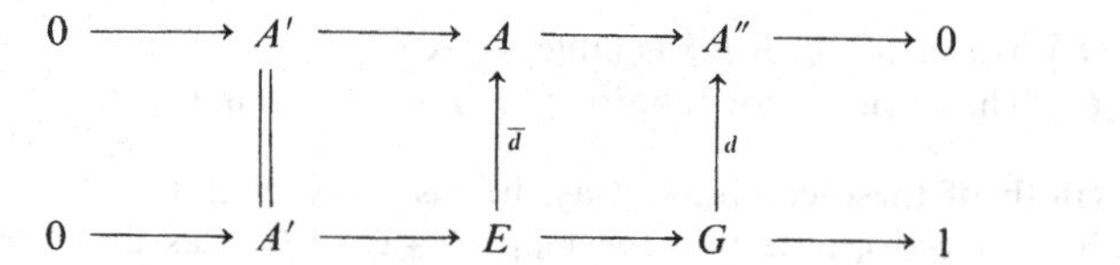

with $\bar{d}$ a derivation. (The latter makes sense because A is an E-module via $E \to G$.) [Hint: Take E to be the set-theoretic pull-back, as in exercise 1a, regarded as a subgroup of $A \rtimes G$.]

(b) The construction in (a) gives a map $\mathrm{Der}(G, A'') \to \mathscr{E}(G, A')$. Show that the diagram

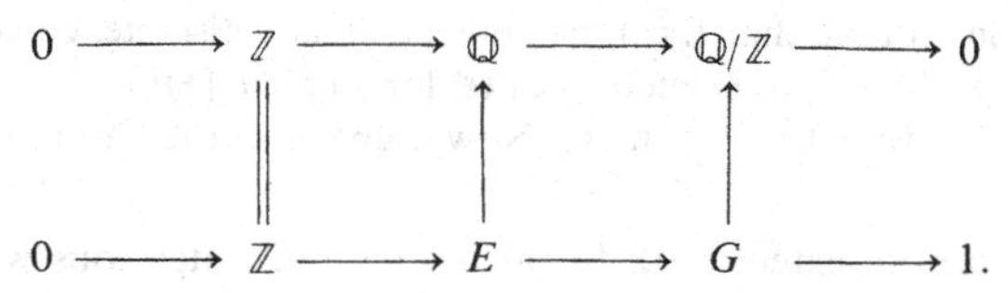

commutes. [Hint: A (set-theoretic) lifting of d to a function $G \to A$ yields a (set-theoretic) cross-section of $E \to G$. Now check definitions.]

3. Let G be a finite group. For any homomorphism $G \to \mathbb{Q}/\mathbb{Z}$ we can construct a central extension of G by $\mathbb{Z}$ by pulling back the canonical extension $0 \to \mathbb{Z} \to \mathbb{Q} \to \mathbb{Q}/\mathbb{Z} \to 0$:

$$\begin{array}{ccccccccc} 0 & \to & \mathbb{Z} & \to & \mathbb{Q} & \to & \mathbb{Q}/\mathbb{Z} & \to & 0 \\ & & \| & & \uparrow & & \uparrow & & \\ 0 & \to & \mathbb{Z} & \to & E & \to & G & \to & 1. \end{array}$$

Thus we have a map $\varphi\colon \mathrm{Hom}(G, \mathbb{Q}/\mathbb{Z}) \to \mathscr{E}(G, \mathbb{Z})$, where G acts trivially on $\mathbb{Z}$. Prove that φ is a bijection. [Hint: By exercise 2, φ can be identified with

$$\delta\colon H^1(G, \mathbb{Q}/\mathbb{Z}) \to H^2(G, \mathbb{Z}).]$$

4. (a) Let E be a group which contains a central subgroup C of finite index n. Prove that there is a homomorphism $E \to C$ whose restriction to C is the n-th power map. [Method 1: We have a central extension $1 \to C \to E \to G \to 1$ with $|G| = n$. Since n annihilates $H^2(G, C)$, the n-th power map $C \to C$ induces the zero-map on $H^2(G, C)$. Exercise 1b now implies that there is a diagram

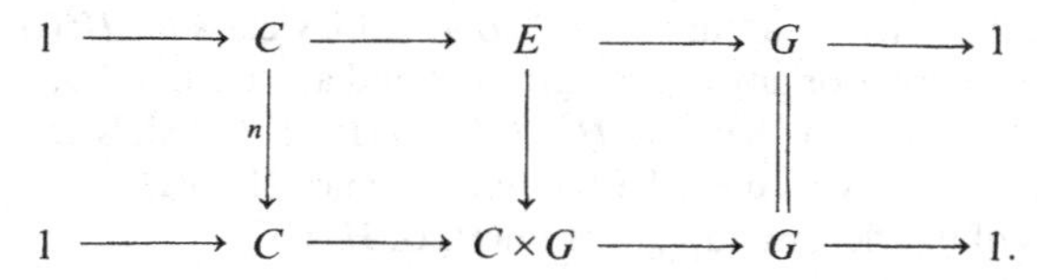

Take the C-component of $E \to C \times G$. Method 2: Use the transfer map $H_1 E \to H_1 C = C$.]

(b) Deduce that the following two conditions on a finitely generated group E are equivalent:

(i) The center C of E has finite index.
(ii) The commutator subgroup $[E, E]$ of E is finite.

(Both of these conditions say, heuristically, that E is "almost abelian.") [Hint: For (i) $\Rightarrow$ (ii), note that the map $E \to C$ of (a) has finite kernel. Conversely, (ii) implies that every generator of E has only finitely many conjugates, hence its centralizer has finite index.]

5. (a) Let E be a group which contains an infinite cyclic central subgroup of finite index. Prove that E is isomorphic to a semi-direct product $F \rtimes \mathbb{Z}$ with F finite. [Hint: Using either exercise 3 or exercise 4a, we can find a surjection $E \twoheadrightarrow \mathbb{Z}$ with finite kernel. This surjection must split since $\mathbb{Z}$ is free.]

(b) Deduce the following theorem of group theory: If E is a torsion-free group which has an infinite cyclic subgroup of finite index, then E is infinite cyclic. [Hint: Let $A \subseteq E$ be an infinite cyclic normal subgroup of finite index. Then E has a subgroup E' of index 1 or 2 such that A is central in E'. It follows from (a) that $E' \approx \mathbb{Z}$, so we have an extension $0 \to \mathbb{Z} \to E \to G \to 1$ with $|G| \leq 2$. If G acts non-trivially on $\mathbb{Z}$ then $H^2(G, \mathbb{Z}) = 0$ by direct computation. But then the extension splits, contradicting the assumption that E is torsion-free. Hence G acts trivially and $E \approx \mathbb{Z}$ by (a).]

*6. Let E be a finitely generated torsion-free group which contains an abelian subgroup of finite index. Prove that E can be embedded as a discrete, co-compact subgroup of the group $\mathbb{R}^n \rtimes O_n$ of isometries of $\mathbb{R}^n$ for some n. [Hint: We have an extension $0 \to \mathbb{Z}^n \to E \to G \to 1$ with G finite. Now argue as in the hint to exercise 4a.]

7. (Universal central extensions). In this exercise all extensions will be assumed to be central and all coefficient modules for cohomology will be assumed to have trivial G-action. We will assume further that G is *perfect*, i.e., that $G = [G, G]$, or, equivalently, that $H_1 G = 0$. All (non-abelian) simple groups are perfect, for example.

(a) Show that there is a universal coefficient isomorphism

$$H^2(G, A) \approx \mathrm{Hom}(H_2 G, A)$$

for any abelian group A.

(b) Deduce that there is a "universal" cohomology class $u \in H^2(G, H_2 G)$, with the following property: For any abelian group A and any $v \in H^2(G, A)$, there is a unique map $f: H_2 G \to A$ such that $v = H^2(G, f)u$. [Hint: Yoneda's lemma (exercise 3a of §I.7) tells you how to describe the natural map $\mathrm{Hom}(H_2 G, —) \xrightarrow{\approx} H^2(G, —)$ of (a) in terms of its effect on $\mathrm{id}_{H_2 G} \in \mathrm{Hom}(H_2 G, H_2 G)$.]

(c) In view of 3.12 and exercise 1b, reinterpret (b) as saying that G admits a "universal central extension"

$$0 \to H_2G \to E \to G \to 1,$$

characterized by the following property: Given any abelian group A and any central extension

$$0 \to A \to E' \to G \to 1,$$

there is a unique map $E \to E'$ making

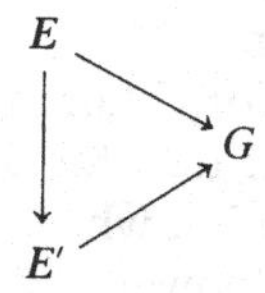

commute. [Hint: To prove uniqueness, note that two such maps $E \rightrightarrows E'$ which induce the same map $H_2G \to A$ must differ by a homomorphism $E \to A$ which factors through G. But G is perfect, so there are no non-trivial maps $G \to A$.]

Remark. In practice one often uses (c) to *compute* H_2G. Namely, one produces a central extension of G with the universal property described in (c), and it then follows that the kernel of this extension is isomorphic to H_2G. See Milnor [1971] for an interesting example of this ($G = SL_n(k)$) and further information about universal central extensions.

8. Let $0 \to A \xrightarrow{i} E \to G \to 1$ be a central extension with G abelian. The *commutator pairing* associated to the extension is the map $c: G \times G \to A$ defined by $c(g, h) = i^{-1}([\tilde{g}, \tilde{h}]) = i^{-1}(\tilde{g}\tilde{h}\tilde{g}^{-1}\tilde{h}^{-1})$, where $\tilde{g}$ and $\tilde{h}$ are lifts of g and h to E.

   (a) Show that c is well-defined, $\mathbb{Z}$-bilinear, and alternating. (The latter means that $c(g, g) = 0$; this implies, in view of bilinearity, that $c(g, h) = -c(h, g)$.) Thus c can be viewed as a map $\bigwedge^2 G \to A$, where $\bigwedge^2 G$, the second exterior power of G, is the quotient of $G \otimes G$ by the subgroup generated by the elements $g \otimes g$.

   (b) If f is a factor set associated to the extension, show that $c(g, h) = f(g, h) - f(h, g)$.

   (c) Let $\theta: H^2(G, A) \to \operatorname{Hom}(\bigwedge^2 G, A)$ be the map which sends the class of a cocycle f to the alternating map $f(g, h) - f(h, g)$. Deduce from (b) that there is a bijection $\ker\theta \approx \mathscr{E}_{ab}(G, A)$, where $\mathscr{E}_{ab}(G, A)$ is the set of equivalence classes of abelian extensions of G by A.

   See exercise 5 of §V.6 below for further results about θ.

4 Application: p-groups with a Cyclic Subgroup of Index p

As an illustration of the theory of group extensions, we will give in this section a classification of p-groups which contain a cyclic subgroup of index p, where p is a prime. (Such a subgroup is necessarily normal; cf. Hall [1959],

§4.2.) We will then use this classification to prove a theorem of Burnside (Theorem 4.3 below) which will be needed in §VI.9.

We begin by listing some examples of p-groups with a cyclic subgroup of index p:

(A) $\mathbb{Z}_q$ ($q = p^n, n \geq 1$).

(B) $\mathbb{Z}_q \times \mathbb{Z}_p$ ($q = p^n, n \geq 1$).

(C) $\mathbb{Z}_q \rtimes \mathbb{Z}_p$ ($q = p^n, n \geq 2$), where the canonical generator of $\mathbb{Z}_p$ acts on $\mathbb{Z}_q$ as multiplication by $1 + p^{n-1}$. (This makes sense because $(1 + p^{n-1})^p \equiv 1 \bmod p^n$ by the binomial theorem.)

If $p = 2$, there are three additional families:

(D) Dihedral 2-groups: Recall that for any integer $m \geq 2$, the dihedral group D_{2m} is defined to be $\mathbb{Z}_m \rtimes \mathbb{Z}_2$, where the generator of $\mathbb{Z}_2$ acts on $\mathbb{Z}_m$ as multiplication by -1. If m is a power of 2, D_{2m} is a 2-group. Note that D_4 is of type (B) and D_8 is of type (C). For $m > 4$, however, D_{2m} is not isomorphic to any of the previous examples; this can be seen by computing abelianizations.

(E) Generalized quaternion 2-groups: Let $\mathbb{H}$ be the quaternion algebra $\mathbb{R} \oplus \mathbb{R}i \oplus \mathbb{R}j \oplus \mathbb{R}k$. For any integer $m \geq 2$, the generalized quaternion group Q_{4m} is defined to be the subgroup of the multiplicative group $\mathbb{H}^*$ generated by $x = e^{\pi i/m}$ and $y = j$. For example, Q_8 is simply the usual quaternion group $\{\pm 1, \pm i, \pm j, \pm k\}$. Note that x is of order $2m$ and that we have the relations $y^2 = x^m$ and $yxy^{-1} = x^{-1}$. It follows that the cyclic subgroup C generated by x is normal and of index 2 and hence that Q_{4m} has order $4m$. If m is a power of 2, Q_{4m} is a 2-group. For future reference we record some properties of Q_{4m}: (a) In the extension $0 \to C \to Q_{4m} \to \mathbb{Z}_2 \to 0$, the generator of $\mathbb{Z}_2$ acts as -1 on C. (b) Every element of $Q_{4m} - C$ is of order 4. (c) Q_8 has exactly three cyclic subgroups of index 2; Q_{4m} for $m > 2$ has a unique cyclic subgroup of index 2. [This follows from (b).] (d) Q_{4m} has a unique element of order 2. (e) The extension $0 \to C \to Q_{4m} \to \mathbb{Z}_2 \to 0$ does not split.

(F) $\mathbb{Z}_q \rtimes \mathbb{Z}_2$ ($q = 2^n$, $n \geq 3$), where the generator of $\mathbb{Z}_2$ acts on $\mathbb{Z}_q$ as multiplication by $-1 + 2^{n-1}$.

We now prove that this list is complete:

(4.1) Theorem. *If G is a p-group with a cyclic subgroup of index p, then G is isomorphic to one of the groups listed in* (A)–(F) *above.*

The proof will use the following lemma from elementary number theory:

(4.2) Lemma. *Let a be an integer such that $a^p \equiv 1 \bmod p^n$ for some $n \geq 2$. If p is odd then $a \equiv 1 \bmod p^{n-1}$. If $p = 2$ then $a \equiv \pm 1 \bmod 2^{n-1}$.*

PROOF. Assuming, as we may, that $a \neq 1$, let $d(a)$ be the largest integer d such that $a \equiv 1 \bmod p^d$. Note that $d(a) \geq 1$ since $a \equiv a^p \equiv 1 \bmod p$, where the first congruence is by Fermat's little theorem. If p is odd or $d(a) \geq 2$, then exercise 3a of §II.4 is applicable and gives $d(a^p) = d(a) + 1$. Thus $d(a) \geq n - 1$ in this case, as required. If $p = 2$ and $d(a) = 1$, then $a \equiv -1 \bmod 4$. We can therefore apply the previous argument to $-a$ to conclude that $-a \equiv 1 \bmod 2^{n-1}$. □

PROOF OF THEOREM 4.1. By hypothesis we have an extension $0 \to \mathbb{Z}_q \to G \to H \to 1$, where $q = p^n$ for some $n \geq 0$ and $|H| = p$. If H acts trivially on $\mathbb{Z}_q$ then G is abelian, being generated by two commuting elements. It then follows from the theory of finite abelian groups that G is of type (A) or (B). Assume now that H acts non-trivially on $\mathbb{Z}_q$. We will use example 2 of §III.1 to compute $H^2(H, \mathbb{Z}_q)$. The action of H on $\mathbb{Z}_q$ is given by an embedding $H \hookrightarrow \mathbb{Z}_q^*$, where $\mathbb{Z}_q^*$ is the group of units of the ring $\mathbb{Z}_q = \mathbb{Z}/p^n\mathbb{Z}$, and clearly we must have $n \geq 2$. Suppose first that p is odd. Then Lemma 4.2 implies that the image of this embedding is $\{1 + b : b \in p^{n-1}\mathbb{Z}/p^n\mathbb{Z}\}$, which is a group of order p, generated by $1 + p^{n-1}$. In particular, H must have a generator which acts as $1 + p^{n-1}$, as in (C). Under this action of H on $\mathbb{Z}_q$, we have $\mathbb{Z}_q^H = \{x \in \mathbb{Z}_q : p^{n-1}x = 0\} = p\mathbb{Z}/p^n\mathbb{Z}$. On the other hand, the norm operator $\sum_{h \in H} h$ on $\mathbb{Z}_q$ is multiplication by $\sum (1 + b)$ $(b \in p^{n-1}\mathbb{Z}/p^n\mathbb{Z})$, which equals p. [$\sum b = 0$ because p is odd.] Thus the image of the norm operator is also equal to $p\mathbb{Z}/p^n\mathbb{Z}$. We conclude that $H^2(H, \mathbb{Z}_q) = 0$, so the extension splits and G is of type (C). Suppose now that $p = 2$. By Lemma 4.2 again, the image of $H \hookrightarrow \mathbb{Z}_q^*$ is a subgroup of $\{\pm 1 + b : b \in 2^{n-1}\mathbb{Z}/2^n\mathbb{Z}\}$. There are therefore three possibilities for the image $a \in \mathbb{Z}_q^*$ of the generator of H:

Case 1: $a = -1$. In this case $\mathbb{Z}_q^H = 2^{n-1}\mathbb{Z}/2^n\mathbb{Z}$ and the norm operator is zero, so $H^2(H, \mathbb{Z}_q) \approx \mathbb{Z}_2$. Thus there are precisely two inequivalent extensions of H by $\mathbb{Z}_q$ corresponding to this action of H on $\mathbb{Z}_q$. Since we have already produced two such extensions in (D) and (E), it follows that G is either of type (D) or type (E).

Case 2: $a = 1 + 2^{n-1}$. Then $\mathbb{Z}_q^H = 2\mathbb{Z}/2^n\mathbb{Z}$ and the norm operator is multiplication by $2 + 2^{n-1} = 2(1 + 2^{n-2})$. We may assume $n \geq 3$, since otherwise we are in case 1. Thus $1 + 2^{n-2} \in \mathbb{Z}_q^*$ and the image of the norm map is $2\mathbb{Z}/2^n\mathbb{Z}$. We therefore have $H^2(H, \mathbb{Z}_q) = 0$, so the extension splits and G is of type (C).

Case 3: $a = -1 + 2^{n-1}$, $n \geq 3$. Then $\mathbb{Z}_q^H = 2^{n-1}\mathbb{Z}/2^n\mathbb{Z}$ and the norm operator is multiplication by 2^{n-1}. Thus $H^2(H, \mathbb{Z}_q) = 0$, the extension splits, and G is of type (F). □

Using this theorem, we can prove the following result (cf. Burnside [1911], §§104 and 105):

(4.3) Theorem. *If G is a p-group which has a unique subgroup of order p, then G is either a cyclic group or a generalized quaternion group.*

PROOF. Arguing by induction on $|G|$, we may assume that every proper subgroup of G is cyclic or generalized quaternion. Choose $H \lhd G$ of index p.

[Such an H always exists, cf. Hall [1959], §4.2.] If H is cyclic, then we are done by Theorem 4.1; for the only groups in the list (A)–(F) which have a unique subgroup of order p are the cyclic groups and the generalized quaternion groups. Suppose, then, that H is generalized quaternion (and hence that $p = 2$). I claim that H has a cyclic subgroup N of index 2 which is normal in G. Indeed, if $\mathscr{S}$ is the set of cyclic subgroups of H of index 2, then we know that card($\mathscr{S}$) is odd (cf. statement (c) in the discussion of type (E) above); the conjugation action of the 2-group G on $\mathscr{S}$ must therefore fix some $N \in \mathscr{S}$, as claimed. Consider now the action of G/N on N. It is given by a homomorphism $G/N \to \mathbb{Z}_q^*$, where $q = |N| \geq 4$. Composing with the canonical projection $\mathbb{Z}_q^* \to \mathbb{Z}_4^* = \{\pm 1\}$, we obtain a surjection $G/N \to \{\pm 1\}$, whose kernel is a group K/N of order 2. Since the generator of K/N does not act as -1 on N, K cannot be generalized quaternion. Thus K is a cyclic subgroup of G of index 2, and we are done as before. □

Remark. It is instructive to look at Burnside's proof of this theorem. Although neither cohomology theory nor the classification of group extensions had yet been developed, Burnside's proof contains, essentially, a direct proof that $\mathscr{E}(G, A) \approx A^G/NA$ for G finite cyclic, where N is the norm operator. [Explicitly, given an extension $0 \to A \to E \to G \to 1$, where G is cyclic of order n with generator t, Burnside lifts t to $\tilde{t} \in E$ and considers $a = \tilde{t}^n \in A$. Then a is in A^G and a mod NA is the element of A^G/NA which classifies the extension.]

Finally, as one further illustration of the use of cohomology in group theory, we will determine all subgroups of a non-dihedral group of type (C):

(4.4) Proposition. *Let* $G = \mathbb{Z}_q \rtimes \mathbb{Z}_p$ $(q = p^n, n \geq 2)$ *as in* (C). *If* $p = 2$, *assume* $n \geq 3$. *Let* $A \subset \mathbb{Z}_q$ *be the subgroup* $p\mathbb{Z}/p^n\mathbb{Z}$ *of index* p.

(a) $A \rtimes \mathbb{Z}_p = A \times \mathbb{Z}_p$ *is a non-cyclic abelian subgroup of* G *of index* p.

(b) *Every element of* G *which is not in* $A \times \mathbb{Z}_p$ *generates a cyclic subgroup of index* p. *There are precisely* p *distinct such cyclic subgroups of index* p.

(c) *Every proper subgroup* H *of* G *is either a subgroup of* $A \times \mathbb{Z}_p$ *or a cyclic subgroup of index* p *as in* (b). *In particular,* H *is abelian.*

Proof. (a) is obvious, since $\mathbb{Z}_p$ acts trivially on A. It is possible to prove (b) by a direct (and tedious) computation, but it is easier to proceed as follows. First, note that any homomorphism $G \to \mathbb{Z}_p$ must factor through the quotient $\mathbb{Z}_p \times \mathbb{Z}_p$ of G. It follows that the subgroups of G of index p correspond to the subgroups of $\mathbb{Z}_p \times \mathbb{Z}_p$ of index p, and there are precisely $p + 1$ of these. [They are the points of the projective line over the field with p elements.] Of these $p + 1$ subgroups of G of index p, one is $A \times \mathbb{Z}_p$ and I claim that the other p are all cyclic. Accepting this for the moment, we can easily prove (b) by a counting argument. For the p cyclic subgroups of index p contain a total of $p(p^n - p^{n-1})$ generators, and none of these generators can be in $A \times \mathbb{Z}_p$. But $\operatorname{card}(G - (A \times \mathbb{Z}_p)) = p^{n+1} - p^n = p(p^n - p^{n-1})$, so (b) follows at once. Since (c) is an immediate consequence of (b), it remains only to prove the claim.

In the proof of 4.1 we calculated the norm operator for the action of $\mathbb{Z}_p$ on $\mathbb{Z}_q$, and it follows from this computation that $H^1(\mathbb{Z}_p, \mathbb{Z}_q) = 0$. In view of Proposition 2.3, this means that the extension

$$0 \to \mathbb{Z}_q \to G \to \mathbb{Z}_p \to 0$$

has a unique splitting (up to conjugacy) and hence that G contains only two conjugacy classes of subgroups of order p. Suppose now that H is a non-cyclic subgroup of G of index p. Then we have an extension

$$0 \to A \to H \to \mathbb{Z}_p \to 0$$

with $\mathbb{Z}_p$ acting trivially on A (because $1 + p^{n-1} \equiv 1 \bmod |A|$). Thus H is abelian and non-cyclic, so H contains at least two subgroups of order p. Since H is normal in G, it follows that H contains both conjugacy classes of subgroups of G of order p. In particular, $0 \times \mathbb{Z}_p \subseteq H$ and hence $H = A \times \mathbb{Z}_p$, as claimed. □

Exercises

1. If G is a generalized quaternion group, show that every subgroup of G is either cyclic or a generalized quaternion group. [This follows from Theorem 4.3, of course, but you can also check it directly.]

2. If G is a dihedral group, show that every subgroup of G is either cyclic or dihedral. [Hint: Every element of $G = \mathbb{Z}_q \rtimes \mathbb{Z}_2$ which is not in $\mathbb{Z}_q$ is of order 2.] If $|G| > 8$, show further that the non-cyclic abelian subgroups of G (i.e., the subgroups isomorphic to $\mathbb{Z}_2 \times \mathbb{Z}_2 = D_4$) are not normal. D_8, on the other hand, contains two non-cyclic abelian normal subgroups.

3. Let $G = \mathbb{Z}_q \rtimes \mathbb{Z}_2$ ($q = 2^n$, $n \geq 3$) as in (F). Let $A \subset \mathbb{Z}_q$ be the subgroup of order 2. Show that the only non-cyclic abelian subgroups of G are $A \times \mathbb{Z}_2$ and its conjugates. Show further that $A \times \mathbb{Z}_2$ is not normal in G. [Hint: If H is a non-cyclic proper subgroup of G, then we have an extension $0 \to H \cap \mathbb{Z}_q \to H \to \mathbb{Z}_2 \to 0$ with the generator of $\mathbb{Z}_2$ acting as -1 on $H \cap \mathbb{Z}_q$. This can only be abelian if $H \cap \mathbb{Z}_q = A$. Now calculate $H^1(\mathbb{Z}_2, \mathbb{Z}_q)$ and use this information as in the proof of 4.4.]

4. If G is a p-group such that every abelian normal subgroup is cyclic, show that G is of type (A), (D), (E), or (F), with $|G| \geq 16$ if G is of type (D). [Hint: Choose a maximal abelian normal subgroup of G, and consider the corresponding extension $0 \to \mathbb{Z}_q \to G \to H \to 1$, $q = p^n$. If $|H| \leq p$ then we are done by Theorem 4.1, since groups of type (B) and (C) have non-cyclic abelian normal subgroups. Suppose, then, that $|H| \geq p^2$, and consider the normal subgroups $H' \subseteq H$ of order p. Such an H' cannot act trivially on $\mathbb{Z}_q$, since the inverse image G' of H' in G would be an abelian normal

subgroup bigger than $\mathbb{Z}_q$. And H' cannot act as in (C) unless $p = 2$ and $n = 2$, since the non-cyclic subgroup of G' of index p (cf. 4.4) would be normal in G. So the only possibility is that $p = 2$ and that the non-trivial element of H' acts as -1 or $-1 + 2^{n-1}$ (with $n \geq 3$ in the latter case). But this is absurd, for the composite $H \to \mathbb{Z}_q^* \to \{\pm 1\}$ has a non-trivial kernel and we could take H' to be contained in this kernel.]

5 Crossed Modules and H^3 (Sketch)

We have seen in §§2 and 3 that H^1 and H^2 have concrete group-theoretic interpretations. It turns out that there are also group-theoretic interpretations of the functors H^n for $n \geq 3$, discovered independently by several people. (Many of the relevant references can be found in MacLane [1979].) We will confine ourselves here to a sketch of the theory in the case of H^3. The essential ideas in this case go back to MacLane [1949], although he did not give the precise classification theorem 5.4 below.

Let E and N be groups. Suppose we are given an action of E on N, denoted $(e, n) \mapsto {}^e n$, as well as a homomorphism $\alpha: N \to E$ satisfying

(5.1) $$^{\alpha(n)}n' = nn'n^{-1} \; (n, n' \in N)$$

and

(5.2) $$\alpha(^e n) = e\alpha(n)e^{-1} \; (e \in E, n \in N)$$

We then say that N (together with α and the action) is a *crossed module* over E.

The canonical example of a crossed module is that where E is the full automorphism group Aut N and $\alpha(n)$ is the inner automorphism associated to n. Thus 5.1 is true by definition, and 5.2 is easily verified. Crossed modules also arise naturally in topology, and it is in this context that they were first introduced (Whitehead [1949]). Namely, if X is a topological space and Y is a subspace, then the relative homotopy group $\pi_2(X, Y)$ is a crossed module over $\pi_1 Y$, with α being the boundary map $\partial: \pi_2(X, Y) \to \pi_1 Y$. For another example, suppose E is the total space of a fibration with fiber F; then one can make $\pi_1 F$ a crossed module over $\pi_1 E$.

Note that any ordinary (abelian) E-module can be viewed as a crossed module, with α being the trivial map. At the other extreme, any normal subgroup of E is a crossed module with E acting by conjugation and α being the inclusion.

Let N be a crossed module over E. We set $A = \ker \alpha$ and $G = \operatorname{coker} \alpha$; the latter makes sense because $\operatorname{im} \alpha$ is normal in E by 5.2. Note that A is central in N (by 5.1) and that the action of E on N induces an action of G on A. Thus we have a 4-term exact sequence

(5.3) $$0 \to A \xrightarrow{i} N \xrightarrow{\alpha} E \xrightarrow{\pi} G \to 1,$$

where A is a G-module. It turns out that exact sequences of this form are classified (up to a suitable equivalence relation) by $H^3(G, A)$.

More precisely, suppose we start with an arbitrary group G and an arbitrary G-module A. Consider all possible exact sequences of the form 5.3, where N is a crossed module over E such that the action of E on N induces the given action of G on A. We impose on these exact sequences the smallest equivalence relation such that

$$0 \to A \to N \to E \to G \to 1$$

is equivalent to

$$0 \to A \to N' \to E' \to G \to 1$$

whenever there is a commutative diagram

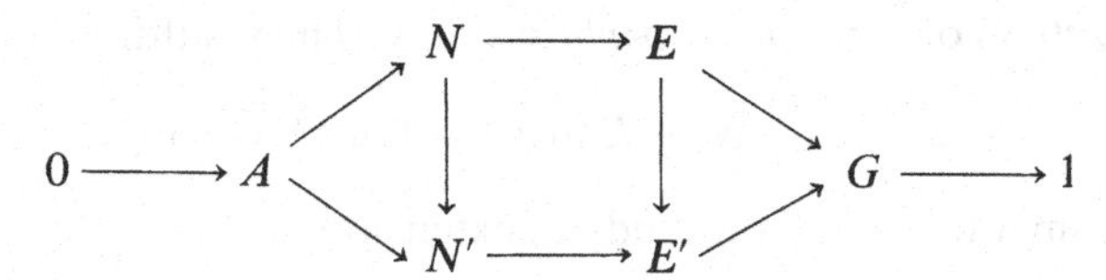

in which the vertical arrows are compatible with the actions of E and E' on N and N'. [Warning: The vertical arrows need not be isomorphisms.] We then have:

(5.4) Theorem. *There is a* 1-1 *correspondence between equivalence classes as above and elements of* $H^3(G, A)$.

We omit the proof, which can be found in the references cited in MacLane [1979]. We will, however, explain how an exact sequence 5.3 gives rise to an element of $H^3(G, A)$. Choose a set-theoretic cross-section $s\colon G \to E$ of π. Its failure to be multiplicative is measured by a function $f\colon G \times G \to \ker \pi$ such that

$$s(g)s(h) = f(g, h)s(gh). \tag{5.5}$$

As in §3, the associativity of the product in E forces a "cocycle condition" on f, which takes the form

$$f(g, h)f(gh, k) = {}^{s(g)}f(h, k)f(g, hk), \tag{5.6}$$

where ${}^{s(g)}f(h, k) = s(g)f(h, k)s(g)^{-1}$. [5.6 is a non-abelian analogue of 3.9; it is proved by computing the triple product $s(g)s(h)s(k)$ in two different ways.]

Since $\ker \pi = \operatorname{im} \alpha$, we can lift f to a function $F\colon G \times G \to N$, and we can ask whether F satisfies the analogue of 5.6 (which now involves the crossed-module action of E on N). The failure of F to do so is measured by a function $c\colon G \times G \times G \to A$ such that

$${}^{s(g)}F(h, k)F(g, hk) = i(c(g, h, k))F(g, h)F(gh, k). \tag{5.7}$$

One then shows that c is a 3-cocycle, whose cohomology class is independent of the choices of s and F, and this cohomology class is the desired element of $H^3(G, A)$.

6 Extensions with Non-Abelian Kernel (Sketch)

The main references for this section are MacLane [1963] and Eilenberg–MacLane [1947].

Let N be a group and recall from §5 that N is an $\mathrm{Aut}(N)$-crossed module via the canonical map $\alpha: N \to \mathrm{Aut}(N)$ and the canonical action of $\mathrm{Aut}(N)$ on N. The kernel of α is the center C of N, and the cokernel of α is, by definition, the group $\mathrm{Out}(N)$ of *outer automorphisms* of N. The resulting exact sequence

(6.1) $$0 \to C \to N \xrightarrow{\alpha} \mathrm{Aut}(N) \to \mathrm{Out}(N) \to 1$$

plays a fundamental role in the study of extensions

(6.2) $$1 \to N \xrightarrow{i} E \xrightarrow{\pi} G \to 1$$

with kernel N.

The first observation to make is that such an extension gives rise not to an action of G on N, but only to an "outer action," i.e., a homomorphism $\psi: G \to \mathrm{Out}(N)$. This is induced by the conjugation action of E on N:

$$\begin{array}{ccccc} N & \longrightarrow & E & \longrightarrow & G \\ \| & & \downarrow & & \downarrow \\ N & \longrightarrow & \mathrm{Aut}(N) & \longrightarrow & \mathrm{Out}(N). \end{array}$$

We therefore fix a homomorphism $\psi: G \to \mathrm{Out}(N)$ and try to understand the set $\mathscr{E}(G, N, \psi)$ of equivalence classes of extensions giving rise to ψ. Note that it is not even obvious, for a given ψ, whether or not $\mathscr{E}(G, N, \psi)$ is non-empty. In particular, we do not have a semi-direct product $N \rtimes G$ in $\mathscr{E}(G, N, \psi)$ unless ψ lifts to a homomorphism $G \to \mathrm{Aut}(N)$.

Suppose now that an extension 6.2 is given. As in §3, choose a set-theoretic cross-section $s: G \to E$ of π. This determines, as usual, a function $f: G \times G \to N$ measuring the failure of s to be a homomorphism. In addition, it determines a set-theoretic lifting $\varphi: G \to \mathrm{Aut}(N)$ of ψ; namely, $\varphi(g)$ is conjugation by $s(g)$. The functions f and φ are related by

(6.3) $$\varphi(g)\varphi(h) = \alpha(f(g, h))\varphi(gh)$$

where α is as in 6.1. Moreover, f satisfies a "cocycle condition" (cf. 5.6)

(6.4) $$f(g, h)f(gh, k) = {}^{\varphi(g)}f(h, k)f(g, hk).$$

Conversely, given f and φ satisfying 6.3 and 6.4, one can construct an extension

$$(6.5) \qquad 1 \to N \to E_{f,\varphi} \to G \to 1,$$

where $E_{f,\varphi}$ is $N \times G$ with a product that can be written down explicitly in terms of f and φ. One sees in this way that extensions are classified by pairs (f, φ) as above, subject to some equivalence relation.

This description of $\mathscr{E}(G, N, \psi)$ can be drastically simplified once we observe that the lifting φ of ψ can be taken to be the same in all extensions under consideration; for by changing the choice of s, we can change our lifting φ to any other lifting. Thus we can fix φ and just consider the extensions $E_{f,\varphi}$, where f ranges over all "cocycles" relative to φ. It follows from 6.3 that any two such f's will differ by a function $G \times G \to C$, and the latter is an honest 2-cocycle of G with coefficients in C, where C is a G-module via ψ. One shows in this way that any two elements of $\mathscr{E}(G, N, \psi)$ have a well-defined "difference" in $H^2(G, C)$. This leads to:

(6.6) Theorem. *The set $\mathscr{E}(G, N, \psi)$ admits a free, transitive action by the abelian group $H^2(G, C)$. Hence either $\mathscr{E}(G, N, \psi) = \varnothing$ or else there is a bijection $\mathscr{E}(G, N, \psi) \approx H^2(G, C)$. This bijection depends on the choice of a particular element of $\mathscr{E}(G, N, \psi)$.*

To complete the classification, we must say when $\mathscr{E}(G, N, \psi) \neq \varnothing$. Recall from §5 that the sequence 6.1 yields an element $u \in H^3(\mathrm{Out}(N), C)$. Applying the cohomology map $\psi^* = H^3(\psi, C)$, we obtain an element $\psi^*u \in H^3(G, C)$. I claim that ψ^*u is the "obstruction" to the existence of an element of $\mathscr{E}(G, N, \psi)$. To see this, we need the following explicit description of ψ^*u, which is easily deduced from §5: Choose a set-theoretic lifting $\varphi: G \to \mathrm{Aut}(N)$ of ψ, and choose a function $f: G \times G \to N$ such that 6.3 holds. Then the failure of f to satisfy 6.4 is measured by a function $G \times G \times G \to C$ which is a 3-cocycle representing ψ^*u.

In particular, if $\mathscr{E}(G, N, \psi) \neq \varnothing$, then we have already seen that φ and f can be chosen to satisfy 6.4, so the 3-cocycle is zero. Conversely, if $\psi^*u = 0$, then we can change the choice of f so that 6.4 holds, and we can then construct an extension 6.5. This proves:

(6.7) Theorem. *A homomorphism $\psi: G \to \mathrm{Out}(N)$ gives rise to an "obstruction" in $H^3(G, C)$, which vanishes if and only if $\mathscr{E}(G, N, \psi) \neq \varnothing$.*

This obstruction, incidentally, does not always vanish. Indeed, for any group G, any G-module C, and any element $v \in H^3(G, C)$, one can construct a group N with center C and a map $\psi: G \to \mathrm{Out}(N)$ whose obstruction is equal to v.

Finally, we mention one special case where the theory of extensions is particularly simple:

(6.8) Corollary. *If N has a trivial center then there is exactly one extension of G by N (up to equivalence) corresponding to any homomorphism $G \to$* Out(N).

This, of course, is an immediate consequence of 6.6 and 6.7, but it also admits an easy direct proof which uses no cohomology theory (cf. exercise 1 below).

Exercises

1. Give a direct proof of 6.8. [Hint: If N has trivial center then any extension of G by N fits into a diagram

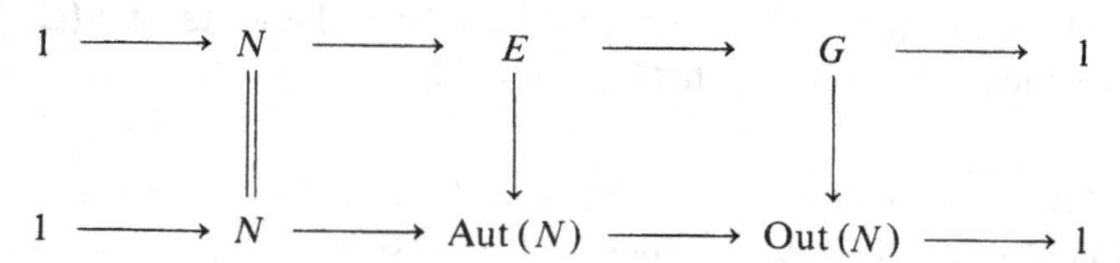

$$\begin{array}{ccccccccc} 1 & \longrightarrow & N & \longrightarrow & E & \longrightarrow & G & \longrightarrow & 1 \\ & & \| & & \downarrow & & \downarrow & & \\ 1 & \longrightarrow & N & \longrightarrow & \operatorname{Aut}(N) & \longrightarrow & \operatorname{Out}(N) & \longrightarrow & 1 \end{array}$$

with exact rows. Such a diagram is necessarily a pull-back diagram (cf. §3, exercise 1a).]

2. Let $1 \to N \to E \to G \to 1$ be an extension of finite groups such that $|N|$ and $|G|$ are relatively prime. We proved in 3.13 that such an extension must split if N if abelian, and we remarked that this result could be generalized to the non-abelian case. One might hope to deduce this generalization directly from 6.6 in view of the vanishing of $H^2(G, C)$. Why doesn't this work?

CHAPTER V
Products

1 The Tensor Product of Resolutions

If G and G' are groups and M (resp. M') is a G-module (resp. G'-module), then $M \otimes M'$ is a $G \times G'$-module in an obvious way: $(g, g') \cdot (m \otimes m') = gm \otimes g'm'$. Note that if M is projective over $\mathbb{Z}G$ and M' is projective over $\mathbb{Z}G'$ then $M \otimes M'$ is projective over $\mathbb{Z}[G \times G']$. In fact, it suffices to verify this in the case where $M = \mathbb{Z}G$ and $M' = \mathbb{Z}G'$, in which case the assertion follows from the obvious isomorphism $\mathbb{Z}G \otimes \mathbb{Z}G' \approx \mathbb{Z}[G \times G']$.

Now let $\varepsilon: F \to \mathbb{Z}$ (resp. $\varepsilon': F' \to \mathbb{Z}$) be a projective resolution of $\mathbb{Z}$ over $\mathbb{Z}G$ (resp. $\mathbb{Z}G'$), and consider the complex $F \otimes F'$. This is a complex of projective $\mathbb{Z}[G \times G']$-modules, and it is augmented over $\mathbb{Z}$ by $\varepsilon \otimes \varepsilon'$: $F \otimes F' \to \mathbb{Z} \otimes \mathbb{Z} = \mathbb{Z}$. Moreover, $\varepsilon \otimes \varepsilon'$ is a weak equivalence. This follows, for instance, from I.7.6; for the latter shows that $\varepsilon: F \to \mathbb{Z}$ and $\varepsilon': F' \to \mathbb{Z}$ are homotopy equivalences if we ignore the G-action, so the same is true of $\varepsilon \otimes \varepsilon'$ (cf. exercise 7c of §I.0). [Alternatively, use the Künneth formula.]

We have therefore established:

(1.1) Proposition. *If $\varepsilon: F \to \mathbb{Z}$ and $\varepsilon': F' \to \mathbb{Z}$ are projective resolutions of $\mathbb{Z}$ over $\mathbb{Z}G$ and $\mathbb{Z}G'$, respectively, then $\varepsilon \otimes \varepsilon': F \otimes F' \to \mathbb{Z}$ is a projective resolution of $\mathbb{Z}$ over $\mathbb{Z}[G \times G']$.*

This result should be thought of as the algebraic analogue of the obvious fact that the cartesian product of a $K(G, 1)$ and a $K(G', 1)$ is a $K(G \times G', 1)$.

(1.2) Corollary. *If $\varepsilon: F \to \mathbb{Z}$ and $\varepsilon': F' \to \mathbb{Z}$ are projective resolutions of $\mathbb{Z}$ over $\mathbb{Z}G$, then so is $\varepsilon \otimes \varepsilon': F \otimes F' \to \mathbb{Z}$, where G acts diagonally on $F \otimes F'$.*

This follows from 1.1 by restriction of scalars with respect to the "diagonal" embedding $G \to G \times G$, given by $g \mapsto (g, g)$.

Note that if $F = F'$ in 1.2 then we have two obvious maps $F \otimes F \to F$ of resolutions, namely, $F \otimes \varepsilon: F \otimes F \to F \otimes \mathbb{Z} = F$ and $\varepsilon \otimes F: F \otimes F \to \mathbb{Z} \otimes F = F$. There is no obvious map in the other direction, but we know from I.7.5 that there exist augmentation-preserving G-chain maps $\Delta: F \to F \otimes F$, and that any two are homotopic. Any such map Δ will be called a *diagonal approximation*. In case F is the standard resolution, for example, there is a well-known diagonal approximation, called the Alexander–Whitney map. In terms of the homogeneous description of F, this map is given by

$$\Delta(g_0, \ldots, g_n) = \sum_{p=0}^{n} (g_0, \ldots, g_p) \otimes (g_p, \ldots, g_n). \tag{1.3}$$

Translating this into the bar notation, we find

$$\Delta[g_1|\cdots|g_n] = \sum_{p=0}^{n} [g_1|\cdots|g_p] \otimes g_1 \cdots g_p[g_{p+1}|\cdots|g_n]. \tag{1.4}$$

EXERCISE

Let G be a finite cyclic group of order m with generator t and let F be the resolution I.6.3. In particular, $F_n = \mathbb{Z}G$ for all $n \geq 0$. Let $\Delta: F \to F \otimes F$ be the map whose (p, q)-component $\Delta_{pq}: F_{p+q} \to F_p \otimes F_q$ is given by

$$\Delta_{pq}(1) = \begin{cases} 1 \otimes 1 & p \text{ even} \\ 1 \otimes t & p \text{ odd}, q \text{ even} \\ \displaystyle\sum_{0 \leq i < j \leq m-1} t^i \otimes t^j & p \text{ odd}, q \text{ odd}. \end{cases}$$

Verify that Δ is a diagonal approximation.

2 Cross-products

Let G, G', F, and F', be as in the previous section. For any G-module M and G'-module M' there is an obvious map

$$(F \otimes_G M) \otimes (F' \otimes_{G'} M') \to (F \otimes F') \otimes_{G \times G'} (M \otimes M'), \tag{2.1}$$

given by $(x \otimes m) \otimes (x' \otimes m') \mapsto (x \otimes x') \otimes (m \otimes m')$. [One justifies this rigorously by noting that

$$(F \otimes_G M) \otimes (F' \otimes_{G'} M') = (F \otimes M)_G \otimes (F' \otimes M')_{G'},$$

which is a certain quotient of $F \otimes M \otimes F' \otimes M'$; making this quotient explicit, one can even see that the map 2.1 is an isomorphism.] If $z \in F \otimes_G M$

and $z' \in F' \otimes_{G'} M'$, then we will denote by $z \times z'$ the image of $z \otimes z'$ under this map. Clearly we have $\partial(z \times z') = \partial z \times z' + (-1)^p z \times \partial z'$ if $\deg z = p$; hence the product of two cycles is a cycle, whose homology class depends only on the classes of the given cycles. We therefore obtain, in view of 1.1, an induced product

$$H_p(G, M) \otimes H_q(G', M') \to H_{p+q}(G \times G', M \otimes M'),$$

called the *homology cross-product* and still denoted $z \otimes z' \mapsto z \times z'$. This product is well-defined, independent of the choice of the resolutions F and F'.

Similarly, given cochains $u \in \mathscr{H}om_G(F, M)$ and $u' \in \mathscr{H}om_{G'}(F', M')$, we define $u \times u' \in \mathscr{H}om_{G \times G'}(F \otimes F', M \otimes M')$ to be the tensor product of the maps u and u' as defined in exercise 7 of §I.0; thus $\langle u \times u', x \otimes x' \rangle = (-1)^{\deg u' \cdot \deg x} \langle u, x \rangle \otimes \langle u', x' \rangle$. It is easy to verify that $\delta(u \times u') = \delta u \times u' + (-1)^p u \times \delta u'$ if $\deg u = p$. [This is a special case of exercise 7a of I.0.] There is therefore an induced *cohomology cross-product*

$$H^p(G, M) \otimes H^q(G', M') \to H^{p+q}(G \times G', M \otimes M').$$

EXERCISES

1. Show that the map 2.1 is an isomorphism. Deduce, under suitable hypotheses, that there is a Künneth formula for computing $H_*(G \times G', M \otimes M')$ in terms of $H_*(G, M)$ and $H_*(G', M')$.

*2. Give hypotheses under which the cochain cross-product

$$\mathscr{H}om_G(F, M) \otimes \mathscr{H}om_{G'}(F', M') \to \mathscr{H}om_{G \times G'}(F \otimes F', M \otimes M')$$

is an isomorphism, and deduce a cohomology Künneth formula.

3 Cup and Cap Products

The previous section dealt with "external" products, involving three groups G, G', and $G \times G'$. In this section we will discuss *internal* products, involving the homology and/or cohomology of a single group G.

Given $u \in H^p(G, M)$ and $v \in H^q(G, N)$, we define the *cup product* of u and v (denoted $u \cup v$ or uv) to be the element $d^*(u \times v) \in H^{p+q}(G, M \otimes N)$, where $d: G \to G \times G$ is the diagonal map. Here $M \otimes N$ has the diagonal G-action (as it must if $d^*: H^*(G \times G, M \otimes N) \to H^*(G, M \otimes N)$ is to make sense).

Checking the definitions, we see that the cup product is induced by the following *cochain cup product*: Let F and F' be projective resolutions of $\mathbb{Z}$ over $\mathbb{Z}G$, and recall (1.2) that $F \otimes F'$ with diagonal G-action is also a projective resolution of $\mathbb{Z}$ over $\mathbb{Z}G$. Given $u \in \mathscr{H}om_G(F, M)$ and $v \in \mathscr{H}om_G(F', N)$, define $u \cup v$ to be $u \times v$ as defined in §2, but now regarded as an element of $\mathcal{H}om_G(F \otimes F', M \otimes N)$. Alternatively, if we prefer to use a single resolution F, then we choose a diagonal approximation $\Delta: F \to F \otimes F$ (cf. §1) and set $u \cup v = (u \times v) \circ \Delta \in \mathcal{H}om_G(F, M \otimes N)$ for $u \in \mathcal{H}om_G(F, M)$ and $v \in \mathcal{H}om_G(F, N)$. For example, if F is the bar resolution and Δ is the Alexander–Whitney map (1.4), then the product of $u \in C^p(G, M)$ and $v \in C^q(G, N)$ is the element $u \cup v \in C^{p+q}(G, M \otimes N)$ given by

$$(u \cup v)(g_1, \ldots, g_{p+q}) = (-1)^{pq} u(g_1, \ldots, g_p) \otimes g_1 \cdots g_p v(g_{p+1}, \ldots, g_{p+q}).$$

We now list some formal properties of the cup product.

(3.1) *Dimension* 0: The cup product $H^0(G, M) \otimes H^0(G, N) \to H^0(G, M \otimes N)$ is the map $M^G \otimes N^G \to (M \otimes N)^G$ induced by the inclusions $M^G \hookrightarrow M$ and $N^G \hookrightarrow N$.

(3.2) *Naturality with respect to coefficient homomorphisms*: Given G-module maps $f: M \to M'$ and $g: N \to N'$ and elements $u \in H^*(G, M)$ and $v \in H^*(G, N)$, we have

$$(f \otimes g)_*(u \cup v) = f_* u \cup g_* v$$

in $H^*(G, M' \otimes N')$, where $f_* = H^*(G, f)$, etc.

3.1 and 3.2 are immediate from the definitions.

(3.3) *Compatibility with* δ: Let $0 \to M' \to M \to M'' \to 0$ be a short exact sequence of G-modules and let N be a G-module such that the sequence $0 \to M' \otimes N \to M \otimes N \to M'' \otimes N \to 0$ is exact. (For example, this holds for any N if $0 \to M' \to M \to M'' \to 0$ is split exact as a sequence of $\mathbb{Z}$-modules.) Then we have $\delta(u \cup v) = \delta u \cup v$ for any $u \in H^p(G, M'')$ and $v \in H^q(G, N)$. In other words, the square

$$\begin{array}{ccc} H^p(G, M'') & \xrightarrow{\delta} & H^{p+1}(G, M') \\ \downarrow{\scriptstyle -\cup v} & & \downarrow{\scriptstyle -\cup v} \\ H^{p+q}(G, M'' \otimes N) & \xrightarrow{\delta} & H^{p+q+1}(G, M' \otimes N). \end{array}$$

commutes.

PROOF. Consider the commutative diagram

$$\begin{array}{ccccccccc} 0 & \longrightarrow & C^*(G, M') & \longrightarrow & C^*(G, M) & \longrightarrow & C^*(G, M'') & \longrightarrow & 0 \\ & & \downarrow & & \downarrow & & \downarrow & & \\ 0 & \longrightarrow & C^*(G, M' \otimes N) & \longrightarrow & C^*(G, M \otimes N) & \longrightarrow & C^*(G, M'' \otimes N) & \longrightarrow & 0, \end{array}$$

where the vertical arrows are given by cup product on the right with a fixed cocycle in $C^q(G, N)$ representing v. These vertical maps commute with the coboundary operators δ in $C^*(G, —)$; for the formula $\delta(a \cup b) = \delta a \cup b + (-1)^{\deg a} a \cup \delta b$ reduces to $\delta(a \cup b) = \delta a \cup b$ if b is a cocycle. The result now follows from the naturality of connecting homomorphisms with respect to maps of short exact sequences (I.0.4). □

Similarly, one proves:

(3.3′) If $0 \to N' \to N \to N'' \to 0$ is a short exact sequence such that $0 \to M \otimes N' \to M \otimes N \to M \otimes N'' \to 0$ is exact, then $\delta(u \cup v) = (-1)^p u \cup \delta v$ in $H^{p+q+1}(G, M \otimes N')$ for any $u \in H^p(G, M)$, $v \in H^q(G, N'')$.

3.3 and 3.3′ allow one to use dimension-shifting arguments in the study of cup products. For we have seen (cf. exercise 3 of §III.7) that we can use a $\mathbb{Z}$-split injection $M \to \overline{\overline{M}}$ for the dimension-shifting argument.

(3.4) *Existence of identity*: The element $1 \in H^0(G, \mathbb{Z}) = \mathbb{Z}$ satisfies $1 \cup u = u = u \cup 1$ for all $u \in H^*(G, M)$, where we make the obvious identifications $\mathbb{Z} \otimes M = M = M \otimes \mathbb{Z}$ of coefficient modules.

This follows from the definitions together with the following two observations: (a) $1 \in H^0(G, \mathbb{Z})$ is represented by the augmentation map ε, regarded as a 0-cocycle in $\mathscr{H}om_G(F, \mathbb{Z})$; and (b) the maps $F \otimes \varepsilon$ and $\varepsilon \otimes F$ are maps of resolutions $F \otimes F \to F$ (cf. §1) and hence induce the "identity map" on cohomology. Alternatively, use the Alexander–Whitney formula.

(3.5) *Associativity*: Given $u_i \in H^*(G, M_i)$ $(i = 1, 2, 3)$, we have $(u_1 u_2) u_3 = u_1(u_2 u_3)$ in $H^*(G, M_1 \otimes M_2 \otimes M_3)$.

Indeed, associativity holds on the cochain level as an identity in $\mathscr{H}om_G(F \otimes F \otimes F,\ M_1 \otimes M_2 \otimes M_3)$. Alternatively, use the Alexander–Whitney formula.

(3.6) *Commutativity*: For any $u \in H^p(G, M)$, $v \in H^q(G, N)$, we have $uv = (-1)^{pq} t_*(vu)$, where $t: N \otimes M \to M \otimes N$ is the canonical isomorphism.

PROOF. Let $\tau: F \otimes F \to F \otimes F$ be the chain automorphism such that $\tau(x \otimes y) = (-1)^{\deg x \cdot \deg y} y \otimes x$, cf. exercise 5 of §I.0. We have a commutative diagram

$$\begin{array}{ccc} \mathscr{H}om_G(F, M) \otimes \mathscr{H}om_G(F, N) & \xrightarrow{\cup} & \mathscr{H}om_G(F \otimes F, M \otimes N) \\ \downarrow & & \downarrow {\scriptstyle \mathscr{H}om_G(\tau, t)} \\ \mathscr{H}om_G(F, N) \otimes \mathscr{H}om_G(F, M) & \xrightarrow{\cup} & \mathscr{H}om_G(F \otimes F, N \otimes M) \end{array}$$

where the vertical arrow on the left is given by $u \otimes v \mapsto (-1)^{\deg u \cdot \deg v} v \otimes u$. [When you check the commutativity of this square, the signs may seem to come out wrong at first; but don't forget that $\langle u, x\rangle = 0$ unless $\deg u = \deg x$.] Since τ is an augmentation-preserving chain map, it induces the identity in cohomology. Hence the vertical arrow on the right induces t_* in cohomology, and this yields 3.6. □

It follows from 3.4–3.6 that $H^*(G, \mathbb{Z})$ is an anti-commutative graded ring and that $H^*(G, M)$ is a graded module over this ring. More generally, if k is an arbitrary commutative ring (with trivial G-action, for simplicity), then we have a product

$$H^*(G, k) \otimes H^*(G, k) \xrightarrow{\cup} H^*(G, k \otimes k) \to H^*(G, k)$$

which makes $H^*(G, k)$ a graded anti-commutative k-algebra. (The unlabelled arrow above is induced by the multiplication map $k \otimes k \to k$.) Similarly, $H^*(G, M)$ is an $H^*(G, k)$-module if M is a kG-module.

(3.7) *Naturality with respect to group homomorphisms*: Given $\alpha: H \to G$, we have $\alpha^*(u \cup v) = \alpha^*u \cup \alpha^*v$ for any $u \in H^*(G, M)$, $v \in H^*(G, N)$.

This is immediate from the definitions.

As a special case of 3.7, $\alpha^*: H^*(G, k) \to H^*(H, k)$ is a ring homomorphism.

(3.8) *Transfer formula*: Suppose $H \subseteq G$ is a subgroup of finite index. For any $u \in H^*(G, M)$ and $v \in H^*(H, N)$, we have

$$\operatorname{cor}_H^G(\operatorname{res}_H^G(u) \cup v) = u \cup \operatorname{cor}_H^G v.$$

This says, in particular, that the transfer map $H^*(H, k) \to H^*(G, k)$ is a homomorphism of $H^*(G, k)$-modules, where $H^*(H, k)$ is regarded as an $H^*(G, k)$-module via the restriction homomorphism $H^*(G, k) \to H^*(H, k)$.

Proof of 3.8. Let F be a projective resolution of $\mathbb{Z}$ over $\mathbb{Z}G$; we will prove the stated formula on the cochain level. Given $u \in \mathcal{H}om(F, M)^G$ and $v \in \mathcal{H}om(F, N)^H$, we have, in $\mathcal{H}om(F \otimes F, M \otimes N)^G$:

$$\begin{aligned}
\operatorname{cor}_H^G(\operatorname{res}_H^G(u) \cup v) &= \sum_{g \in G/H} g \cdot (u \otimes v) \\
&= \sum_{g \in G/H} u \otimes gv \qquad \text{[because } u \text{ is } G\text{-invariant]} \\
&= u \otimes \sum_{g \in G/H} gv \\
&= u \cup \operatorname{cor}_H^G v.
\end{aligned}$$

□

There is a second internal product, called the *cap product*, which is useful in the study of duality (cf. §§VI.7 and VIII.10). If F is a projective resolution of $\mathbb{Z}$ over $\mathbb{Z}G$, then there is a chain map

$$\gamma: \mathcal{H}om_G(F, M) \otimes ((F \otimes F) \otimes_G N) \to F \otimes_G (M \otimes N)$$

given by $u \otimes (x \otimes y \otimes n) \mapsto (-1)^{\deg u \cdot \deg x} x \otimes u(y) \otimes n$. [The reader can check directly that γ is well-defined and is a chain map. Alternatively, we can appeal to the exercises of §I.0: According to exercise 7b of the latter, we have a chain map $\mathcal{H}om_G(F, M) \to \mathcal{H}om((F \otimes N) \otimes_G F, (F \otimes N) \otimes_G M)$ given by $u \mapsto \mathrm{id}_{F \otimes N} \otimes u$. This corresponds by exercise 6b of §I.0 to a chain map

$$\mathcal{H}om_G(F, M) \otimes ((F \otimes N) \otimes_G F) \to (F \otimes N) \otimes_G M,$$

which is precisely γ, modulo the identifications $(F \otimes N) \otimes_G F = (F \otimes F) \otimes_G N$ and $(F \otimes N) \otimes_G M = F \otimes_G (M \otimes N)$. These identifications can be derived by manipulating triple tensor products as in the proof of III.2.2.]

For $u \in \mathcal{H}om_G(F, M)^p = \mathcal{H}om_G(F, M)_{-p}$ and $z \in (F \otimes F)_q \otimes_G N$, $\gamma(u \otimes z)$ is an element of $F_{q-p} \otimes_G (M \otimes N)$, denoted $u \cap z$ and called the *cap product* of u and z. The same notation and terminology are used for the induced product

$$H^p(G, M) \otimes H_q(G, N) \to H_{q-p}(G, M \otimes N).$$

As with the cup product, one can use a diagonal approximation $\Delta: F \to F \otimes F$ to compute the cap product in terms of a single resolution F. Namely, one composes γ with the map

$$\mathrm{id} \otimes (\Delta \otimes \mathrm{id}): \mathcal{H}om_G(F, M) \otimes (F \otimes_G N) \to \mathcal{H}om_G(F, M) \otimes ((F \otimes F) \otimes_G N).$$

The cap product, which may seem strange at first, can be motivated by the fact that it is *adjoint* to the cup product, in a sense which we now explain. Consider the "evaluation map"

$$\mathcal{H}om_G(F, M) \otimes (F \otimes_G N) \to M \otimes_G N,$$

given by $u \otimes (x \otimes n) \mapsto u(x) \otimes n$. We denote by $\langle u, z \rangle$ the image of $u \otimes z$ under this map. [Except for the fact that we are carrying along the factor N, this is the same as the evaluation map introduced in exercise 3 of §I.0.] The evaluation map is a chain map, i.e., $\langle \delta u, z \rangle + (-1)^{\deg u} \langle u, \partial z \rangle = 0$; so there is an induced pairing

$$H^p(G, M) \otimes H_p(G, N) \to M \otimes_G N,$$

still denoted $\langle \cdot, \cdot \rangle$, which is independent of the choice of resolution. One now checks, directly from the definitions, that the following adjunction formula holds for any $u \in H^p(G, M_1)$, $v \in H^q(G, M_2)$, $z \in H_{p+q}(G, M_3)$:

$$\langle u \cup v, z \rangle = \langle u, v \cap z \rangle. \tag{3.9}$$

(Note: Both sides of this equation are in $(M_1 \otimes M_2 \otimes M_3)_G$.) In particular, taking $u = 1 \in H^0(G, \mathbb{Z})$, we find:

(3.10) For any $v \in H^q(G, M)$ and $z \in H_q(G, N)$, we have $v \cap z = \langle v, z \rangle$ in $H_0(G, M \otimes N) = M \otimes_G N$.

The cap product has properties analogous to 3.1–3.8. We leave it to the interested reader to write these down.

EXERCISES

1. Given $m \in H^0(G, M) = M^G$ and $u \in H^q(G, N)$, show that $m \cup u = f_m(u)$, where $f_m: H^*(G, N) \to H^*(G, M \otimes N)$ is induced by the coefficient homomorphism $n \mapsto m \otimes n$. [Hint: Use 3.2 and 3.4.] State and prove a similar interpretation of the cap product $H^0(G, M) \otimes H_q(G, N) \to H_q(G, M \otimes N)$.

2. Using the diagonal approximation given in the exercise of §1, compute all cup products in $H^*(G, —)$ if G is finite cyclic.

3. Let G be a finite group which acts freely on S^{2k-1} as in §I.6.

   (a) For any G-module M, show that there is an iterated coboundary map d: $H^i(G, M) \to H^{i+2k}(G, M)$ which is an isomorphism for $i > 0$ and an epimorphism for $i = 0$. [Hint: Tensor the sequence I.6.1 with M, break it up into short exact sequences, and use the dimension-shifting argument.]

   (b) Show that there is an element $u \in H^{2k}(G, \mathbb{Z})$ such that the "periodicity map" d of (a) is given by $d(v) = u \cup v$ for all $v \in H^*(G, M)$. [Hint: By 3.3, $d(w \cup v) = d(w) \cup v$ for any $w \in H^*(G, \mathbb{Z})$, $v \in H^*(G, M)$; now set $w = 1$.]

   (c) Using (b), calculate the ring structure on $H^*(G, \mathbb{Z})$ for G finite cyclic.

4. Let G be cyclic of order n. For any $m \in \mathbb{Z}_n$ there is an endormorphism $\alpha(m)$ of G, given by $\alpha(m)g = g^m$. Calculate $\alpha(m)^*: H^*(G, \mathbb{Z}) \to H^*(G, \mathbb{Z})$. [Hint: In view of what you know about the ring structure on $H^*(G, \mathbb{Z})$ from exercise 2 and/or 3c, it suffices to calculate $\alpha(m)^*$ on $H^2(G, \mathbb{Z})$. This can be done non-computationally by using the universal coefficient isomorphism $H^2(G, \mathbb{Z}) \approx \operatorname{Ext}(H_1 G, \mathbb{Z})$ or by using the interpretation of H^2 in terms of group extensions.]

5. Using exercise 4 and Theorem III.10.3, calculate the integral cohomology of the symmetric group on 3 letters.

4 Composition Products

We observed several times in Chapter III that there are a number of chain complexes that can be used to compute $H_*(G, —)$ and $H^*(G, —)$. Here is one more example of this:

(4.1) Lemma. *Let $\varepsilon: F \to \mathbb{Z}$ and $\varepsilon': F' \to \mathbb{Z}$ be resolutions of $\mathbb{Z}$ over $\mathbb{Z}G$ such that F is projective and F' is $\mathbb{Z}$-free. For any G-module M, the map $\varepsilon' \otimes M$: $F' \otimes M \to \mathbb{Z} \otimes M = M$ induces weak equivalences $F \otimes_G (F' \otimes M) \to F \otimes_G M$ and $\mathscr{H}om_G(F, F' \otimes M) \to \mathscr{H}om_G(F, M)$. Hence*

$$H_*(G, M) \approx H_*(F \otimes_G (F' \otimes M))$$

and

$$H^*(G, M) \approx H^*(\mathcal{H}om_G(F, F' \otimes M)).$$

PROOF. We already noted in §III.2 that $\varepsilon' \otimes M: F' \otimes M \to M$ is a weak equivalence. The lemma therefore follows from I.8.5 and I.8.6. □

Note that the last isomorphism in 4.1 can be written in the form $H^*(G, M) \approx [F, F' \otimes M]^*$, where $[F, F' \otimes M]^n = [F, F' \otimes M]_{-n}$ is the group of homotopy classes of chain maps of degree $-n$ from F' to $F' \otimes M$ (cf. §I.0). The significance of this is that it allows us to construct products by composition of chain maps. Taking $M = \mathbb{Z}$ and $F = F'$, for instance, the isomorphism takes the form $H^*(G, \mathbb{Z}) \approx [F, F]^*$. Since homotopy classes of chain maps can be composed, we obtain a *composition product*

$$H^*(G, \mathbb{Z}) \otimes H^*(G, \mathbb{Z}) \to H^*(G, \mathbb{Z}).$$

More generally, let $\varepsilon: F \to \mathbb{Z}$, $\varepsilon': F' \to \mathbb{Z}$, and $\varepsilon'': F'' \to \mathbb{Z}$ be resolutions of $\mathbb{Z}$ over $\mathbb{Z}G$ such that F and F' are projective and F'' is $\mathbb{Z}$-free. [In practice, we will either take F'' projective or $F'' = \mathbb{Z}$.] For any G-modules M, N, there is a cochain product

(4.2) $$\mathcal{H}om_G(F', F'' \otimes M) \otimes \mathcal{H}om_G(F, F' \otimes N) \to \mathcal{H}om_G(F, F'' \otimes M \otimes N)$$

given by $u \otimes v \mapsto (u \otimes \mathrm{id}_N) \circ v$ for $u \in \mathcal{H}om_G(F', F'' \otimes M)$, $v \in \mathcal{H}om_G(F, F' \otimes N)$:

$$F \xrightarrow{v} F' \otimes N \xrightarrow{u \otimes \mathrm{id}} F'' \otimes M \otimes N.$$

One easily verifies that 4.2 is a chain map (cf. exercise 4 of §I.0); in view of 4.1, there is an induced cohomology product

(4.3) $$H^*(G, M) \otimes H^*(G, N) \to H^*(G, M \otimes N),$$

called the *composition product*. It is well-defined, independent of the choices of resolution.

Similarly, there is a product

(4.4) $$\mathcal{H}om_G(F', F'' \otimes M) \otimes (F \otimes_G (F' \otimes N)) \to F \otimes_G (F'' \otimes M \otimes N)$$

given by

$$u \otimes x \otimes x' \otimes n \mapsto (-1)^{\deg u \cdot \deg x} x \otimes u(x') \otimes n$$

for $u \in \mathcal{H}om_G(F', F'' \otimes M)$, $x \in F$, $x' \in F'$, $n \in N$. One verifies as in the definition of the cap product that 4.4 is chain map; in view of 4.1, there is an induced product

(4.5) $$H^*(G, M) \otimes H_*(G, N) \to H_*(G, M \otimes N).$$

The following result, although not unexpected, is useful:

(4.6) Theorem. *The products* 4.3 *and* 4.5 *coincide with the cup and cap products.*

PROOF. Consider the cochain composition product 4.2 with $F = F'$ and $F'' = \mathbb{Z}$:

$$(*) \qquad \mathcal{H}om_G(F, M) \otimes \mathcal{H}om_G(F, F \otimes N) \to \mathcal{H}om_G(F, M \otimes N).$$

This induces the product 4.3 via the weak equivalence $\alpha = \mathcal{H}om_G(F, \varepsilon \otimes N)$: $\mathcal{H}om_G(F, F \otimes N) \to \mathcal{H}om_G(F, N)$ of 4.1. In order to compare 4.3 to the cup product, we would like to find a weak inverse of α, i.e., a map $\alpha' : \mathcal{H}om_G(F, N) \to \mathcal{H}om_G(F, F \otimes N)$ which induces in cohomology the isomorphism inverse to that induced by α. This would allow us to convert $(*)$ to a cochain product

$$(**) \qquad \mathcal{H}om_G(F, M) \otimes \mathcal{H}om_G(F, N) \to \mathcal{H}om_G(F, M \otimes N)$$

by composing $(*)$ with

$$\mathrm{id} \otimes \alpha' : \mathcal{H}om_G(F, M) \otimes \mathcal{H}om_G(F, N) \to \mathcal{H}om_G(F, M) \otimes \mathcal{H}om_G(F, F \otimes N).$$

It turns out that a diagonal approximation $\Delta: F \to F \otimes F$ gives rise to such an α'. Namely, we take α' to be the composite

$$\mathcal{H}om_G(F, N) \xrightarrow{\mathrm{id}_F \otimes -} \mathcal{H}om_G(F \otimes F, F \otimes N) \xrightarrow{- \circ \Delta} \mathcal{H}om_G(F, F \otimes N),$$

i.e., $\alpha'(u) = (\mathrm{id}_F \otimes u) \circ \Delta$. Computing $\alpha\alpha' : \mathcal{H}om_G(F, N) \to \mathcal{H}om_G(F, N)$, one finds that it is the map

$$u \mapsto (\varepsilon \otimes \mathrm{id}_N) \circ (\mathrm{id}_F \otimes u) \circ \Delta = (\varepsilon \otimes u) \circ \Delta.$$

But this is simply $u \mapsto \varepsilon \cup u$; since the cocycle ε represents $1 \in H^0(G, \mathbb{Z})$, it follows that α' is indeed a weak inverse of α.

We now have a cochain product $(**)$ which induces the composition product in cohomology, and it is given explicitly by

$$\begin{aligned} u \otimes v \mapsto (u \otimes \mathrm{id}_N) \circ \alpha'(v) &= (u \otimes \mathrm{id}_N)(\mathrm{id}_F \otimes v)\Delta \\ &= (u \otimes v)\Delta \\ &= u \cup v, \end{aligned}$$

where the latter is the cochain cup product defined via Δ. Thus the composition product in cohomology equals the cup product.

The proof that 4.5 equals the cap product is similar, and even easier. This time we consider the map 4.4, with $F = F'$ and $F'' = \mathbb{Z}$, and we seek a weak inverse $\beta' : F \otimes_G N \to F \otimes_G (F \otimes N)$ of the weak equivalence $\beta = F \otimes \varepsilon \otimes N : F \otimes_G (F \otimes N) \to F \otimes_G N$ of 4.1. I claim we can take $\beta' = \Delta \otimes N$. For the composite $(F \otimes \varepsilon \otimes N) \circ (\Delta \otimes N) : F \otimes_G N \to F \otimes_G N$ is a map of the form $\tau \otimes N$, where $\tau : F \to F$ is an augmentation-preserving chain map, hence it is homotopic to the identity. We now use β' to convert 4.4 to a product

$$\mathcal{H}om_G(F, M) \otimes (F \otimes_G N) \to F \otimes_G (M \otimes N),$$

and it is transparent that this map coincides with the chain level cap product defined by Δ. □

We close by remarking that the methods of this section can be used to define composition products

(4.7) $$\mathrm{Ext}_G^*(M', M'') \otimes \mathrm{Ext}_G^*(M, M') \to \mathrm{Ext}_G^*(M, M'')$$

and

(4.8) $$\mathrm{Ext}_G^*(M', M'') \otimes \mathrm{Tor}_*^G(M, M') \to \mathrm{Tor}_*^G(M, M'')$$

for any G-modules M, M', M''. One uses suitable resolutions F, F', F'' of the three modules and one defines maps $\mathscr{H}om_G(F', F'') \otimes \mathscr{H}om_G(F, F') \to \mathscr{H}om_G(F, F'')$ and $\mathscr{H}om_G(F', F'') \otimes (F \otimes_G F') \to F \otimes_G F''$ similar to 4.2 and 4.4. The products 4.7 and 4.8 are obviously closely related to 4.3 and 4.5, but it does not seem that either type is a special case of the other. For instance 4.3, which we now know is the same as the cup product, can be written in the form

$$\mathrm{Ext}_G^*(\mathbb{Z}, M) \otimes \mathrm{Ext}_G^*(\mathbb{Z}, N) \to \mathrm{Ext}_G^*(\mathbb{Z}, M \otimes N).$$

This fits into the framework of 4.7 in the important special case where $N = \mathbb{Z}$, but not in general. In the other direction, we know that the Ext groups in 4.7 can be expressed as cohomology groups if M and M' are $\mathbb{Z}$-free (cf. III.2.2), so we expect 4.7 to be describable in terms of cup products in this case, but not in general.

Note, incidentally, that 4.7 and 4.8 can be defined for Ext and Tor over arbitrary rings, not just group rings. One need only be careful about which modules are left modules and which are right modules.

Exercises

1. Carry out the details of the definitions of 4.7 and 4.8.

2. Describe 4.7 in terms of cup products if M and M' are $\mathbb{Z}$-free. Describe 4.8 in terms of cap products if M' is $\mathbb{Z}$-free and M is $\mathbb{Z}$-torsion-free. [Hint: Use Theorem 4.6.]

5 The Pontryagin Product

If G is an *abelian* group then the multiplication map $G \times G \to G$ is a group homomorphism. Using this map, we will define an internal homology product, called the Pontryagin product. For simplicity, we will take the coefficient module to be a commutative ring k, with trivial G-action. See exercise 1 below for a more general product.

The *Pontryagin product* on $H_*(G, k)$ for G abelian is defined to be the composite

$$H_*(G, k) \otimes H_*(G, k) \xrightarrow{\times} H_*(G \times G, k \otimes k) \xrightarrow{\mu_*} H_*(G, k),$$

where $\mu: (G \times G, k \otimes k) \to (G, k)$ is the multiplication map $((g, g') \mapsto gg', \lambda \otimes \lambda' \mapsto \lambda\lambda')$. To compute this product in terms of a projective resolution F of $\mathbb{Z}$ over $\mathbb{Z}G$, we need to choose an augmentation-preserving chain map $\tau: F \otimes F \to F$ compatible with the multiplication map $G \times G \to G$, i.e., satisfying $\tau(gx \otimes g'y) = gg'\tau(x \otimes y)$. In other words, writing $xy = \tau(x \otimes y)$ for $x, y \in F$, what we have is a $\mathbb{Z}G$-bilinear product on F such that (a) $\varepsilon(xy) = \varepsilon(x)\varepsilon(y)$ and (b) $\partial(xy) = \partial x \circ y + (-1)^{\deg x} x \circ \partial y$. (For brevity, we will call a $\mathbb{Z}G$-bilinear product on F satisfying (a) and (b) an *admissible product.*) Such a product induces a k-bilinear product on $F \otimes_G k$, which in turn induces the Pontryagin product on $H_*(G, k)$. As in 3.4–3.6, one verifies that this product makes $H_*(G, k)$ an associative, anti-commutative, graded k-algebra with identity.

In case F is the standard resolution, there is a canonical product on F, called the *shuffle product*. This has its origins in the Eilenberg–Zilber theorem of algebraic topology (cf. MacLane [1963], VIII.8, X.12), and in the present context it takes the following form. Let $\mathscr{S}_n$ be the group of permutations of $\{1, \ldots, n\}$, and let $\mathscr{S}_n$ act $\mathbb{Z}G$-linearly on F_n by

$$\sigma[g_1|\cdots|g_n] = (-1)^{\operatorname{sgn}\sigma}[g_{\sigma^{-1}(1)}|\cdots|g_{\sigma^{-1}(n)}].$$

If $n = p + q$ then an element $\sigma \in \mathscr{S}_n$ is said to be a (p, q)-*shuffle* if $\sigma(i) < \sigma(j)$ for $1 \leq i < j \leq p$ and for $p + 1 \leq i < j \leq p + q$. [Such a permutation can be viewed as a way of shuffling a deck of p cards with a deck of q cards.] The shuffle product on F is now defined to be the $\mathbb{Z}G$-bilinear product such that

$$[g_1|\cdots|g_p] \cdot [g_{p+1}|\cdots|g_{p+q}] = \sum_{\sigma} \sigma[g_1|\cdots|g_{p+q}],$$

where σ ranges over all (p, q)-shuffles. One can verify that, in addition to being admissible, this product is associative, anti-commutative, and has an identity. Since F is $\mathbb{Z}$-torsion-free, anti-commutativity implies *strict* anti-commutativity, i.e., $x^2 = 0$ if $\deg x$ is odd. It follows easily that $F \otimes_G k$ is strictly anti-commutative. Consequently:

(5.1) Proposition. *For any abelian group G and commutative ring k, the ring $H_*(G, k)$ is strictly anti-commutative.*

(By contrast, the cohomology ring $H^*(G, k)$ is not strictly anti-commutative in general. An attempt to prove strict anti-commutativity by the method used above fails because the Alexander–Whitney cochain product is only anti-commutative up to homotopy.)

We now give some other examples of resolutions with admissible products.

(5.2) Let G be infinite cyclic and let F be the resolution of I.4.5. Let 1 and x denote the $\mathbb{Z}G$-basis elements in dimensions 0 and 1, respectively. We set $1 \cdot 1 = 1$, $1 \cdot x = x \cdot 1 = x$, and $x^2 = 0$, and it is trivial to verify admissibility.

(Note that F, as a $\mathbb{Z}G$-algebra, is the *exterior algebra* $\bigwedge(x)$ with one generator x of degree 1.)

Our next example will use the notion of *divided polynomial algebra*, which is defined as follows: Consider the elements $y^{(i)} = y^i/i!$ $(i \geq 0)$ in the polynomial ring $\mathbb{Q}[y]$. We have $y^{(i)}y^{(j)} = (i, j)y^{(i+j)}$, where (i, j) is the binomial coefficient $\binom{i+j}{i} = (i + j)!/i!j!$, so the $\mathbb{Z}$-submodule of $\mathbb{Q}[y]$ generated by the $y^{(i)}$ is a subring. This ring is denoted $\Gamma(y)$ or $\Gamma_{\mathbb{Z}}(y)$ and is called the divided polynomial ring (over $\mathbb{Z}$) in one variable y. More generally, if R is an arbitrary ring, then the R-algebra $R \otimes \Gamma(y)$ is called the divided polynomial algebra over R in one variable y and is denoted $\Gamma(y)$ or $\Gamma_R(y)$. We will regard $\Gamma(y)$ as a *graded ring*, with $\deg y = 2$.

(5.3) Let G be a finite cyclic group and let F be the resolution I.6.3 with one $\mathbb{Z}G$-basis element e_i in dimension i, $\partial e_{2i} = Ne_{2i-1}$ $(i > 0)$, and $\partial e_{2i-1} = (t - 1)e_{2i-2}$. Define a $\mathbb{Z}G$-bilinear product on F by:

(i) $$e_{2i}e_{2j} = (i, j)e_{2i+2j}$$

(ii) $$e_{2i}e_{2j+1} = (i, j)e_{2i+2j+1} = e_{2j+1}e_{2i}$$

(iii) $$e_{2i+1}e_{2j+1} = 0.$$

Thus F, as a $\mathbb{Z}G$-algebra, is simply $\bigwedge(e_1) \otimes_{\mathbb{Z}G} \Gamma(e_2)$. [Recall that if A and B are graded algebras over a commutative ring R, then their graded tensor product $A \otimes_R B$ is a graded algebra, with

$$(a \otimes b)(a' \otimes b') = (-1)^{\deg b \cdot \deg a'} aa' \otimes bb'.]$$

One can check that this product on F is admissible; it is also associative and anti-commutative and has an identity.

Remark. The motivation for the definition above of the product in F is as follows: Let $x = e_1$ and $y = e_2$. Suppose we want our product to be associative and anti-commutative, to have e_0 as identity, and to satisfy (iv) $xe_{2i} = e_{2i+1}$. Let $y^i = c_i e_{2i}$, where c_i is an element of $\mathbb{Z}G$ to be determined. Then $\partial(y^i) = c_i \partial(e_{2i}) = c_i N e_{2i-1}$. On the other hand, admissibility and (iv) imply that

$$\partial(y^i) = iy^{i-1}\partial y = ic_{i-1}e_{2i-2}Nx = ic_{i-1}Ne_{2i-1}.$$

So we must have $c_i N = ic_{i-1}N$. The most obvious way to satisfy this is to take $c_i = i!$, so that $e_{2i} = y^i/i!$. Formulas (i)–(iii) now follow at once.

(5.4) Suppose F and F' are resolutions with admissible product for G and G'. Then it is easily verified that $F \otimes F'$ is a resolution with admissible product for $G \times G'$, where, as usual,

$$(x \otimes x')(y \otimes y') = (-1)^{\deg x' \cdot \deg y} xy \otimes x'y'.$$

Combining examples 5.2–5.4, one can write down a resolution with admissible product for any finitely generated abelian group G.

Finally, we use 5.4 to relate the cross-product to the Pontryagin product. If F and F' are as in 5.4, there is a chain isomorphism

$$(F \otimes_G k) \otimes_k (F' \otimes_{G'} k) \to (F \otimes F') \otimes_{G \times G'} k, \tag{5.5}$$

given by $(x \otimes \lambda) \otimes (x' \otimes \lambda') \mapsto (x \otimes x') \otimes \lambda\lambda'$. For $z \in F \otimes_G k$ and $z' \in F' \otimes_{G'} k$ we denote by $z \times z'$ the image of $z \otimes z'$ under this map. As in §2, there is an induced homology cross-product

$$H_*(G, k) \otimes_k H_*(G', k) \xrightarrow{\times} H_*(G \times G', k). \tag{5.6}$$

(5.7) Proposition. *If G and G' are abelian, then the cross-product 5.6 is a k-algebra homomorphism. Moreover, $z \times z' = i_* z \cdot i'_* z'$ for any $z \in H_*(G, k)$ and $z' \in H_*(G', k)$, where $i: G \to G \times G'$ and $i': G' \to G \times G'$ are the inclusions.*

Proof. The first assertion is an easy consequence of 5.4. Since $z \otimes z' = (z \otimes 1) \cdot (1 \otimes z')$, it now follows that $z \times z' = (z \times 1) \cdot (1 \times z')$. I claim that $z \times 1 = i_* z$; indeed, by naturality of the cross-product it suffices to check this for $G' = 1$, in which case it is obvious. Similarly, $1 \times z' = i'_* z'$, whence the proposition. □

We can now state the following Künneth formula:

(5.8) Corollary. *If G and G' are abelian and k is a principal ideal domain, then there is a split-exact sequence*

$$0 \to \bigoplus_{p+q=n} H_p(G, k) \otimes_k H_q(G', k) \xrightarrow{\mu} H_n(G \times G', k)$$

$$\to \bigoplus_{p+q=n-1} \operatorname{Tor}_1^k(H_p(G, k), H_q(G', k)) \to 0,$$

where $\mu(z \otimes z') = i_* z \cdot i'_* z'$.

Proof. In view of the isomorphism 5.5 and the Künneth theorem for chain complexes, we have a split-exact sequence as above with μ equal to the cross-product map 5.6. [This is true even if G and G' are non-abelian.] Now apply 5.7. □

Exercises

1. Recall that if R is a *commutative* ring, then $M \otimes_R N$ has an R-module structure for any two R-modules M and N, defined by $r \cdot (m \otimes n) = rm \otimes n = m \otimes rn$. This applies in particular if $R = \mathbb{Z}G$ with G abelian. The resulting tensor product $M \otimes_{\mathbb{Z}G} N$ should not be confused with the tensor product $M \otimes_G N = (M \otimes N)_G$ which we introduced in §III.0 for arbitrary G. (By contrast, the present tensor product $M \otimes_{\mathbb{Z}G} N$, as an abelian group, is equal to $(M \otimes N)_G$ where G acts *anti-diagonally* on $M \otimes N$: $g \cdot (m \otimes n) = g^{-1}m \otimes gn$.)

(a) If G is abelian, show that there is a Pontryagin product $H_*(G, M) \otimes H_*(G, N) \to H_*(G, M \otimes_{\mathbb{Z}G} N)$.

(b) Give an example where $M \otimes_{\mathbb{Z}G} N \not\approx M \otimes_G N$. [Hint: Take G cyclic of order 3 and $M = N = \mathbb{Z}_7$, with non-trivial G-action.]

2. State and prove a result analogous to 5.7, expressing the cohomology cross-product in terms of the cup product.

3. (a) Prove that homology commutes with direct limits. More precisely, let $(G_\alpha)_{\alpha \in D}$ be a direct system of groups, where D is a directed set, and let $G = \varinjlim G_\alpha$. For any G-module M, we have a compatible family of maps $H_*(G_\alpha, M) \to H_*(G, M)$ $(\alpha \in D)$, hence a map $\varphi: \varinjlim H_*(G_\alpha, M) \to H_*(G, M)$. Prove that φ is an isomorphism. [Hint: Use the standard resolution.]

   (b) Prove 5.1 without using the shuffle product. [Hint: Use (a) to reduce to the case where G is finitely generated.]

4. Let $n = p + q + r$. Define the notion of (p, q, r)-shuffle and prove that the 3-fold shuffle product in the bar resolution is given by

$$[g_1|\cdots|g_p] \cdot [g_{p+1}|\cdots|g_{p+q}] \cdot [g_{p+q+1}|\cdots|g_{p+q+r}] = \sum_\sigma \sigma[g_1|\cdots|g_{p+q+r}],$$

where σ ranges over the (p, q, r)-shuffles. Generalize.

6 Application: Calculation of the Homology of an Abelian Group

For any abelian group G we have $H_1(G, k) = H_1(G) \otimes k = G \otimes k$. We can therefore construct elements of $H_*(G, k)$ by taking products in the sense of §5 of elements of $G \otimes k$. (If $k = \mathbb{Z}$, for example, the reader has already seen this construction in exercise 1 of §II.3.) Our first goal in this section is to show (at least if k is a principal ideal domain) that there are no relations among these products other than the relations imposed by the fact that $H_*(G, k)$ is a strictly anti-commutative k-algebra. We begin by reviewing the notion of exterior algebra, so that we can state this result precisely.

Let k be an arbitrary commutative ring, let V be a k-module, and let $T^p(V) = V \otimes \cdots \otimes V$ (p copies of V), where $\otimes = \otimes_k$. (By convention, $T^0(V) = k$.) There is an obvious k-bilinear product on $T^*(V)$ given by juxtaposition of tensors, making $T^*(V)$ a graded k-algebra. We define the *exterior algebra* of V, denoted $\bigwedge^*(V)$ or $\bigwedge_k^*(V)$, to be the quotient of $T^*(V)$ by the two-sided ideal generated by the elements $v \otimes v \in T^2(V)$. Explicitly, then, $\bigwedge^p(V)$ is the quotient of $T^p(V)$ by the k-submodule generated by the elements $v_1 \otimes \cdots \otimes v_p$ such that $v_i = v_{i+1}$ for some i. The product in $\bigwedge^*(V)$ is sometimes denoted $x \wedge y$.

We have $\bigwedge^0(V) = k$ and $\bigwedge^1(V) = V$; moreover, $\bigwedge^1(V)$ generates $\bigwedge^*(V)$ as a k-algebra. Since the generators $v \in V = \bigwedge^1(V)$ satisfy $v^2 = 0$ by construction, it follows easily that $\bigwedge^*(V)$ is *strictly anti-commutative.* Moreover, it is immediate from the definition that $\bigwedge^*(V)$ has the following universal property:

(6.1) If A^* is a strictly anti-commutative graded k-algebra, then any k-module map $V \to A^1$ extends uniquely to a k-algebra map $\bigwedge^*(V) \to A^*$:

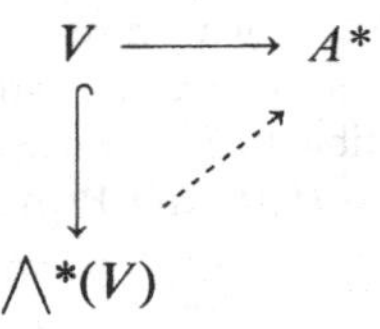

Thus $\bigwedge^*(V)$ is the strictly anti-commutative k-algebra "freely generated" by V. The following two properties of the exterior algebra functor will be crucial in our study of homology:

(6.2) $\bigwedge^*(V_1 \oplus V_2) \approx \bigwedge^*(V_1) \otimes \bigwedge^*(V_2)$. More precisely, let $i_1: \bigwedge^*(V_1) \to \bigwedge^*(V_1 \oplus V_2)$ and $i_2: \bigwedge^*(V_2) \to \bigwedge^*(V_1 \oplus V_2)$ be induced by the inclusions $V_1 \hookrightarrow V_1 \oplus V_2$ and $V_2 \hookrightarrow V_1 \oplus V_2$, and let $\varphi: \bigwedge^*(V_1) \otimes \bigwedge^*(V_2) \to \bigwedge^*(V_1 \oplus V_2)$ be defined by $\varphi(x \otimes y) = i_1 x \cdot i_2 y$. Then φ is an isomorphism.

PROOF. There is a k-map $V_1 \oplus V_2 \to \bigwedge^*(V_1) \otimes \bigwedge^*(V_2)$ given by $v_1 \mapsto v_1 \otimes 1$, $v_2 \mapsto 1 \otimes v_2$ $(v_i \in V_i)$. By 6.1 this extends to an algebra map

$$\bigwedge^*(V_1 \oplus V_2) \to \bigwedge^*(V_1) \otimes \bigwedge^*(V_2),$$

which is the inverse of φ. □

(6.3) Let $(V_\alpha)_{\alpha \in D}$ be a direct system of k-modules, where D is a directed set. Then $\bigwedge^*(\varinjlim V_\alpha) \approx \varinjlim \bigwedge^*(V_\alpha)$. More precisely, the canonical maps $V_\alpha \to \varinjlim V_\alpha$ induce a compatible family of maps $\bigwedge^*(V_\alpha) \to \bigwedge^*(\varinjlim V_\alpha)$ and hence a map $\varphi : \varinjlim \bigwedge^*(V_\alpha) \to \bigwedge^*(\varinjlim V_\alpha)$; this map φ is an isomorphism.

PROOF. The inclusions $V_\alpha \hookrightarrow \bigwedge^*(V_\alpha)$ induce a k-map $\varinjlim V_\alpha \to \varinjlim \bigwedge^*(V_\alpha)$. This extends by 6.1 to an algebra map $\bigwedge^*(\varinjlim V_\alpha) \to \varinjlim \bigwedge^*(V_\alpha)$, which is the inverse of φ. □

Using 6.2 and 6.3, one can give the following concrete description of $\bigwedge^*(V)$ if V is a free k-module: Choose a basis $(x_i)_{i \in I}$ with I simply ordered; then $\bigwedge^p(V)$ has a k-basis consisting of the monomials $x_{i_1} \cdots x_{i_p}$ with $i_1 < \cdots < i_p$. One often writes $\bigwedge^*(V) = \bigwedge^*(x_i)_{i \in I}$.

Returning now to the study of $H_*(G, k)$ for G abelian, the isomorphism $G \otimes k \to H_1(G, k)$ extends by 6.1 to a k-algebra map $\psi: \bigwedge^*(G \otimes k) \to H_*(G, k)$, which is an isomorphism in dimensions 0 and 1. It is an obvious but important fact that ψ is a *natural* map of functors of G.

(6.4) Theorem. *Assume that k is a principal ideal domain.*

(i) *The map $\psi: \bigwedge^*(G \otimes k) \to H_*(G, k)$ is injective for every abelian group G and is a split injection if G is finitely generated.*
(ii) *Suppose that every prime p such that G has p-torsion is invertible in k; then ψ is an isomorphism.*
(iii) *If k has characteristic zero (e.g., if $k = \mathbb{Z}$) then ψ is an isomorphism in dimension 2.*

Note that the hypothesis of (ii) holds, in particular, if $k = \mathbb{Q}$, or if $k = \mathbb{Z}_p$ and G is p-torsion-free, or if $k = \mathbb{Z}$ and G is torsion-free.

Proof of 6.4. Suppose first that G is cyclic. Then $\bigwedge^p(G \otimes k) = 0$ for $p > 1$, so (i) holds trivially. Under the hypothesis of (ii) we also have $H_p(G, k) = 0$ for $p > 1$ by the computation in §III.1, so ψ is an isomorphism. The same computation shows that $H_2(G, k) = 0$ if char $k = 0$, so (iii) holds.

Now suppose G is finitely generated and hence a finite direct product of cyclic groups. Arguing by induction on the number of cyclic factors, we may assume that $G = G_1 \times G_2$ and that the theorem is known for G_1 and G_2. Consider the diagram

$$\begin{array}{ccc} \bigwedge^*(G_1 \otimes k) \otimes_k \bigwedge^*(G_2 \otimes k) & \xrightarrow[\approx]{\varphi} & \bigwedge^*(G \otimes k) \\ \Big\downarrow{\scriptstyle \psi(G_1) \otimes \psi(G_2)} & & \Big\downarrow{\scriptstyle \psi(G)} \\ H_*(G_1, k) \otimes_k H_*(G_2, k) & \xrightarrow{\mu} & H_*(G, k), \end{array}$$

where φ is the isomorphism of 6.2 and μ is the split injection of 5.8. Looking at the definitions of φ and μ and using the naturality of ψ, we see that this diagram commutes. Since $\psi(G_1)$ and $\psi(G_2)$ are split injections by hypothesis, it now follows that $\psi(G)$ is a split injection, whence (i). Now assume that the hypothesis of (ii) holds. Then it also holds for G_1 and G_2, so $\psi(G_1)$ and $\psi(G_2)$ are isomorphisms. Since $G_i \otimes k$ is a free k-module (with one basis element for every infinite cyclic factor of G_i), it follows that $H_*(G_i, k)$ is k-free. Therefore the Tor term in the Künneth formula vanishes, so μ is an isomorphism and (ii) follows at once. Similarly, μ is always an isomorphism in dimension 2 because the Tor term involves $H_0(-, k) = k$. Assuming now that char $k = 0$, we know that $\psi(G_1)$ and $\psi(G_2)$ are isomorphisms in dimensions ≤ 2, whence (iii).

Finally, any group G is the direct limit of its finitely generated subgroups G_α. We therefore have a diagram

$$\begin{array}{ccc} \varinjlim \bigwedge^*(G_\alpha \otimes k) & \xrightarrow[\approx]{\varphi} & \bigwedge^*(G \otimes k) \\ \Big\downarrow{\scriptstyle \varinjlim \psi(G_\alpha)} & & \Big\downarrow{\scriptstyle \psi(G)} \\ \varinjlim H_*(G_\alpha, k) & \xrightarrow[\approx]{} & H_*(G, k), \end{array}$$

where φ is the isomorphism of 6.3 and the unlabelled isomorphism is that of exercise 3a of §5. As above, this diagram commutes because ψ is natural, and the theorem follows at once. □

If G has p-torsion and p is not invertible in k, then the situation is more complicated. For simplicity we will confine ourselves to the case $k = \mathbb{Z}_p$, and even in this case we will be somewhat sketchy. See Cartan [1954/55] for more details and for the calculation of $H_*(G, \mathbb{Z})$.

Recall from 5.3 that if G is a finite cyclic group with generator t then there is a free resolution F of the form $\bigwedge(x) \otimes_{\mathbb{Z}G} \Gamma(y)$, where $\deg x = 1, \deg y = 2$, $\partial x = (t - 1) \cdot 1$, and $\partial y = Nx$. If p is a prime dividing $|G|$, then $F \otimes_G \mathbb{Z}_p = \bigwedge(x) \otimes \Gamma(y)$ with $\partial x = 0$ and $\partial y = 0$, where $\bigwedge$, $\otimes$, and Γ are now over $\mathbb{Z}_p$; hence $H_*(G, \mathbb{Z}_p) \approx \bigwedge(x) \otimes \Gamma(y)$. Thus we have, in addition to the exterior algebra part of $H_*(G, \mathbb{Z}_p)$ which we understand via 6.4, a 2-dimensional generator y and its divided powers $y^{(i)}$. This suggests that the homology of an abelian group should have, in addition to its ring structure, a "divided power" structure. We now make this precise.

Let $A = (A_n)_{n \geq 0}$ be a strictly anti-commutative graded ring. By a *system of divided powers* on A we mean a family of functions $A_{2n} \to A_{2ni}$ for $n > 0$ and $i \geq 0$, denoted $x \mapsto x^{(i)}$, with the following properties:

(a) $x^{(0)} = 1, x^{(1)} = x$.

(b) $x^{(i)}x^{(j)} = (i, j)x^{(i+j)}$, where $(i, j) = (i + j)!/i!j!$.

(c) $(x + y)^{(i)} = \sum_{j+k=i} x^{(j)}y^{(k)}$.

(d) $$(xy)^{(i)} = \begin{cases} 0 & \text{if } \deg x \text{ and } \deg y \text{ are odd and } i \geq 2 \\ x^i y^{(i)} & \text{if } \deg x \text{ and } \deg y \text{ are even and } \deg y > 0. \end{cases}$$

(e) $(x^{(i)})^{(j)} = \varepsilon_{ij} x^{(ij)}$ if $i, j > 0$, where

$$\varepsilon_{ij} = (i, i - 1)(2i, i - 1) \cdots ((j - 1)i, i - 1).$$

If A has a differential ∂ (which is always assumed to satisfy $\partial(xy) = \partial x \cdot y + (-1)^{\deg x} x \cdot \partial y$), then we require also:

(f) $\partial x^{(i)} = x^{(i-1)} \partial x$ for $i > 0$.

Note that this implies that $x^{(i)}$ is a cycle if x is a cycle. But there is not necessarily an induced system of divided powers on $H_* A$; see exercise 3 below.

In working with divided powers it is often convenient to introduce the formal power series $e^{tx} = \sum_{i \geq 0} x^{(i)} t^i$, where t is an indeterminate. Formula (c) then takes the form $e^{t(x+y)} = e^{tx} e^{ty}$, so $x \mapsto e^{tx}$ is a homomorphism from the additive group A_{2n} $(n > 0)$ to the multiplicative group of formal power series $1 + y_1 t + y_2 t^2 + \cdots$ with $y_i \in A_{2ni}$.

(6.5) EXAMPLES

1. If A is a strictly anti-commutative graded $\mathbb{Q}$-algebra, then A admits a system of divided powers with $x^{(i)} = x^i/i!$. Moreover, this is the unique system of divided powers on A; for (a) and (b) imply that, in any algebra with divided powers, $xx^{(j-1)} = jx^{(j)}$ and hence $x^i = i!x^{(i)}$.

2. Let $\alpha: R \to S$ be a homomorphism of commutative rings and let A be a (strictly anti-commutative) graded R-algebra with a system of divided powers. Then the S-algebra $S \otimes_R A$ obtained by extension of scalars admits a unique system of divided powers such that $(1 \otimes x)^{(i)} = 1 \otimes x^{(i)}$. Indeed, we must have $(s \otimes x)^{(i)} = s^i \otimes x^{(i)}$ by (d), so uniqueness follows from (c). To prove existence, one can use the universal mapping property of the tensor product to define e^{tz} for $z \in S \otimes_R A_{2n}$ by $e^{t(s \otimes x)} = \sum_i (s^i \otimes x^{(i)})t^i$; details are left to the reader.

3. The divided polynomial algebra $\Gamma_R(y)$ defined in §5 admits a system of divided powers. Indeed, in view of example 2, it suffices to check this if $R = \mathbb{Z}$, in which case the assertion follows from the fact that $\Gamma(y) \subset \mathbb{Q}[y]$ is closed under the divided power operations [cf. identity (e)]. In exactly the same way we can construct a divided polynomial algebra $\Gamma_R(y_i)_{i \in I}$ in several variables; if I is simply ordered, then this has an R-basis consisting of the "monomials" $y_{i_1}^{(n_1)} \cdots y_{i_k}^{(n_k)}$ with $i_1 < \cdots < i_k$ and $n_j \geq 0$. If V is a free R-module with basis $(y_i)_{i \in I}$, then we set $\Gamma(V) = \Gamma_R(y_i)$; this has V in dimension 2 and has a universal mapping property (in the category of R-algebras with divided powers) analogous to 6.1. It follows, in particular, that $\Gamma(V)$ is well-defined up to canonical isomorphism, independent of the choice of basis.

4. The Pontryagin ring $H_*(G, k)$ of an abelian group G admits a natural system of divided powers. To see this, consider the bar resolution F with the shuffle product. If $x = [g_1|\cdots|g_{2n}]$ $(n > 0)$, then we have (cf. §5, exercise 4)

$$x^i = \sum_{\sigma \in \mathscr{S}\!h} \sigma[g_1|\cdots|g_{2n}|\cdots|g_1|\cdots|g_{2n}],$$

where the "block" $g_1|\cdots|g_{2n}$ is repeated i times and $\mathscr{S}\!h \subset \mathscr{S}_{2ni}$ is the set of $(2n, \ldots, 2n)$-shuffles. Now there is an obvious embedding $\mathscr{S}_i \hookrightarrow \mathscr{S}_{2ni}$, obtained by letting $\mathscr{S}_i$ permute i sets of $2n$ elements blockwise. Since $\mathscr{S}_i \subset \mathscr{S}_{2ni}$ is a group of *even* permutations, it is clear that $\mathscr{S}_i$ fixes

$$[g_1|\cdots|g_{2n}|\cdots|g_1|\cdots|g_{2n}].$$

Moreover, $\mathscr{S}\!h$ is closed under right multiplication by elements of $\mathscr{S}_i$; the above formula for x^i can therefore be written

$$x^i = i! \sum_{\sigma \in \mathscr{S}\!h/\mathscr{S}_i} \sigma[g_1|\cdots|g_{2n}|\cdots|g_1|\cdots|g_{2n}].$$

It follows at once (via identity (c)) that $F \subset \mathbb{Q} \otimes F$ is closed under the divided power operations and hence admits a system of divided powers.

This yields (cf. example 2) a system of divided powers on $k \otimes_G F$, and I claim that there is an induced system of divided powers on $H_*(G, k)$. To prove this, it suffices (cf. exercise 3 below) to show that $z^{(i)}$ is a boundary if $z \in k \otimes_G F_{2n}$ is a boundary. For this purpose we view $k \otimes_G F$ as $(k \otimes F)_G$ and note that $k \otimes F$ is acyclic in positive dimensions (because $\varepsilon: F \to \mathbb{Z}$ is a homotopy equivalence if we forget the G-action). Now any boundary $z \in k \otimes_G F$ is the image of a boundary $w \in k \otimes F$, and $w^{(i)}$ is a cycle by identity (f). But then $w^{(i)}$ is a boundary by acyclicity, and hence its image $z^{(i)}$ is also a boundary.

Returning now to the study of $H_*(G, \mathbb{Z}_p)$, consider the split-exact universal coefficient sequence

$$0 \to H_2 G \otimes \mathbb{Z}_p \to H_2(G, \mathbb{Z}_p) \to \operatorname{Tor}(H_1 G, \mathbb{Z}_p) \to 0.$$

We have $H_2 G \approx \bigwedge^2(G)$ by 6.4(iii), and it is easily seen that $\bigwedge^2(G) \otimes \mathbb{Z}_p \approx \bigwedge^2_{\mathbb{Z}_p}(G_p)$, where $G_p = G \otimes \mathbb{Z}_p = G/pG$. Also, $\operatorname{Tor}(H_1 G, \mathbb{Z}_p) = \operatorname{Tor}(G, \mathbb{Z}_p) = {}_pG$, where the latter denotes $\{g \in G: pg = 0\}$. The sequence above therefore takes the form

$$0 \to \bigwedge^2(G_p) \to H_2(G, \mathbb{Z}_p) \to {}_pG \to 0.$$

Choose a splitting ${}_pG \to H_2(G, \mathbb{Z}_p)$ of this sequence. (If p is odd, we may use the *canonical splitting* given in exercise 4b below.) This extends to a $\mathbb{Z}_p$-algebra homomorphism $\varphi: \Gamma({}_pG) \to H_*(G, \mathbb{Z}_p)$ compatible with divided powers, cf. example 3 above. Combining φ with the map $\psi: \bigwedge(G_p) \to H_*(G, \mathbb{Z}_p)$ studied in 6.4, we obtain an algebra map $\rho: \bigwedge(G_p) \otimes \Gamma({}_pG) \to H_*(G, \mathbb{Z}_p)$ given by $\rho(x \otimes y) = \psi(x)\varphi(y)$. We can now state:

(6.6) Theorem. *The map $\rho: \bigwedge(G_p) \otimes \Gamma({}_pG) \to H_*(G, \mathbb{Z}_p)$ is an isomorphism. It is natural if $p \neq 2$.*

If G is cyclic, this follows easily from our earlier computations. The general case can now be deduced by using the Künneth formula and direct limits as in the proof of 6.4. See Cartan [1954/55] for more details.

Exercises

1. If G is an abelian group (written additively) and $n \in \mathbb{Z}$, compute the endomorphism of $H_*(G, \mathbb{Q})$ induced by the endomorphism $g \mapsto ng$ of G.

2. Let A and B be strictly anti-commutative graded k-algebras. Show that $A \otimes_k B$ is the *sum* of A and B in the category of strictly anti-commutative graded k-algebras, via the maps $a \mapsto a \otimes 1$, $b \mapsto 1 \otimes b$ ($a \in A$, $b \in B$). Thus 6.2 can be interpreted as saying that the exterior algebra functor preserves sums.

3. Let A be a strictly anti-commutative graded ring with a differential ∂ and a system of divided powers.

   (a) Suppose that $x^{(i)}$ is a boundary whenever x is a boundary. Show that there is an induced system of divided powers on H_*A.

   (b) Give an example to show that the hypothesis of (a) need not hold.

4. (a) Show that the injection $\psi: \bigwedge^2(G \otimes k) \to H_2(G, k)$ of 6.4 takes $(g \otimes 1) \wedge (h \otimes 1)$ to the class of $[g|h] - [h|g]$.

   (b) If 2 is invertible in k, show that the map $C_2(G, k) \to \bigwedge^2(G \otimes k)$ given by $[g|h] \mapsto (g \otimes 1) \wedge (h \otimes 1)/2$ induces a map $H_2(G, k) \to \bigwedge^2(G \otimes k)$ which is a left inverse of ψ.

5. Let G be abelian and let A be a G-module with trivial G-action. In view of the isomorphism $H_2 G \approx \bigwedge^2 G$, the universal coefficient theorem (cf. exercise 3 of §III.1) gives us a split exact sequence

$$0 \to \operatorname{Ext}(G, A) \to H^2(G, A) \xrightarrow{\theta} \operatorname{Hom}(\textstyle\bigwedge^2 G, A) \to 0.$$

Deduce from exercise 4a that θ coincides with the map called θ in exercise 8 of §IV.3, hence the latter is a (split) surjection. Thus every alternating map comes from a 2-cocycle. (A non-cohomological proof of this surprising fact can be found in Hughes [1951], proof of Theorem 2.) Incidentally, it also follows that $\operatorname{Ext}(G, A) \approx \mathscr{E}_{ab}(G, A)$. [More generally, it is known for any ring R that $\operatorname{Ext}_R^1(M, N)$ is isomorphic to the set of equivalence classes of R-module extensions of M by N, whence the name "Ext." A proof of this can be found in almost any book on homological algebra.]

CHAPTER VI

Cohomology Theory of Finite Groups

1 Introduction

Homology and cohomology are usually thought of as dual to one another. We have seen in Chapter III, for example, that homology has a number of formal properties and that cohomology has "dual" properties. If G is finite, however, then homology and cohomology seem to have *similar* properties rather than dual ones. For example, since every subgroup H of a finite group G has finite index, we have restriction and corestriction maps for *arbitrary* subgroups, in both homology and cohomology. For another example, the distinction between induced modules and co-induced modules disappears, so we have a single class $\mathscr{I}$ of G-modules (namely, the induced modules $\mathbb{Z}G \otimes A$) with the following properties: (a) Every $M \in \mathscr{I}$ is acyclic for both homology and cohomology. (b) For every G-module M there is a module $\overline{M} \in \mathscr{I}$ such that M is a quotient of $\overline{M}$ and M can be embedded in $\overline{M}$.

Tate discovered an ingenious way to exploit these similarities between H_* and H^* for G finite. Namely, he showed that there is a quotient $\tilde{H}^0$ of H^0 and a sub-functor $\tilde{H}_0 \subseteq H_0$ such that the functors $\ldots, H_2, H_1, \tilde{H}_0, \tilde{H}^0, H^1, H^2, \ldots$ fit together to form a "cohomology theory" $\hat{H}^*$, involving functors $\hat{H}^i$ for all $i \in \mathbb{Z}$:

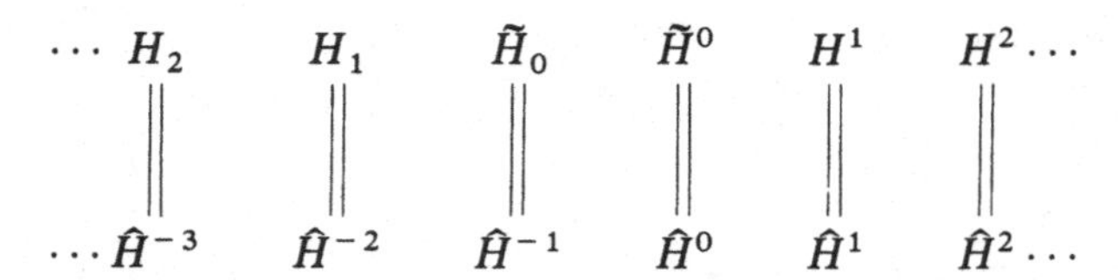

$$\begin{matrix} \cdots H_2 & H_1 & \tilde{H}_0 & \tilde{H}^0 & H^1 & H^2 \cdots \\ \| & \| & \| & \| & \| & \| \\ \cdots \hat{H}^{-3} & \hat{H}^{-2} & \hat{H}^{-1} & \hat{H}^0 & \hat{H}^1 & \hat{H}^2 \cdots \end{matrix}$$

The purpose of this chapter is to develop this *Tate cohomology theory* and to illustrate its usefulness by discussing (a) the Nakayama–Rim theory of cohomologically trivial modules (§8) and (b) the theory of groups with periodic cohomology (§9).

2 Relative Homological Algebra

We indicated in exercise 2 of §I.7 that the fundamental lemma I.7.4 and its corollaries could be generalized in various ways. In particular, we will need in this chapter a "relative-dual" version of those results, which we work out in the present section.

Throughout the section, G will be a fixed group (not necessarily finite) and H a fixed subgroup. For the purposes of the present chapter, the case of interest will be that where G is finite and $H = \{1\}$. But the general case will be needed later, in Chapter X.

An injection $i: M' \hookrightarrow M$ of G-modules will be called *admissible* if it is a split injection when regarded as an injection of H-modules, i.e., if there is an H-map $\pi: M \to M'$ such that $\pi i = \mathrm{id}_{M'}$. An exact sequence $M' \xrightarrow{i} M \xrightarrow{j} M''$ will be called *admissible* if the inclusion $\mathrm{im}\, j \hookrightarrow M''$ is admissible. An acyclic chain complex C of G-modules will be called *admissible* if each exact sequence $C_{i+1} \to C_i \to C_{i-1}$ is admissible; in view of I.0.3, this is equivalent to saying that C is contractible when regarded as a complex of H-modules. Finally, a G-module Q is *relatively injective* if it satisfies the following equivalent conditions:

(i) Every mapping problem

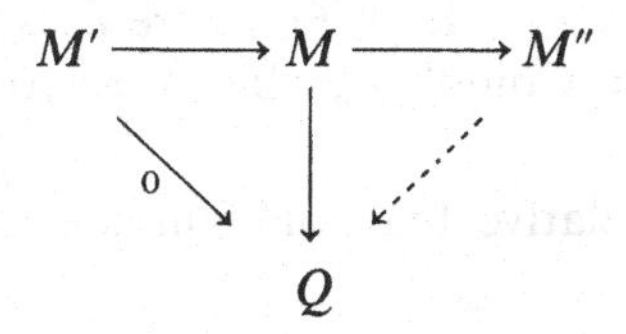

with admissible exact row can be solved.

(ii) Every mapping problem

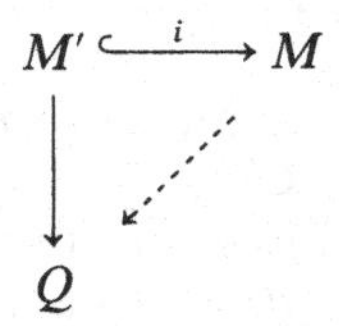

with i an admissible injection can be solved.

(iii) The contravariant functor $\mathrm{Hom}_G(-, Q)$ takes admissible injections of G-modules to surjections of abelian groups.

In particular, every injective module is certainly relatively injective. But there are many relative injectives which are not injective:

(2.1) Proposition. *For any H-module N, the G-module* $\operatorname{Coind}_H^G N$ *is relatively injective.*

Proof. By the universal property of co-induction (III.3.6), we have

$$\operatorname{Hom}_G(-, \operatorname{Coind}_H^G N) \approx \operatorname{Hom}_H(\operatorname{Res}_H^G(-), N).$$

The functor on the right clearly takes any admissible injection of G-modules to a surjection of abelian groups. □

(2.2) Corollary. *For any G-module M there is a canonical admissible injection $M \hookrightarrow \overline{M}$, where $\overline{M}$ is relatively injective. If $(G: H) < \infty$, then this construction has the following properties*: (a) *If M is free (resp. projective) as a $\mathbb{Z}H$-module, then $\overline{M}$ is free (resp. projective) over $\mathbb{Z}G$.* (b) *If M is finitely generated as a G-module, then so is $\overline{M}$.*

Proof. Take $\overline{M} = \operatorname{Coind}_H^G \operatorname{Res}_H^G M$ and use the canonical H-split injection $M \hookrightarrow \overline{M}$ of III.3.7. If $(G: H) < \infty$, then we have $\operatorname{Coind}_H^G(-) \approx \operatorname{Ind}_H^G(-)$, and (a) and (b) follow easily. □

(2.3) Corollary. *Suppose $(G: H) < \infty$. Then any $\mathbb{Z}G$-projective module is relatively injective.*

Proof. It is easy to see that a direct summand of a relative injective is relatively injective, so it suffices to consider free modules. Now if F is a free $\mathbb{Z}G$-module, then clearly $F \approx \mathbb{Z}G \otimes_{\mathbb{Z}H} F' = \operatorname{Ind}_H^G F'$, where F' is a free $\mathbb{Z}H$-module of the same rank. But $\operatorname{Ind}_H^G F' \approx \operatorname{Coind}_H^G F'$, which is relatively injective by 2.1. □

We now record the "relative dual" of Lemma I.7.4, in the form in which we will need it:

(2.4) Proposition. *Let C and C' be chain complexes of G-modules and let r be an integer. Suppose that C_i' is relatively injective for $i < r$ and that $C_{i+1} \to C_i \to C_{i-1}$ is exact and admissible for $i \le r$.*

(a) *Any family $(f_i: C_i \to C_i')_{i \ge r}$ of maps commuting with boundary operators extends to a chain map $C \to C'$.*
(b) *Let $f, g: C \to C'$ be chain maps and let $(h_i: C_i \to C_{i+1}')_{i \ge r-1}$ be a family of maps such that $\partial_{i+1}' h_i + h_{i-1}\partial_i = f_i - g_i$ for $i \ge r$. Then $(h_i)_{i \ge r-1}$ extends to a homotopy from f to g.*

The proof is virtually identical to that of I.7.4, except that all arrows are reversed. □

By a *relative injective resolution* of a G-module M we mean a non-negative cochain complex Q of relative injectives, together with a weak equivalence $\eta: M \to Q$ such that the augmented complex

$$0 \to M \to Q^0 \to Q^1 \to \cdots$$

is admissible. As an immediate corollary of 2.4, we have:

(2.5) Corollary. *Any two relative injective resolutions of M are canonically homotopy equivalent.*

It is clear that relative injective resolutions exist for any M; indeed, we can take $\eta: M \to Q^0$ to be the canonical admissible injection of 2.2, then apply 2.2 again to get an admissible injection $\operatorname{coker} \eta \hookrightarrow Q^1$, etc. Moreover, if $(G: H) < \infty$ and M is projective as a $\mathbb{Z}H$-module, then I claim that the modules Q^n which occur in this resolution will be projective over $\mathbb{Z}G$. This is clear for Q^0, by 2.2a. Since η is H-split, it follows that $\operatorname{coker} \eta$ is projective as a $\mathbb{Z}H$-module, hence 2.2 implies that Q^1 is projective over $\mathbb{Z}G$. Continuing in this way, one proves the claim. Similarly, if M is finitely generated, then so is each Q^n. Summarizing, we have shown:

(2.6) Proposition. *Suppose $(G: H) < \infty$. If M is a $\mathbb{Z}G$-module which is projective (resp. finitely generated and projective) as a $\mathbb{Z}H$-module, then M admits a relative injective resolution $\eta: M \to Q$ such that each Q^n is a projective (resp. finitely generated projective) $\mathbb{Z}G$-module.*

EXERCISE

Show that the resolution Q obtained in the proof of 2.6 is a complex of free $\mathbb{Z}G$-modules if M is free as a $\mathbb{Z}H$-module.

3 Complete Resolutions

We now specialize the relative homological algebra of §2 to the case where G is finite and $H = \{1\}$. As a consequence of 2.6 we obtain the perhaps surprising result that the G-module $\mathbb{Z}$ admits a "backwards" resolution $0 \to \mathbb{Z} \to Q^0 \to Q^1 \to \cdots$, where each Q^i is finitely generated and projective. [This becomes less surprising, however, when one realizes that such a backwards resolution can be obtained by taking the *dual* of an ordinary projective resolution of finite type; cf. 3.5 below.] Setting $F_i = Q^{-i-1}$, this takes the form

(3.1) $$0 \to \mathbb{Z} \xrightarrow{\eta} F_{-1} \to F_{-2} \to \cdots.$$

Now splice 3.1 onto an ordinary projective resolution $\varepsilon\colon (F_n)_{n\geq 0} \to \mathbb{Z}$, to obtain an acyclic complex

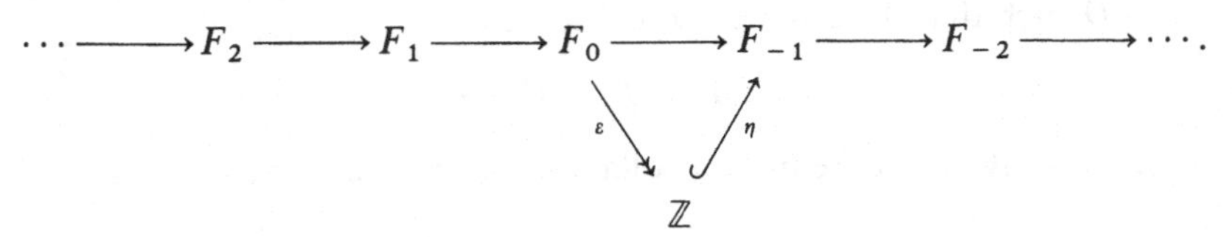

A complex of this type is called a complete resolution for G. More precisely, a *complete resolution* is an acyclic complex $F = (F_i)_{i\in\mathbb{Z}}$ of projective $\mathbb{Z}G$-modules, together with a map $\varepsilon\colon F_0 \to \mathbb{Z}$ such that $\varepsilon\colon F_+ \to \mathbb{Z}$ is a resolution in the usual sense, where $F_+ = (F_i)_{i\geq 0}$. It follows from the definition that $\partial\colon F_0 \to F_{-1}$ factors uniquely as $\partial = \eta\varepsilon$, where $\eta\colon \mathbb{Z} \to F_{-1}$ is a monomorphism. It also follows that the resulting complex 3.1 is a relative injective resolution of $\mathbb{Z}$. In fact, we know from 2.3 that each F_i is relatively injective, and admissibility follows from exercise 3b of §I.8, which we repeat here:

(3.2) Lemma. *Any acyclic chain complex C of free abelian groups is contractible.*

PROOF. The abelian group Z_n of n-cycles is free, being a subgroup of a free abelian group, so the sequence $0 \to Z_{n+1} \to C_{n+1} \to Z_n \to 0$ splits. □

We do not require in the definition of complete resolution that F be of finite type, i.e., that each F_i be finitely generated, but we saw at the beginning of this section that there do exist complete resolutions of finite type.

As with ordinary resolutions, we may view the map ε in a complete resolution as a chain map $F \to \mathbb{Z}$. (Note, however, that it is *not* a weak equivalence.) Given complete resolutions $\varepsilon\colon F \to \mathbb{Z}$ and $\varepsilon'\colon F' \to \mathbb{Z}$, a chain map $\tau\colon F \to F'$ is *augmentation-preserving* if $\varepsilon'\tau = \varepsilon$.

(3.3) Proposition. *If $\varepsilon\colon F \to \mathbb{Z}$ and $\varepsilon'\colon F' \to \mathbb{Z}$ are complete resolutions, then there exists a unique homotopy class of augmentation-preserving maps from F to F'. These maps are homotopy equivalences.*

PROOF. By I.7.4 we can find an augmentation-preserving chain map $\tau_+\colon F_+ \to F'_+$. Since F is acyclic and admissible (by 3.2) and F' is dimension-wise relatively injective, τ_+ can be extended to negative dimensions by 2.4. Similarly, given two augmentation-preserving maps $\tau, \tau'\colon F \to F'$, I.7.4 gives us a homotopy between τ_+ and τ'_+, which can then be extended to negative dimensions by 2.4. It is clear from the uniqueness that any map of complete resolutions is a homotopy equivalence. □

Finally, we want to show that the negative part of a complete resolution of finite type is the dual of an ordinary projective resolution of $\mathbb{Z}$ over $\mathbb{Z}G$. Recall that a G-module M has a dual $M^* = \operatorname{Hom}_G(M, \mathbb{Z}G)$, which we studied in I.8.3 for finitely generated projective modules. [Note that M^* is naturally a *right* G-module if M is a left G-module, but we can convert it to a left module as in §III.0. Thus $(gu)(m) = u(m)g^{-1}$ for $g \in G, u \in M^*, m \in M$.] On the other hand, one can also consider the dual $\text{Hom}(M, \mathbb{Z})$ of M as a $\mathbb{Z}$-module. This is a G-module via the diagonal action (cf. §III.0), where G acts trivially on $\mathbb{Z}$. Thus $(gu)(m) = u(g^{-1}m)$ for $g \in G$, $u \in \text{Hom}(M, \mathbb{Z})$, $m \in M$. It turns out that $\text{Hom}(M, \mathbb{Z})$ coincides with the dual M^* when G is finite:

(3.4) Proposition. *For any finite group G and any (left) G-module M, there is a G-module isomorphism*

$$\psi : \text{Hom}(M, \mathbb{Z}) \to M^* = \text{Hom}_G(M, \mathbb{Z}G),$$

given by $\psi(u)(m) = \sum_{g \in G} u(g^{-1}m)g$ *for* $u \in \text{Hom}(M, \mathbb{Z})$, $m \in M$.

PROOF. One can simply verify this by straightforward definition-checking, but in fact it follows from things we have already done. Namely, $\mathbb{Z}G$ is the induced module $\mathbb{Z}G \otimes \mathbb{Z}$, so it is isomorphic to the co-induced module $\text{Hom}(\mathbb{Z}G, \mathbb{Z})$ by III.5.9. The universal property III.3.6 of co-induction therefore implies that $\text{Hom}_G(M, \mathbb{Z}G) \approx \text{Hom}(M, \mathbb{Z})$ (as abelian groups), and an examination of the proofs of III.3.6 and III.5.9 yields the specific isomorphism ψ. It is easy to verify that ψ is compatible with the G-action. □

As an application of this, we will give a concrete interpretation of the "backwards projective resolution" $\mathbb{Z} \to Q$ discussed at the beginning of this section. By the *dual* of a chain complex F (over an arbitrary ring R) we mean the complex $\bar{F} = \mathcal{H}om_R(F, R)$. Thus $\bar{F}^n = \bar{F}_{-n} = \text{Hom}_R(F_n, R) = (F_n)^*$.

(3.5) Proposition. *If $\varepsilon: P \to \mathbb{Z}$ is a finite type projective resolution of $\mathbb{Z}$ over $\mathbb{Z}G$ (G finite), then $\varepsilon^*: \mathbb{Z}^* = \mathbb{Z} \to \bar{P}$ is a backwards projective resolution; moreover, up to isomorphism every finite type backwards projective resolution is obtained in this way. Consequently, any finite type complete resolution is obtained from two ordinary finite type projective resolutions $P' \to \mathbb{Z}$ and $P \to \mathbb{Z}$ by splicing together P' and $\overline{\Sigma P}$, where ΣP is the suspension of P (§I.0).*

PROOF. The augmented chain complex associated to $\varepsilon: P \to \mathbb{Z}$ is contractible as a complex of abelian groups (by 3.2, for instance). It therefore remains contractible when the duality functor $(\ \ )^* \approx \text{Hom}(-, \mathbb{Z})$ is applied, so $\varepsilon^*: \mathbb{Z}^* = \mathbb{Z} \to \bar{P}$ is a backwards projective resolution. This proves the first assertion. Similarly, given any finite type backwards projective resolution $\mathbb{Z} \to Q$, its dual $\bar{Q} \to \mathbb{Z}$ is an ordinary projective resolution, and Q can be identified with the dual of $\bar{Q}$ by I.8.3d. The remaining assertion is clear. □

EXERCISE

If G acts freely on S^{2k-1} as in §I.6, show that there is a complete resolution F for G which is periodic of period $2k$. Write this out explicitly if G is finite cyclic.

4 Definition of $\hat{H}^*$

The *Tate cohomology* of a finite group G with coefficients in a G-module M is defined by

$$\hat{H}^i(G, M) = H^i(\mathcal{H}om_G(F, M))$$

for all $i \in \mathbb{Z}$, where F is a complete resolution for G. By 3.3, $\hat{H}^i$ is well-defined up to canonical isomorphism.

Let $F_+ = (F_i)_{i \geq 0}$ and $F_- = (F_i)_{i < 0}$, and let C^+ (resp. C^-) be the complex $\mathcal{H}om_G(F_+, M)$ (resp. $\mathcal{H}om_G(F_-, M)$). Then we have (cf. proof of II.5.1)

$$\hat{H}^i(G, M) = \begin{cases} H^i(C^+) & i > 0 \\ H^i(C^-) & i < -1 \end{cases}$$

and there is an exact sequence

$$0 \to \hat{H}^{-1}(G, M) \to H^{-1}(C^-) \xrightarrow{\alpha} H^0(C^+) \to \hat{H}^0(G, M) \to 0,$$

where α is induced by the coboundary operator $\delta\colon C^{-1} \to C^0$. More precisely, α is determined by the diagram

$$(*) \qquad \begin{array}{ccc} C^{-1} & \xrightarrow{\delta} & C^0 \\ \downarrow & & \uparrow \\ H^{-1}(C^-) & \xrightarrow{\alpha} & H^0(C^+). \end{array}$$

Now $\varepsilon\colon F_+ \to \mathbb{Z}$ is a projective resolution of $\mathbb{Z}$, so $H^i(C^+) \approx H^i(G, M)$. And if we assume (as we may) that F is of finite type, then 3.5 implies that $F_- = \overline{\Sigma P}$ for some projective resolution $P \to \mathbb{Z}$ of finite type. The duality isomorphism I.8.3c now yields

$$C^- = \mathcal{H}om_G(F_-, M) = \mathcal{H}om_G(\overline{\Sigma P}, M) \approx \Sigma P \otimes_G M,$$

so $H^i(C^-) \approx H_{-i}(\Sigma P \otimes_G M) = H_{-i-1}(P \otimes_G M) = H_{-i-1}(G, M)$. We therefore have

$$\hat{H}^i(G, M) = \begin{cases} H^i(G, M) & i > 0 \\ H_{-i-1}(G, M) & i < -1 \end{cases}$$

and there is an exact sequence

$$0 \to \hat{H}^{-1}(G, M) \to H_0(G, M) \xrightarrow{\alpha} H^0(G, M) \to \hat{H}^0(G, M) \to 0.$$

I claim that α is the norm map $\overline{N}: M_G \to M^G$ defined in §III.1. To prove this, it is convenient to assume that the projective resolutions F_+ and P above both have $\mathbb{Z}G$ in dimension 0 and both start with the canonical augmentation $\mathbb{Z}G \to \mathbb{Z}$. [We can certainly assume this, since the resolutions can be taken to be arbitrary finite type projective resolutions.] Then $\eta = \varepsilon^*: \mathbb{Z} \to \mathbb{Z}G$ is given by $\eta(1) = \sum_{g \in G} g$. Since $\operatorname{Hom}_G(\mathbb{Z}G, M) = M$, the diagram $(*)$ becomes

$$\begin{array}{ccc} M & \xrightarrow{N} & M \\ \big\downarrow & & \big\uparrow \\ H_0(G, M) & \xrightarrow{\alpha} & H^0(G, M), \end{array}$$

and it is easy to check that the vertical maps are the canonical maps $M \twoheadrightarrow M_G$ and $M^G \hookrightarrow M$. This proves the claim.

We have now proved

$$\hat{H}^{-1}(G, M) = \ker \bar{N} \subseteq H_0(G, M)$$

and

$$\hat{H}^0(G, M) = \operatorname{coker} \bar{N} \twoheadleftarrow H^0(G, M).$$

For example, $\hat{H}^{-1}(G, \mathbb{Z}) = 0$ and $\hat{H}^0(G, \mathbb{Z}) = \mathbb{Z}/|G| \cdot \mathbb{Z}$.

We can summarize the results above by means of the following diagram:

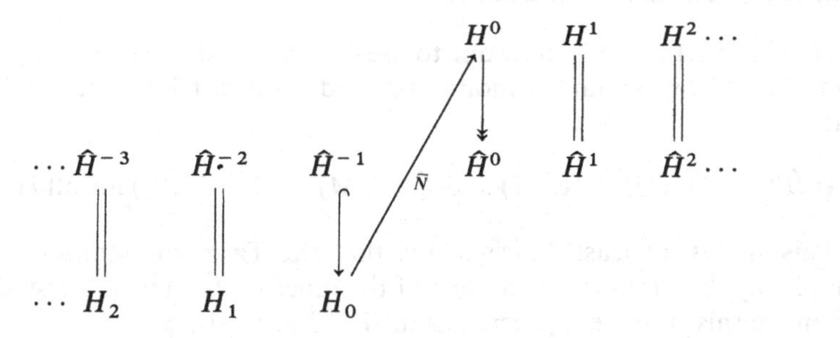

One can also define *Tate homology groups*, by $\hat{H}_*(G, M) = H_*(F \otimes_G M)$, where F is a complete resolution. Arguments analogous to those above show that

$$\hat{H}_i = \begin{cases} H_i & i > 0 \\ \ker \bar{N} & i = 0 \\ \operatorname{coker} \bar{N} & i = -1 \\ H^{-i-1} & i < -1. \end{cases}$$

In other words, $\hat{H}_i = \hat{H}^{-i-1}$. In view of this, it may seem pointless to introduce $\hat{H}_*$ since it, like $\hat{H}^*$, just consists of the functors H_i and H^i for $i > 0$ together with modifications of H^0 and H_0. As we will see in §7, however, the "equality" $\hat{H}_i = \hat{H}^{-i-1}$ should really be viewed as a duality theorem, and, as such, it has important consequences.

5 Properties of $\hat{H}^*$

Because Tate cohomology was defined in terms of resolutions, it is easy to prove that many of the formal properties of H^* hold also for $\hat{H}^*$. For example, one proves as in III.6.1(ii′):

(5.1) A short exact sequence $0 \to M' \to M \to M'' \to 0$ of G-modules gives rise to a long exact sequence

$$\cdots \xrightarrow{\delta} \hat{H}^i(G, M') \to \hat{H}^i(G, M) \to \hat{H}^i(G, M'') \xrightarrow{\delta} \hat{H}^{i+1}(G, M') \to \cdots.$$

Similarly, the proof of Shapiro's lemma (III.6.2) goes through in the present context, since a complete resolution for G can also be regarded as a complete resolution for any $H \subseteq G$. Since co-induction is the same as induction (because G is finite), Shapiro's lemma takes the form:

(5.2) If $H \subseteq G$ and M is an H-module, then $\hat{H}^*(H, M) \approx \hat{H}^*(G, \mathbb{Z}G \otimes_{\mathbb{Z}H} M)$.

Taking $H = \{1\}$ and noting that $\hat{H}^*(\{1\}, -) = 0$, we obtain:

(5.3) $\hat{H}^*(G, \mathbb{Z}G \otimes A) = 0$ for any abelian group A. Consequently, each $\hat{H}^i$ is both effaceable and co-effaceable.

Note that 5.1 and 5.3 allow one to use dimension-shifting. Namely, given a G-module M we can find G-modules K and C (as in III.7.1 and III.7.2) such that:

(5.4) $\hat{H}^i(G, M) \approx \hat{H}^{i+1}(G, K)$ and $\hat{H}^i(G, M) \approx \hat{H}^{i-1}(G, C)$ for all $i \in \mathbb{Z}$.

This shows, at least heuristically, that the Tate cohomology theory is completely determined by any one of the functors $\hat{H}^i$. The interested reader can make this statement precise, as in III.7.3 and III.7.5.

Given $H \subseteq G$, a complete resolution F for G, and a G-module M, we have cochain maps $\mathscr{H}om_G(F, M) \hookrightarrow \mathscr{H}om_H(F, M)$ and $\mathscr{H}om_H(F, M) \to \mathscr{H}om_G(F, M)$, where the second map is a transfer map, defined as in definition (C) of §III.9. Consequently:

(5.5) For any $H \subseteq G$ and any G-module M, one has a restriction map $\hat{H}^*(G, M) \to \hat{H}^*(H, M)$ and a corestriction map $\hat{H}^*(H, M) \to \hat{H}^*(G, M)$. These maps have formal properties analogous to those stated in III.9.5.

It follows, in particular, that the analogues of III.10.2 and III.10.3 hold for $\hat{H}^*$.

Finally, we will show that the theory of cup products extends to $\hat{H}^*$:

(5.6) There is a cup product $\hat{H}^p(G, M) \otimes \hat{H}^q(G, N) \to \hat{H}^{p+q}(G, M \otimes N)$, with formal properties analogous to 3.1–3.8 of Chapter V.

(Note: To state the analogue of V.3.7, one has to assume that $\alpha: H \to G$ is an inclusion, since that is the only case where we have a map $\alpha^*: \hat{H}^*(G) \to \hat{H}^*(H)$.)

In particular, $\hat{H}^*(G, \mathbb{Z})$ is an anti-commutative graded ring with identity element $1 \in \mathbb{Z}/|G| \cdot \mathbb{Z} = \hat{H}^0(G, \mathbb{Z})$, and $\hat{H}^*(G, M)$ is a module over $\hat{H}^*(G, \mathbb{Z})$ for any M.

The construction of the cup product requires some work. Recall that we based our treatment of cup products in ordinary cohomology on the fact that the tensor product of resolutions is a resolution. On the cochain level, then, the cup product was simply the map

$$(5.7) \qquad \mathcal{H}om_G(F, M) \otimes \mathcal{H}om_G(F, N) \to \mathcal{H}om_G(F \otimes F, M \otimes N)$$

given by tensor product of graded maps. We pointed out that one could also define the cup product in terms of a single resolution F by choosing a diagonal approximation $\Delta: F \to F \otimes F$ and composing the product above with

$$\mathcal{H}om_G(\Delta, M \otimes N): \mathcal{H}om_G(F \otimes F, M \otimes N) \to \mathcal{H}om_G(F, M \otimes N).$$

Note that, from this point of view, the cup product

$$\mathcal{H}om_G(F, M)^p \otimes \mathcal{H}om_G(F, N)^q \to \mathcal{H}om_G(F, M \otimes N)^{p+q}$$

depends only on the (p, q)-component $\Delta_{pq}: F_{p+q} \to F_p \otimes F_q$ of Δ.

Now suppose F is a complete resolution. The first difficulty in trying to imitate the procedure above is that $F \otimes F$ does not appear to be a complete resolution. In particular, $(F \otimes F)_+$ is *not* equal to the resolution $F_+ \otimes F_+$. Consequently, a definition of the form (5.7) would not in any obvious way induce a cohomology product $\hat{H}^p \otimes \hat{H}^q \to \hat{H}^{p+q}$, so a diagonal approximation now appears to be crucial, rather than a mere convenience. Secondly, a moment's thought shows that $F \otimes F$ is not really the appropriate target for a diagonal approximation. Indeed, for any $n \in \mathbb{Z}$ there are infinitely many (p, q) such that $p + q = n$, and dimension-shifting considerations suggest that the corresponding cup products should all be non-trivial. Thus Δ should have a non-trivial component Δ_{pq} for all (p, q), so the target of Δ should be the graded module which in dimension n is $\prod_{p+q=n} F_p \otimes F_q$ rather than $\bigoplus_{p+q=n} F_p \otimes F_q$. This discussion motivates the following definitions.

If C and C' are graded modules, their *completed tensor product* $C \hat{\otimes} C'$ is defined by

$$(C \hat{\otimes} C')_n = \prod_{p+q=n} C_p \otimes C'_q.$$

Given two other graded modules D, D' and maps $u: C \to D$ of degree r and $v: C' \to D'$ of degree s, there is a map $u \hat{\otimes} v: C \hat{\otimes} C' \to D \hat{\otimes} D'$ of degree $r + s$ defined by

$$(u \hat{\otimes} v)_n = \prod_{p+q=n} (-1)^{ps} u_p \otimes v_q: \prod_{p+q=n} C_p \otimes C'_q \to \prod_{p+q=n} D_{p+r} \otimes D'_{q+s}.$$

Now let $\varepsilon\colon F \to \mathbb{Z}$ be a complete resolution, and let d be the differential in F. Then $F \hat{\otimes} F$ has the "partial differentials" $d \hat{\otimes} \mathrm{id}_F$ and $\mathrm{id}_F \hat{\otimes} d$; these are of square zero and anti-commute, since $(d \hat{\otimes} \mathrm{id}_F)(\mathrm{id}_F \hat{\otimes} d) = d \hat{\otimes} d$ whereas $(\mathrm{id}_F \hat{\otimes} d)(d \hat{\otimes} \mathrm{id}_F) = -d \hat{\otimes} d$ (cf. exercise 8 of §I.0). So the "total differential" $\partial = d \hat{\otimes} \mathrm{id}_F + \mathrm{id}_F \hat{\otimes} d$ is of square zero and makes $F \hat{\otimes} F$ a chain complex. Moreover, we have the "augmentation" $\varepsilon \hat{\otimes} \varepsilon\colon F \hat{\otimes} F \to \mathbb{Z} \hat{\otimes} \mathbb{Z} = \mathbb{Z}$. By a *complete diagonal approximation* we mean an augmentation-preserving chain map $\Delta\colon F \to F \hat{\otimes} F$. It is by no means obvious that complete diagonal approximations exist, but we will prove below that they do. Accepting this for the moment, we can define a cochain cup product

$$\mathcal{H}om_G(F, M) \otimes \mathcal{H}om_G(F, N) \xrightarrow{\cup} \mathcal{H}om_G(F, M \otimes N)$$

by $u \cup v = (u \hat{\otimes} v) \circ \Delta$.

One verifies the usual coboundary formula

$$\delta(u \cup v) = \delta u \cup v + (-1)^p u \cup \delta v,$$

and it follows that there is an induced product $\hat{H}^p(G, M) \otimes \hat{H}^q(G, N) \to \hat{H}^{p+q}(G, M \otimes N)$. It is immediate that this product is natural with respect to coefficient homomorphisms (as in V.3.2), and one proves exactly as in V.3.3 and V.3.3′ that it is compatible with connecting homomorphisms δ in long exact cohomology sequences. Moreover, the fact that Δ is augmentation-preserving allows us to calculate the cup product $\hat{H}^0 \otimes \hat{H}^0 \to \hat{H}^0$, and we find as in V.3.1 that it is induced by the obvious map $M^G \otimes N^G \to (M \otimes N)^G$. [This makes sense because $\hat{H}^0(G, -)$ is a quotient of $(-)^G$.]

We can now use dimension-shifting to prove that the cup product is independent of the choice of F and Δ. More precisely:

(5.8) Lemma. *There is at most one cup product on* $\hat{H}^*(G, -)$ *satisfying the analogues of* V.3.1, V.3.3, and V.3.3′.

PROOF. Given a module M, let $\overline{M}$ be the induced module $\mathbb{Z}G \otimes M$ and let $0 \to K \to \overline{M} \to M \to 0$ be the canonical $\mathbb{Z}$-split exact sequence. For any G-module N the sequence $0 \to K \otimes N \to \overline{M} \otimes N \to M \otimes N \to 0$ is exact, and the module $\overline{M} \otimes N$ is induced (cf. §III.5, exercise 2a). We therefore have dimension-shifting isomorphisms

$$\delta\colon \hat{H}^i(G, M) \xrightarrow{\approx} \hat{H}^{i+1}(G, K) \quad \text{and} \quad \delta\colon \hat{H}^i(G, M \otimes N) \xrightarrow{\approx} \hat{H}^{i+1}(G, K \otimes N);$$

moreover, any cup product satisfying the analogue of V.3.3 is compatible with these isomorphisms, in the sense that

$$\begin{array}{ccc} \hat{H}^p(G, M) & \approx & \hat{H}^{p+1}(G, K) \\ \Big\downarrow{\scriptstyle - \cup v} & & \Big\downarrow{\scriptstyle - \cup v} \\ \hat{H}^{p+q}(G, M \otimes N) & \approx & \hat{H}^{p+q+1}(G, K \otimes N) \end{array}$$

commutes for any $v \in \hat{H}^q(G, N)$. Now suppose we have two cup products $\hat{H}^p \otimes \hat{H}^q \to \hat{H}^{p+q}$ as in the statement of the lemma. By hypothesis they agree when $p = q = 0$, so the square above allows us to prove by descending induction on p that they agree for $p \leq 0$ and $q = 0$. Similarly, writing N as a quotient of an induced module, we can extend this by descending induction on q to the case $p \leq 0$, $q \leq 0$. Next we embed M in an induced module and prove inductively that the cup products agree for p arbitrary and $q \leq 0$, and finally we embed N in an induced module to prove that they agree for all p and q. □

Still assuming the existence of Δ, we can now easily prove the remaining properties of the cup product: The fact that $1 \in \mathbb{Z}/|G| \cdot \mathbb{Z} = \hat{H}^0(G, \mathbb{Z})$ is an identity is proved as in V.3.4 or by dimension-shifting [it is obviously true in dimension 0]; associativity is proved by dimension-shifting [it is obviously true in dimension 0]; commutativity can be deduced from 5.8 [$u \cup v$ and $(-1)^{pq} t_*(v \cup u)$ are two cup products with the required properties] or, alternatively, there is a direct proof analogous to that of V.3.6; finally, the fact that the cup product behaves properly with respect to restriction and co-restriction follows directly from the definition as in V.3.7 and V.3.8, or by dimension-shifting.

To complete the construction of the cup product, we must prove the existence of a complete diagonal approximation Δ. We will use the following three lemmas:

(5.9) Lemma. *$F \hat{\otimes} F$ is acyclic and dimension-wise relatively injective.*

Proof. Relative injectivity follows from the fact that an arbitrary direct product of relative injectives is relatively injective. To prove acyclicity, let $h: F \to F$ be a contracting homotopy for F, regarded as a complex of abelian groups (cf. 3.2), and let $H = h \hat{\otimes} \mathrm{id}_F$. I claim that H is a contracting homotopy for $F \hat{\otimes} F$ (regarded as a complex of abelian groups). Indeed, recalling the definition of the differential ∂ in $F \hat{\otimes} F$, we have

$$\begin{aligned}\partial H + H\partial &= (d \hat{\otimes} \mathrm{id}_F + \mathrm{id}_F \hat{\otimes} d) \circ (h \hat{\otimes} \mathrm{id}_F) \\ &\quad + (h \hat{\otimes} \mathrm{id}_F) \circ (d \hat{\otimes} \mathrm{id}_F + \mathrm{id}_F \hat{\otimes} d) \\ &= dh \hat{\otimes} \mathrm{id}_F - h \hat{\otimes} d + hd \hat{\otimes} \mathrm{id}_F + h \hat{\otimes} d \qquad \text{(cf. exercise 8 of §I.0)} \\ &= (dh + hd) \hat{\otimes} \mathrm{id}_F \\ &= \mathrm{id}_F \hat{\otimes} \mathrm{id}_F \\ &= \mathrm{id}_{F \hat{\otimes} F},\end{aligned}$$

whence the claim. □

(5.10) Lemma. *Let (C, ∂) and (C', ∂') be two acyclic chain complexes of $\mathbb{Z}G$-modules. Assume that each C_i is projective and each C'_i is relatively injective. If*

$\tau_0: C_0 \to C'_0$ is a map such that $\partial'_0 \tau_0 \partial_1 = 0$, then τ_0 extends to a chain map $\tau: C \to C'$.

PROOF. Since C is projective and C' is acyclic, we can construct $(\tau_i)_{i \geq 0}$ as in the proof of the fundamental lemma I.7.4 so as to commute with boundary operators; indeed, the condition $\partial'_0 \tau_0 \partial_1 = 0$ is exactly what is needed to start the inductive construction, cf. I.7.3a. Similarly, since C is acyclic and admissible (by 3.2) and C' is relatively injective, $(\tau_i)_{i \geq 0}$ can be extended to negative dimensions by 2.4. □

(5.11) Lemma. *Let C be an admissible acyclic complex of $\mathbb{Z}G$-modules. For any projective $\mathbb{Z}G$-module P, the complex $P \otimes C$ (with diagonal G-action) is contractible as a complex of $\mathbb{Z}G$-modules.*

PROOF. It suffices to prove this for $P = \mathbb{Z}G$. By III.5.7, we have $\mathbb{Z}G \otimes C$ isomorphic to the induced complex $\mathbb{Z}G \otimes C'$, where C' is C regarded as a complex of abelian groups. Since C' is contractible by hypothesis, it follows that $\mathbb{Z}G \otimes C$ is contractible. □

In view of 5.9 and 5.10, the construction of $\Delta: F \to F \hat{\otimes} F$ is reduced to the construction of a map $\alpha = (\alpha_p): F_0 \to \prod_{p \in \mathbb{Z}} F_p \otimes F_{-p}$ such that (i) $\partial\alpha | B_0 = 0$ and (ii) $(\varepsilon \otimes \varepsilon)\alpha_0 = \varepsilon$, where $B_0 \subset F_0$ is the module of boundaries. Let $\partial'_{pq} = d_p \otimes F_q: F_p \otimes F_q \to F_{p-1} \otimes F_q$ and let $\partial''_{pq} = (-1)^p F_p \otimes d_q: F_p \otimes F_q \to F_p \otimes F_{q-1}$. Then $\partial\alpha: F_0 \to \prod_{p \in \mathbb{Z}} F_{p-1} \otimes F_{-p}$ has components $\partial'_{p,-p}\alpha_p + \partial''_{p-1,1-p}\alpha_{p-1}$. So (i) is equivalent to (i') $(\partial'\alpha_p + \partial''\alpha_{p-1}) | B_0 = 0$, where we have omitted the subscripts on ∂' and ∂'' to simplify the notation. We now start constructing α by taking $\alpha_0: F_0 \to F_0 \otimes F_0$ to be any map satisfying (ii); this is possible because F_0 is projective. Assuming that $p > 0$ and that α_{p-1} has been defined, we wish to define $\alpha_p: F_0 \to F_p \otimes F_{-p}$ so that the diagram

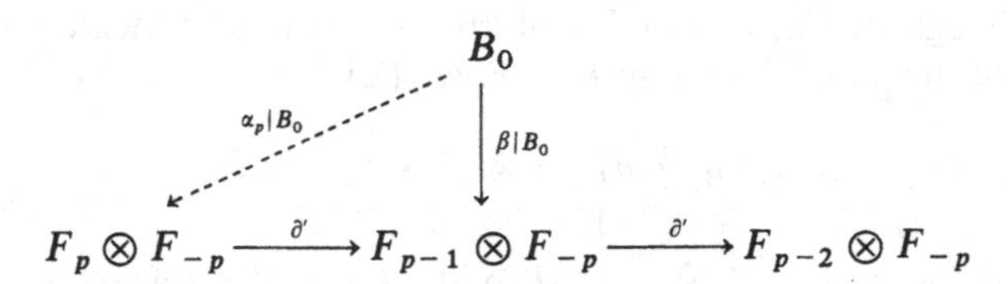

commutes, where $\beta = -\partial''\alpha_{p-1}$. I claim that $\partial'\beta | B_0 = 0$. In fact, if $p > 1$ then we can assume inductively that $(\partial'\alpha_{p-1} + \partial''\alpha_{p-2}) | B_0 = 0$, so that on B_0 we have

$$\begin{aligned} \partial'\beta &= -\partial'\partial''\alpha_{p-1} && \text{by definition of } \beta \\ &= \partial''\partial'\alpha_{p-1} && \text{because } \partial' \text{ and } \partial'' \text{ anti-commute} \\ &= -\partial''\partial''\alpha_{p-2} && \text{by the inductive hypothesis} \\ &= 0; \end{aligned}$$

and if $p = 1$ then on B_0 we have

$$\begin{aligned} \partial'\beta &= -\partial'\partial''\alpha_0 && \text{by definition of } \beta \\ &= -(d_0 \otimes d_0)\alpha_0 && \text{by definition of } \partial' \text{ and } \partial'' \\ &= -(\eta \otimes \eta)(\varepsilon \otimes \varepsilon)\alpha_0 && \text{because } d_0 = \eta\varepsilon \\ &= -(\eta \otimes \eta)\varepsilon && \text{by (ii)} \\ &= 0 && \text{because } \varepsilon | B_0 = 0. \end{aligned}$$

This proves the claim. By 5.11 the complex $(F_* \otimes F_{-p}, \partial')$ is contractible. We can therefore choose a contracting homotopy h and set $\alpha_p = h\beta$ to complete the inductive step. A similar argument constructs α_p for $p < 0$ by descending induction, and the proof of 5.6 is complete. □

Finally, we remark that there are also *cap products* in the Tate theory. No new difficulties arise here, so we confine ourselves to a brief discussion. There is a map

$$\mathscr{H}om_G(F, M) \to \mathscr{H}om((F \hat{\otimes} F) \otimes_G N, F \otimes_G (M \otimes N))$$

given by $u \mapsto \text{id} \hat{\otimes} u \otimes \text{id}_N$. This corresponds to a map

$$\gamma\colon \mathscr{H}om_G(F, M) \otimes ((F \hat{\otimes} F) \otimes_G N) \to F \otimes_G (M \otimes N).$$

We now define the cap product

$$\mathscr{H}om_G(F, M) \otimes (F \otimes_G N) \xrightarrow{\frown} F \otimes_G (M \otimes N)$$

to be γ composed with

$$\text{id} \otimes (\Delta \otimes \text{id})\colon \mathscr{H}om_G(F, M) \otimes (F \otimes_G N) \to \mathscr{H}om_G(F, M) \otimes ((F \hat{\otimes} F) \otimes_G N),$$

where $\Delta\colon F \to F \hat{\otimes} F$ is any complete diagonal approximation. This induces a well-defined cap product

$$\hat{H}^* \otimes \hat{H}_* \to \hat{H}_*$$

with the usual properties.

Exercises

1. Show that the cup product on $\hat{H}^*$ is compatible with that defined in Chapter V on H^*. More precisely, we have a natural map $H^* \to \hat{H}^*$ which is an isomorphism in positive dimensions and an epimorphism in dimension 0; show that this map preserves products.

2. Let $\hat{H}^*(G)_{(p)}$ be the p-primary component of $\hat{H}^*(G) = \hat{H}^*(G, \mathbb{Z})$, so that we have $\hat{H}^*(G) = \bigoplus_{p \| G|} \hat{H}^*(G)_{(p)}$ by the analogue of III.10.2.

   (a) Show that each $\hat{H}^*(G)_{(p)}$ is an ideal in $\hat{H}^*(G)$, hence so is $\bigoplus_{q \neq p} H^*(G)_{(q)}$. Consequently, $\hat{H}^*(G)_{(p)}$ is a quotient ring of $\hat{H}^*(G)$ via the projection $\hat{H}^*(G) \twoheadrightarrow \hat{H}^*(G)_{(p)}$. [Note: $\hat{H}^*(G)_{(p)}$ is *not* a subring of $\hat{H}^*(G)$; indeed, although the inclusion $\hat{H}^*(G)_{(p)} \hookrightarrow \hat{H}^*(G)$ does preserve products, it does not preserve identity elements.]

(b) Show that there is a *ring* isomorphism $\hat{H}^*(G) \xrightarrow{\approx} \prod_{p \mid |G|} \hat{H}^*(G)_{(p)}$, where each factor on the right is a ring via (a), and multiplication in the product is done componentwise.

6 Composition Products

As in ordinary cohomology (§V.4) there is a second method of defining products, based on composition of chain maps. We begin with the analogue of V.4.1:

(6.1) Proposition. *Let G be a finite group, let F be an acyclic chain complex of projective $\mathbb{Z}G$-modules, and let $\varepsilon': F' \to \mathbb{Z}$ be a complete resolution.*

(a) *For any G-module M, the map $\varepsilon' \otimes M: F' \otimes M \to M$ induces a weak equivalence $\mathcal{H}om_G(F, F' \otimes M) \to \mathcal{H}om_G(F, M)$. In particular, if F is a complete resolution then $\hat{H}^*(G, M) \approx H^*(\mathcal{H}om_G(F, F' \otimes M)) = [F, F' \otimes M]^*$.*

(b) *If F is of finite type then $\varepsilon' \otimes M$ induces a weak equivalence*

$$F \hat{\otimes}_G (F' \otimes M) \to F \hat{\otimes}_G M = F \otimes_G M.$$

In particular, if F is a complete resolution then

$$\hat{H}_*(G, M) \approx H_*(F \hat{\otimes}_G (F' \otimes M)).$$

Proof. For (a) we must prove $[F, F' \otimes M]_n \xrightarrow{\approx} [F, M]_n$ for all $n \in \mathbb{Z}$. Replacing F by its n-fold suspension $\Sigma^n F$, we reduce to the case $n = 0$; thus it suffices to show $[F, F' \otimes M] \xrightarrow{\approx} [F, M]$. Let $u: F \to M$ be a chain map. This means that u is a map $F_0 \to M$ such that $ud_1 = 0$, where d is the boundary operator in F. Since F_0 is projective, we can lift u to a map $\tau_0: F_0 \to F'_0 \otimes M$ such that $(\varepsilon' \otimes M)\tau_0 = u$. Recall that the boundary operator $d'_0: F'_0 \to F'_{-1}$ admits a factorization $d'_0 = \eta'\varepsilon'$, and consider $(d'_0 \otimes M)\tau_0 d_1$:

$$\begin{array}{ccccc} F_1 \xrightarrow{d_1} F_0 & & & & \\ \downarrow{\scriptstyle \tau_0} \quad \searrow{\scriptstyle u} & & & & \\ F'_0 \otimes M \xrightarrow{\varepsilon' \otimes M} M \xrightarrow{\eta' \otimes M} F'_{-1} \otimes M. & & & & \end{array}$$

(with the bottom arrow $d'_0 \otimes M: F'_0 \otimes M \to F'_{-1} \otimes M$)

We have

$$\begin{aligned} (d'_0 \otimes M)\tau_0 d_1 &= (\eta' \otimes M)(\varepsilon' \otimes M)\tau_0 d_1 \\ &= (\eta' \otimes M)ud_1 \\ &= 0; \end{aligned}$$

so we may apply lemma 5.10 to conclude that τ_0 extends to a chain map $\tau: F \to F' \otimes M$. [Note that $F'_i \otimes M$ is relatively injective; indeed, it suffices to observe that if L is a free $\mathbb{Z}G$-module then $L \otimes M$ is an induced module by

III.5.7.] This proves the surjectivity of $[F, F' \otimes M] \to [F, M]$. Now suppose $\tau: F \to F' \otimes M$ is a chain map such that $(\varepsilon' \otimes M)\tau: F \to M$ is null-homotopic, i.e., such that $(\varepsilon' \otimes M)\tau_0 = sd_0$ for some $s: F_{-1} \to M$. Lift s to a map h_{-1}: $F_{-1} \to F'_0 \otimes M$ satisfying $(\varepsilon' \otimes M)h_{-1} = s$, and consider $(d'_0 \otimes M)h_{-1}d_0$:

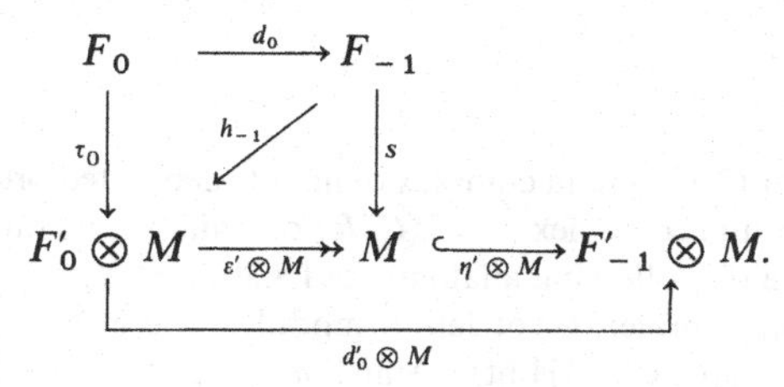

We have

$$\begin{aligned}(d'_0 \otimes M)h_{-1}d_0 &= (\eta' \otimes M)(\varepsilon' \otimes M)h_{-1}d_0 \\ &= (\eta' \otimes M)sd_0 \\ &= (\eta' \otimes M)(\varepsilon' \otimes M)\tau_0 \\ &= (d'_0 \otimes M)\tau_0.\end{aligned}$$

As in the proof of the fundamental lemma I.7.4, this is precisely what is needed to extend h_{-1} to $(h_i)_{i \geq -1}$ satisfying $hd + (d' \otimes M)h = \tau$, cf. I.7.3b. We can now use Proposition 2.4b to extend $(h_i)_{i \geq -1}$ to a null-homotopy h of τ. Thus $[F, F' \otimes M] \to [F, M]$ is injective, whence (a). To prove (b), consider the dual $\bar{F} = \mathcal{H}om_G(F, \mathbb{Z}G)$. This is again projective and acyclic (cf. proof of 3.5), and we have a duality isomorphism $F \hat{\otimes}_G - \approx \mathcal{H}om_G(\bar{F}, -)$ by exercise 1 below. So (b) follows from (a). □

Remark. It is perhaps surprising at first that we obtain the same conclusion here as in I.8.5 and I.8.6, since our map $\varepsilon' \otimes M$ is *not* a weak equivalence. On the other hand, we have a very strong hypothesis on F (acyclicity), and this compensates for the failure of $\varepsilon' \otimes M$ to be a weak equivalence.

It is now a routine matter to imitate the definitions in §V.4: Let $\varepsilon: F \to \mathbb{Z}$ and $\varepsilon': F' \to \mathbb{Z}$ be complete resolutions, and let $\varepsilon'': F'' \to \mathbb{Z}$ be either a complete resolution or the identity map $\mathbb{Z} \to \mathbb{Z}$. Then there are chain maps

$$\mathcal{H}om_G(F', F'' \otimes M) \otimes \mathcal{H}om_G(F, F' \otimes N) \to \mathcal{H}om_G(F, F'' \otimes M \otimes N)$$

and

$$\mathcal{H}om_G(F', F'' \otimes M) \otimes (F \hat{\otimes}_G (F' \otimes N)) \to F \hat{\otimes}_G (F'' \otimes M \otimes N),$$

defined as in V.4.2 and V.4.4. These induce *composition products*

$$\hat{H}^*(G, M) \otimes \hat{H}^*(G, N) \to \hat{H}^*(G, M \otimes N)$$

and

$$\hat{H}^*(G, M) \otimes \hat{H}_*(G, N) \to \hat{H}_*(G, M \otimes N).$$

The proof of V.4.6 now goes through, essentially verbatim, and yields:

(6.2) Theorem. *The composition products defined above coincide with the cup and cap products.*

EXERCISES

1. Let R be a ring, let C be a chain complex of finitely generated projective R-modules, and let $\bar{C}$ be the dual complex $\mathscr{H}om_R(C, R)$ of finitely generated projective right R-modules. Prove the following analogue of I.8.3b:

   For any chain complex C' of left R-modules, there is a chain isomorphism $\varphi: \bar{C} \hat{\otimes}_R C' \xrightarrow{\approx} \mathscr{H}om_R(C, C')$. [Hint: Define $\varphi_{pq}: \bar{C}_p \otimes_R C'_q \to \mathrm{Hom}_R(C_{-p}, C'_q)$ by $\langle \varphi_{pq}(u \otimes x'), x \rangle = (-1)^{pq} \langle u, x \rangle x'$ for $u \in \bar{C}_p = \mathrm{Hom}_R(C_{-p}, R)$, $x' \in C'_q$, $x \in C_{-p}$. It is an isomorphism by I.8.3b. Now set $\varphi_n = \prod_{p+q=n} \varphi_{pq}: \prod_{p+q=n} \bar{C}_p \otimes_R C'_q \to \prod_{p+q=n} \mathrm{Hom}_R(C_{-p}, C'_q) = \mathscr{H}om_R(C, C')_n$.] Similarly prove analogues of I.8.3c and I.8.3d.

2. Let C and $\bar{C}$ be as in exercise 1. For any $z \in (\bar{C} \hat{\otimes}_R C)_n$ and any chain complex C', there is a graded map $\psi_z: \mathscr{H}om_R(C, C') \to \bar{C} \hat{\otimes}_R C'$ of degree n, given by $\psi_z(u) = (\mathrm{id}_{\bar{C}} \hat{\otimes} u)(z)$. Show that there is a cycle $z \in (\bar{C} \hat{\otimes}_R C)_0$ such that for any C', ψ_z is the inverse of the isomorphism $\varphi: \bar{C} \hat{\otimes}_R C' \xrightarrow{\approx} \mathscr{H}om_R(C, C')$ of exercise 1. [Hint: Let $z \in (\bar{C} \hat{\otimes}_R C)_0$ correspond to id_C under $\varphi: \bar{C} \hat{\otimes}_R C \xrightarrow{\approx} \mathscr{H}om_R(C, C)$. Then $z = (z_p)_{p \in \mathbb{Z}}$, where $z_p \in \bar{C}_{-p} \otimes_R C_p = (C_p)^* \otimes_R C_p$ corresponds to $(-1)^p \mathrm{id}_{C_p}$ under the canonical isomorphism $(C_p)^* \otimes_R C_p \approx \mathrm{Hom}_R(C_p, C_p)$. Checking the definition of ψ_z, you should find that ψ_z is induced by maps $\psi_{pq}: \mathrm{Hom}_R(C_{-p}, C'_q) \to \bar{C}_p \otimes_R C'_q$ given by $\psi_{pq}(u) = (-1)^{(q+p)p}(\mathrm{id}_{\bar{C}_p} \otimes u)(z_{-p})$. Exercise 7 of §I.8 now shows that $\psi_{pq} = \varphi_{pq}^{-1}$.]

*3. Is the finiteness hypothesis in 6.1b necessary?

*4. Let F be a complete resolution, let n be an arbitrary integer, and let Z be the module $Z_n(F)$ of n-cycles. Show that there is a ring homomorphism $\mathrm{Hom}_G(Z, Z) \to [F, F]$ which is surjective and has as its kernel the group I of maps $Z \to Z$ which extend to maps $F_n \to Z$. Deduce that I is a 2-sided ideal in $\mathrm{Hom}_G(Z, Z)$ and that there is a ring isomorphism $\mathrm{Hom}_G(Z, Z)/I \approx \hat{H}^0(G, \mathbb{Z}) = \mathbb{Z}/|G| \cdot \mathbb{Z}$. [Hint: Use the techniques of the proof of 6.1.]

7 A Duality Theorem

It follows from the existence of finite type complete resolutions that the groups $\hat{H}^i(G, \mathbb{Z})$ are finitely generated. On the other hand, they are annihilated by $|G|$, so they are finite. The main purpose of this section is to show that $\hat{H}^i(G, \mathbb{Z})$ and $\hat{H}^{-i}(G, \mathbb{Z})$ are *dual* finite abelian groups; more precisely, we will show that the duality between them is given by the cup product

$$\hat{H}^i(G, \mathbb{Z}) \otimes \hat{H}^{-i}(G, \mathbb{Z}) \to \hat{H}^0(G, \mathbb{Z}) = \mathbb{Z}/|G| \cdot \mathbb{Z}.$$

We will give a proof which uses Tate homology theory and the cap product; a different proof, which uses only cup products and dimension-shifting, can be found in Cartan–Eilenberg [1956], XII.6. We begin by reviewing the relevant duality theory for abelian groups.

For any abelian group A we set $A' = \mathrm{Hom}(A, \mathbb{Q}/\mathbb{Z})$. Since $\mathbb{Q}/\mathbb{Z}$ is injective, the (contravariant) functor $(\quad)'$ is *exact*. In case $nA = 0$ for some $n > 0$, we have $\mathrm{Hom}(A, \mathbb{Q}/\mathbb{Z}) = \mathrm{Hom}(A, n^{-1}\mathbb{Z}/\mathbb{Z}) \approx \mathrm{Hom}(A, \mathbb{Z}/n\mathbb{Z})$; hence we can identify A' with $\mathrm{Hom}(A, \mathbb{Z}_n)$. In particular, if A is cyclic of order n then so is A'. Consequently, $A' \approx A$ (non-canonically) for any finite abelian group A. As usual, we must pass to the double dual to get a canonical isomorphism: There is a natural map $A \to A''$ given by $a \mapsto (f \mapsto f(a))$, which is an isomorphism if A is finite.

A map $\rho: A \otimes B \to \mathbb{Q}/\mathbb{Z}$ gives rise to a map $\bar{\rho}: A \to B'$; we will say that ρ is a *duality pairing* if $\bar{\rho}$ is an isomorphism. Of course ρ also gives rise to a map $\bar{\bar{\rho}}: B \to A'$, which is equal to the composite $B \to B'' \xrightarrow{\bar{\rho}'} A'$. It follows easily that, if A and B are finite, ρ is a duality pairing if and only if $\bar{\bar{\rho}}$ is an isomorphism. Thus a duality pairing between *finite* abelian groups provides an isomorphism of each one with the dual of the other. The canonical example of a duality pairing is, of course, the evaluation map $A' \otimes A \to \mathbb{Q}/\mathbb{Z}$, where A is arbitrary. Finally, if $nA = 0$ and $nB = 0$, then we can similarly speak of duality pairings $A \otimes B \to \mathbb{Z}_n$.

If G is a group and M is a G-module, then $M' = \mathrm{Hom}(M, \mathbb{Q}/\mathbb{Z})$ inherits a G-action in the usual way: $(gu)(m) = u(g^{-1}m)$ for $g \in G, u \in M', m \in M$. There is an *evaluation pairing* $\rho: H^i(G, M') \otimes H_i(G, M) \to \mathbb{Q}/\mathbb{Z}$, obtained by composing the pairing $\langle \cdot, \cdot \rangle$ of §V.3 with the evaluation map $M' \otimes_G M \to \mathbb{Q}/\mathbb{Z}$. Similarly, if G is finite, there is a pairing $\hat{\rho}: \hat{H}^i(G, M') \otimes \hat{H}_i(G, M) \to \mathbb{Q}/\mathbb{Z}$.

(7.1) Proposition. *The pairings ρ and $\hat{\rho}$ are duality pairings. Thus $H^i(G, M') \approx H_i(G, M)'$ for any G and M, and $\hat{H}^i(G, M') \approx \hat{H}_i(G, M)'$ if G is finite.*

PROOF. Let F be a projective resolution of $\mathbb{Z}$ over $\mathbb{Z}G$. Then $\mathcal{H}om_G(F, M') = \mathcal{H}om_G(F, \mathrm{Hom}(M, \mathbb{Q}/\mathbb{Z})) \approx \mathcal{H}om_G(F \otimes M, \mathbb{Q}/\mathbb{Z}) = \mathcal{H}om(F \otimes_G M, \mathbb{Q}/\mathbb{Z}) = (F \otimes_G M)'$. Since $(\quad)'$ is exact, we can pass to homology to obtain $H^*(G, M') \approx H_*(G, M)'$. It is easy to check that this isomorphism corresponds to the pairing ρ. The argument for $\hat{\rho}$ is the same. □

Assume now that G is finite. As we remarked briefly at the end of §4, there is an isomorphism $\hat{H}^i(G, M) \approx \hat{H}_{-1-i}(G, M)$. We wish to make this more precise by showing that there is an isomorphism of this type given by cap product with a "fundamental class":

(7.2) Proposition. *There is an element $z \in \hat{H}_{-1}(G, \mathbb{Z})$ such that the cap product map $\frown z: \hat{H}^i(G, M) \to \hat{H}_{-1-i}(G, M)$ is an isomorphism for any G-module M.*

PROOF. Let (F, d) be a complete resolution of finite type, with $d_0 = \eta\varepsilon$ as in §3, and let $\bar{F}$ be the dual complex $\mathcal{H}om_G(F, \mathbb{Z}G)$. Thus $\bar{F}_i = (F_{-i})^*$. Then $\bar{F}$ is still projective and acyclic (cf. proof of 3.5), and the boundary operator $\bar{F}_1 \to \bar{F}_0$ is the composite

$$(F_{-1})^* \xrightarrow{\eta^*} \mathbb{Z} \xrightarrow{\varepsilon^*} (F_0)^*.$$

Except for indexing, then, $\bar{F}$ is a complete resolution; more precisely, $\bar{F}$ is the suspension ΣE of a complete resolution E. We now apply the duality isomorphism $\mathcal{H}om_G(F, M) \approx \bar{F} \otimes_G M$ of I.8.3b. This yields

$$\hat{H}^i(G, M) \approx H_{-i}(\bar{F} \otimes_G M) = H_{-i-1}(E \otimes_G M) = \hat{H}_{-i-1}(G, M).$$

I claim that this isomorphism is given, up to sign, by cap product with a universal element $z \in \hat{H}_{-1}(G, \mathbb{Z})$. There is an easy abstract proof of this, outlined in exercise 1 below. We will give here a direct but somewhat tedious proof, based on the composition product and the exercises of §6.

Let w_0 be the 0-cycle in $\bar{F} \hat{\otimes}_G F$ which corresponds to id_F under the canonical isomorphism $\bar{F} \hat{\otimes}_G F \approx \mathcal{H}om_G(F, F)$ of exercise 1 of §6. Using the notation and result of exercise 2 of §6, we find that the isomorphism

$$\mathcal{H}om_G(F, M) \xrightarrow{\approx} \bar{F} \otimes_G M$$

used above is the map ψ_{w_0}. Now $\bar{F} \hat{\otimes}_G F = \Sigma E \hat{\otimes}_G F$, so w_0 can also be regarded as a (-1)-cycle w_{-1} in $E \hat{\otimes}_G F$. Moreover, one discovers by checking definitions that $\psi_{w_{-1}}: \mathcal{H}om_G(F, M)_i \to (E \otimes_G M)_{i-1}$ is equal to $(-1)^i \psi_{w_0}: \mathcal{H}om_G(F, M)_i \to (\bar{F} \otimes_G M)_i$. Thus the isomorphism $\hat{H}^i(G, M) \to \hat{H}_{-i-1}(G, M)$ of the previous paragraph is induced, up to sign, by $\psi_{w_{-1}}$. On the other hand, $\psi_{w_{-1}}$ is precisely the composition product with w_{-1} as defined in §6; it therefore induces the cap product with the class $z \in \hat{H}_{-1}(G, \mathbb{Z})$ represented by w_{-1}. This proves the claim and the proposition. ☐

Remark. The element z above is necessarily a generator of the cyclic group $\hat{H}_{-1}(G, \mathbb{Z}) = \mathbb{Z}/|G|\mathbb{Z}$; for the cap product isomorphism $\frown z: \hat{H}^0(G, \mathbb{Z}) \to \hat{H}_{-1}(G, \mathbb{Z})$ takes $1 \in \hat{H}^0(G, \mathbb{Z}) = \mathbb{Z}/|G|\mathbb{Z}$ to $1 \frown z = z$.

(7.3) Corollary. *For any G-module M, the composite*

$$\hat{H}^i(G, M') \otimes \hat{H}^{-1-i}(G, M) \xrightarrow{\smile} H^{-1}(G, M' \otimes M) \xrightarrow{\alpha_*} \hat{H}^{-1}(G, \mathbb{Q}/\mathbb{Z})$$

is a duality pairing, where α_* is induced by the canonical pairing α: $M' \otimes M \to \mathbb{Q}/\mathbb{Z}$.

(Note that the statement makes sense because

$$\hat{H}^{-1}(G, \mathbb{Q}/\mathbb{Z}) = \ker\{\bar{N}: (\mathbb{Q}/\mathbb{Z})_G \to (\mathbb{Q}/\mathbb{Z})^G\} = |G|^{-1}\mathbb{Z}/\mathbb{Z}$$

and $\hat{H}^*(G, -)$ is annihilated by $|G|$.)

Proof. By 7.1 and 7.2 we have $\hat{H}^i(G, M') \xrightarrow{\approx} \hat{H}_i(G, M)' \xrightarrow{\approx} \hat{H}^{-1-i}(G, M)'$. The composite isomorphism is given explicitly by $u \mapsto (v \mapsto \bar{\alpha}\langle u, v \frown z\rangle)$, where

$\bar{\alpha}: M' \otimes_G M \to \mathbb{Q}/\mathbb{Z}$ is induced by α. Since $\bar{\alpha}\langle u, v \cap z\rangle = \bar{\alpha}\langle u \cup v, z\rangle = \langle \alpha_*(u \cup v), z\rangle$, it follows that the composite

$$\hat{H}^i(G, M') \otimes \hat{H}^{-1-i}(G, M) \xrightarrow{\cup} \hat{H}^{-1}(G, M' \otimes M) \xrightarrow{\alpha_*} \hat{H}^{-1}(G, \mathbb{Q}/\mathbb{Z}) \xrightarrow{\langle -, z\rangle} \mathbb{Q}/\mathbb{Z}$$

is a duality pairing. To complete the proof, we need only note that $\langle -, z\rangle$ maps $\hat{H}^{-1}(G, \mathbb{Q}/\mathbb{Z})$ isomorphically onto $|G|^{-1}\mathbb{Z}/\mathbb{Z}$. Indeed, $\langle -, z\rangle$ is the composite

$$\hat{H}^{-1}(G, \mathbb{Q}/\mathbb{Z}) \xrightarrow{\approx} \hat{H}_{-1}(G, \mathbb{Z})' \to \mathbb{Q}/\mathbb{Z},$$

where the first map is given by 7.1 and the second is evaluation at the generator $z \in \hat{H}_{-1}(G, \mathbb{Z})$. And it is obvious that if A is a finite cyclic group, then evaluation at a generator of A gives an isomorphism $A' \xrightarrow{\approx} |A|^{-1}\mathbb{Z}/\mathbb{Z}$. □

We can now prove our main result:

(7.4) Theorem. *The cup product* $\hat{H}^i(G, \mathbb{Z}) \otimes \hat{H}^{-i}(G, \mathbb{Z}) \to \hat{H}^0(G, \mathbb{Z}) = \mathbb{Z}/|G| \cdot \mathbb{Z}$ *is a duality pairing.*

Proof. Since $\hat{H}^*(G, \mathbb{Q}) = 0$, the coefficient sequence $0 \to \mathbb{Z} \to \mathbb{Q} \to \mathbb{Q}/\mathbb{Z} \to 0$ yields an isomorphism $\delta: \hat{H}^j(G, \mathbb{Q}/\mathbb{Z}) \xrightarrow{\approx} \hat{H}^{j+1}(G, \mathbb{Z})$ for all j. Moreover, we have (cf. V.3.3) a commutative diagram

$$\begin{array}{ccc}
\hat{H}^{i-1}(G, \mathbb{Q}/\mathbb{Z}) \otimes \hat{H}^{-i}(G, \mathbb{Z}) & \xrightarrow{\cup} & \hat{H}^{-1}(G, \mathbb{Q}/\mathbb{Z}) \\
\downarrow{\scriptstyle \delta \otimes \mathrm{id}\,\approx} & & \downarrow{\scriptstyle \delta\,\approx} \\
\hat{H}^i(G, \mathbb{Z}) \otimes \hat{H}^{-i}(G, \mathbb{Z}) & \xrightarrow{\cup} & \hat{H}^0(G, \mathbb{Z}).
\end{array}$$

Since the top row is a duality pairing by 7.3 (with $M = \mathbb{Z}$), so is the bottom row. □

Remark. The fact that $\hat{H}^i(G, \mathbb{Z})$ and $\hat{H}^{-i}(G, \mathbb{Z})$ are dual to one another can be proved quite easily by means of the universal coefficient theorem and the interpretation of $\hat{H}^*$ in terms of H^* and H_*. [Recall that $\mathrm{Ext}(A, \mathbb{Z}) \approx A'$ if A is finite, cf. exercise 2 below.] What is not obvious, however, is that the *cup product* provides a duality pairing. This is the essential content of 7.4.

Exercises

1. The purpose of this exercise is to give an alternate proof that the isomorphism $\hat{H}^i(G, M) \xrightarrow{\approx} \hat{H}_{-1-i}(G, M)$ established in the first part of the proof of 7.2 is given, up to sign, by cap product with a fixed element $z \in H_{-1}(G, \mathbb{Z})$. Call that isomorphism φ and let $z = \varphi(1)$.

   (a) Show that φ is natural and that it is compatible with connecting homomorphisms in long exact sequences.

(b) Show that $\varphi: \hat{H}^0(G, M) \to \hat{H}_{-1}(G, M)$ is given by cap product with z. [By definition of z, φ and $\cap z$ agree on $1 \in \hat{H}^0(G, \mathbb{Z})$. If now M and $u \in \hat{H}^0(G, M)$ are arbitrary, there is a coefficient homomorphism $\mathbb{Z} \to M$ such that $1 \mapsto u$ under the induced map $\hat{H}^0(G, \mathbb{Z}) \to \hat{H}^0(G, M)$, so $\varphi(u) = u \cap z$ by naturality.]

(c) Now use dimension-shifting to show that φ and $\cap z$ agree in all dimensions, up to sign.

2. For any abelian torsion group A, prove that $\operatorname{Ext}(A, \mathbb{Z}) \approx A'$. [Hint: Look at the long exact $\operatorname{Ext}^*(A, -)$-sequence associated to the short exact sequence $0 \to \mathbb{Z} \to \mathbb{Q} \to \mathbb{Q}/\mathbb{Z} \to 0$. Equivalently, compute $\operatorname{Ext}(A, \mathbb{Z})$ by means of the injective resolution $0 \to \mathbb{Z} \to \mathbb{Q} \to \mathbb{Q}/\mathbb{Z} \to 0$ of $\mathbb{Z}$.]

3. If M is a G-module which is free as an abelian group, show that

$$\hat{H}^i(G, M^*) \otimes \hat{H}^{-i}(G, M) \xrightarrow{\smile} \hat{H}^0(G, M^* \otimes M) \to \hat{H}^0(G, \mathbb{Z})$$

is a duality pairing.

4. Let k be an arbitrary commutative ring and Q an injective k-module. Let $A' = \operatorname{Hom}_k(A, Q)$ for any k-module A. If M is a kG-module, show that the pairing

$$\hat{H}^i(G, M') \otimes \hat{H}^{-1-i}(G, M) \xrightarrow{\smile} \hat{H}^{-1}(G, M' \otimes M) \to \hat{H}^{-1}(G, Q) \hookrightarrow Q$$

induces an isomorphism $\hat{H}^i(G, M') \xrightarrow{\approx} \hat{H}^{-1-i}(G, M)'$.

8 Cohomologically Trivial Modules

The theory to be presented in this section is due to Nakayama and Rim. Our treatment is based on Serre [1968].

A G-module M (where G is finite) is said to be *cohomologically trivial* if $\hat{H}^i(H, M) = 0$ for all $i \in \mathbb{Z}$ and all $H \subseteq G$. For example, any induced module $\mathbb{Z}G \otimes A$ is cohomologically trivial by 5.3, since $\mathbb{Z}G \otimes A$ is still induced when regarded as an H-module for any $H \subseteq G$. [This is an easy special case of the double-coset formula III.5.6b.] In particular, if k is an arbitrary commutative ring, it follows that any free kG-module is cohomologically trivial and hence that any projective kG-module is cohomologically trivial. Our goals in this section are (a) to prove that, conversely, under suitable hypotheses cohomologically trivial kG-modules are projective; and (b) to find simple criteria for checking that a module is cohomologically trivial. These results have applications to class field theory (cf. Serre [1968]) as well as to algebraic K-theory and homotopy theory.

We begin with the case where G is a p-group for some prime p. In this case everything will be deduced from the following simple result:

(8.1) Proposition. *If G is a p-group and M is a G-module in which every element has order a power of p, then $M^G \neq 0$.*

PROOF. We may assume that M is finitely generated as a G-module, hence also as an abelian group. But then M is finite, of order a power of p. Since every G-orbit in M has order p^a for some $a \geq 0$, it follows that $|M^G| \equiv |M| \equiv 0 \pmod{p}$, so that $M^G \neq 0$. □

(8.2) Corollary. *If k is a field of characteristic p, G is a p-group, and M is a simple kG-module, then $M \approx k$, with trivial G-action.*

(Recall that a *simple* module is a non-zero module which contains no proper non-zero submodule.)

PROOF. Since $M^G \neq 0$ by 8.1, we must have $M^G = M$, so G acts trivially on M. Simplicity now implies that $\dim_k M = 1$. □

Let I be the *augmentation ideal* of kG, i.e., the kernel of the k-algebra homomorphism $\varepsilon: kG \to k$ such that $\varepsilon(g) = 1$ for all $g \in G$. We can also describe I as the annihilator of the kG-module k.

(8.3) Corollary. *If G is a p-group and k is a field of characteristic p, then the augmentation ideal I of kG is nilpotent.*

PROOF. Choose a composition series $0 = J_0 \subset J_1 \subset \cdots \subset J_n = kG$ for kG, regarded as a left kG-module. Thus each J_i is a left ideal and J_i/J_{i-1} is simple for $1 \leq i \leq n$. [The existence of a composition series follows from the finite-dimensionality of kG over k.] By 8.2, I annihilates J_i/J_{i-1}, hence $IJ_i \subseteq J_{i-1}$. Consequently, $I^n = I^n J_n \subseteq J_0 = 0$. □

If G is cyclic of order $q = p^a$, for example, we can easily see directly that I is nilpotent; for I is generated by $t - 1$ where t is a generator of G (cf. §I.2, exercise 1b), and $(t-1)^q = t^q - 1 = 0$ since char $k = p$.

(8.4) Corollary. *Let G be a p-group and k a field of characteristic p. If M is a kG-module such that $H_0(G, M) = 0$, then $M = 0$.*

PROOF. It is easy to see that $H_0(G, M) = M/IM$ (cf. first paragraph of §II.2). Thus $H_0(G, M) = 0 \Rightarrow M = IM \Rightarrow M = I^n M$ for all n. In view of 8.3, this implies that $M = 0$. □

We can now achieve, for kG-modules as above, the two goals (a) and (b) stated at the beginning of this section:

(8.5) Theorem. *Let G be a p-group and k a field of characteristic p. The following conditions on a kG-module M are equivalent:*

(i) *M is free.*
(ii) *M is projective.*
(iii) *M is cohomologically trivial.*
(iv) *$H_1(G, M) = 0$.*
(v) *$\hat{H}^i(G, M) = 0$ for some $i \in \mathbb{Z}$.*

PROOF. Clearly (i) $\Rightarrow$ (ii) $\Rightarrow$ (iii) $\Rightarrow$ (iv) $\Rightarrow$ (v). To complete the proof we will show (iv) $\Rightarrow$ (i) and (v) $\Rightarrow$ (iv). Choose a k-basis for the vector space M_G, and lift the basis elements to elements m_j $(j \in J)$ in M. Let F be a free kG-module with basis $(e_j)_{j \in J}$ and define $f: F \to M$ by $f(e_j) = m_j$. By construction, then, $H_0(G, f): H_0(G, F) \to H_0(G, M)$ is an isomorphism. By right exactness of $H_0(G, -)$, it follows that $H_0(G, \operatorname{coker} f) = 0$, and 8.4 now implies that $\operatorname{coker} f = 0$. Consider now the long exact homology sequence associated to $0 \to \ker f \to F \to M \to 0$. If $H_1(G, M) = 0$, then the fact that $H_0(G, f)$ is an isomorphism implies that $H_0(G, \ker f) = 0$. Applying 8.4 again, we conclude that $\ker f = 0$, so f is an isomorphism. Thus M is free and we have proven (iv) $\Rightarrow$ (i).

Assume now that (v) holds. By dimension-shifting we can find a kG-module N such that $\hat{H}^n(G, N) \approx \hat{H}^{n+i+2}(G, M)$ for all $n \in \mathbb{Z}$. [Note that the dimension-shifting argument does not take us out of the category of kG-modules, because $\mathbb{Z}G \otimes A$ is a kG-module if A is a k-module.] In particular, $H_1(G, N) = \hat{H}^{-2}(G, N) = \hat{H}^i(G, M) = 0$, so N is free by the previous paragraph. But then $\hat{H}^*(G, N) = 0$, hence $\hat{H}^*(G, M) = 0$, hence $H_1(G, M) = 0$. Thus $(v) \Rightarrow$ (iv). □

Note that the theorem implies, in particular, that $H^i(G, \mathbb{Z}_p) \neq 0$ for all $i > 0$ if G is a (non-trivial) p-group.

Remarks

1. Corollary 8.2 can be interpreted as saying that I is the Jacobson radical of kG. From this point of view, 8.3 is simply the well-known result that the Jacobson radical of an Artin ring is nilpotent, and 8.4 is a special case of "Nakayama's lemma." The equivalence of (i), (ii), and (iv) in 8.5 could also be stated in a more general context, as a theorem about modules over a local ring. [It follows from 8.3 that kG is a (non-commutative) local ring, with I as its unique maximal ideal.]

2. The proof above yields the following more precise version of the implication (iv) $\Rightarrow$ (i): For any kG-module M, $\dim_k H_0(G, M)$ is the minimal number of generators of M and $\dim_k H_1(G, M)$ is the minimal number of defining relations among any minimal set of generators.

Next we will consider more general G-modules M, still assuming that G is a p-group. Let $M_p = M \otimes \mathbb{Z}_p = M/pM$.

(8.6) Lemma. *Let G be a p-group and M a G-module which is p-torsion-free. For any $i \in \mathbb{Z}$, $\hat{H}^i(G, M_p) = 0$ if and only if $\hat{H}^i(G, M) = \hat{H}^{i+1}(G, M) = 0$.*

PROOF. From the short exact sequence $0 \to M \xrightarrow{p} M \to M_p \to 0$ we obtain an exact sequence

$$\hat{H}^i(G, M) \xrightarrow{p} \hat{H}^i(G, M) \to \hat{H}^i(G, M_p) \to \hat{H}^{i+1}(G, M) \xrightarrow{p} \hat{H}^{i+1}(G, M).$$

If $\hat{H}^i(G, M) = \hat{H}^{i+1}(G, M) = 0$, then clearly this implies that $\hat{H}^i(G, M_p) = 0$. Conversely, if $\hat{H}^i(G, M_p) = 0$, then we conclude that $\hat{H}^i(G, M)$ is p-divisible and $\hat{H}^{i+1}(G, M)$ is p-torsion-free. But $\hat{H}^*(G, M)$ is annihilated by $|G|$, which is a power of p, so $\hat{H}^i(G, M) = \hat{H}^{i+1}(G, M) = 0$. □

(8.7) Theorem. *If G is a p-group, then the following two conditions are equivalent for any G-module M:*

(i) *M is cohomologically trivial.*
(ii) *$\hat{H}^i(G, M) = 0$ for two consecutive integers i.*

If M is p-torsion-free, then (i) *and* (ii) *are also equivalent to:*

(iii) *$\hat{H}^i(G, M_p) = 0$ for some i.*
(iv) *M_p is cohomologically trivial.*
(v) *M_p is free over $\mathbb{Z}_p[G]$.*

PROOF. Suppose first that M is p-torsion-free. Then (i) ⇒ (ii) trivially; (ii) ⇒ (iii) by 8.6; (iii) ⇔ (iv) ⇔ (v) by 8.5; and (iv) ⇒ (i) by 8.6 applied to G and all of its subgroups. If M is now arbitrary, then we can prove the equivalence of (i) and (ii) by using dimension-shifting to reduce to the previous case; for we can choose an exact sequence $0 \to M' \to F \to M \to 0$ with F free over $\mathbb{Z}G$ and then apply the previous case to M'. □

An important consequence of 8.7 is that cohomological triviality can be checked (if G is a p-group) by looking only at $\hat{H}^*(G, M)$, even though the definition involves $\hat{H}^*(H, M)$ for arbitrary subgroups $H \subseteq G$.

We turn now to the case where G is an arbitrary finite group. For each prime p dividing $|G|$, choose a p-Sylow subgroup $G(p)$.

(8.8) Proposition. *A G-module M is cohomologically trivial if and only if its restriction to $G(p)$ is cohomologically trivial for each p.*

PROOF. The "only if" part is trivial. To prove the "if" part, note first that if the restriction of M to $G(p)$ is cohomologically trivial, then $\hat{H}^*(P, M) = 0$ for any p-group $P \subseteq G$. For we know that $gPg^{-1} \subseteq G(p)$ for some $g \in G$, so there is a conjugation isomorphism $\hat{H}^*(P, M) \approx \hat{H}^*(gPg^{-1}, M) = 0$. If $H \subseteq G$ is now an arbitrary subgroup, then we know from transfer theory (cf. §III.10) that $\hat{H}^*(H, M) = \bigoplus_p \hat{H}^*(H, M)_{(p)} \hookrightarrow \bigoplus_p \hat{H}^*(H(p), M) = 0$, where p ranges over the primes dividing $|H|$ and $H(p)$ is a p-Sylow subgroup of H. Thus M is cohomologically trivial. □

Combining 8.8 with our previous results on cohomological triviality over p-groups, we obtain a solution to problem (b) stated at the beginning of the section. In particular, we have:

(8.9) Theorem. *A G-module M is cohomologically trivial if and only if for each prime p there are two consecutive integers i such that $\hat{H}^i(G(p), M) = 0$.*

The solution to problem (a) (with $k = \mathbb{Z}$) is given by the following theorem of Rim:

(8.10) Theorem. *If M is a G-module which is $\mathbb{Z}$-free and cohomologically trivial, then M is $\mathbb{Z}G$-projective.*

PROOF. Choose a short exact sequence $0 \to K \to F \to M \to 0$ with F $\mathbb{Z}G$-free. We will prove that this sequence splits, so that M is a direct summand of F and hence is projective. I claim that the obstruction to splitting the sequence lies in $H^1(G, \operatorname{Hom}(M, K))$. More precisely, we have a short exact sequence of G-modules $0 \to \operatorname{Hom}(M, K) \to \operatorname{Hom}(M, F) \to \operatorname{Hom}(M, M) \to 0$ since M is $\mathbb{Z}$-free, and this yields an exact sequence $\operatorname{Hom}_G(M, F) \to \operatorname{Hom}_G(M, M) \xrightarrow{\delta} H^1(G, \operatorname{Hom}(M, K))$. [Recall that

$$\operatorname{Hom}_G(-, -) = \operatorname{Hom}(-, -)^G = H^0(G, \operatorname{Hom}(-, -)).]$$

Hence the extension splits if and only if $\delta(\mathrm{id}_M) = 0$. It will therefore suffice to prove:

(8.11) Lemma. *Let M and K be G-modules which are $\mathbb{Z}$-free. If M is cohomologically trivial then so is* $\operatorname{Hom}(M, K)$.

PROOF. By 8.8 we need only consider the case where G is a p-group, in which case it suffices by 8.7 to prove that $\operatorname{Hom}(M, K)_p$ is cohomologically trivial. From the exact sequence $0 \to K \xrightarrow{p} K \to K_p \to 0$ we get (since M is $\mathbb{Z}$-free) an exact sequence $0 \to \operatorname{Hom}(M, K) \xrightarrow{p} \operatorname{Hom}(M, K) \to \operatorname{Hom}(M, K_p) \to 0$. Thus $\operatorname{Hom}(M, K)_p \approx \operatorname{Hom}(M, K_p) = \operatorname{Hom}(M_p, K_p)$. But M_p is $\mathbb{Z}_p[G]$-free by 8.7, so $\operatorname{Hom}(M_p, K_p)$ is induced (cf. §III.5, exercise 2b) and hence cohomologically trivial. □

Remark. The group $H^1(G, \operatorname{Hom}(M, K))$ which arose in the proof of 8.10 is isomorphic to $\operatorname{Ext}^1_{\mathbb{Z}G}(M, K)$ by III.2.2. The reader familiar with the theory of extensions of modules will therefore not be surprised that this group contains the obstruction to splitting the extension $0 \to K \to F \to M \to 0$.

Finally, we will deduce from 8.10 a characterization (also due to Rim) of cohomologically trivial modules which are not necessarily $\mathbb{Z}$-free. If R is a ring and M is an R-module, the *projective dimension* of M, denoted proj dim M or $\operatorname{proj\,dim}_R M$, is defined to be the infimum of the set of integers n such that M admits a projective resolution $0 \to P_n \to \cdots \to P_0 \to M \to 0$ of length n. (In particular, proj dim $M = \infty$ if M does not admit a projective resolution of finite length.)

(8.12) Theorem. *The following conditions on a $\mathbb{Z}G$-module M are equivalent*:

(i) *M is cohomologically trivial.*
(ii) proj dim $M \leq 1$.
(iii) proj dim $M < \infty$.

PROOF. (i) $\Rightarrow$ (ii): Suppose M is cohomologically trivial, and choose a short exact sequence $0 \to P_1 \to P_0 \to M \to 0$ with P_0 $\mathbb{Z}G$-projective. Then P_1 is cohomologically trivial and $\mathbb{Z}$-free, so P_1 is projective by 8.10 and (ii) holds. (ii) $\Rightarrow$ (iii) trivially. (iii) $\Rightarrow$ (i): Suppose proj dim $M < \infty$ and let $0 \to P_n \to \cdots \to P_0 \to M \to 0$ be a projective resolution of finite length. Breaking up this resolution into short exact sequences $0 \to Z_i \to P_i \to Z_{i-1} \to 0$, one sees by descending induction on i that Z_i is cohomologically trivial. In particular, $M = Z_{-1}$ is cohomologically trivial. □

Rim's results can be used to provide examples of projective modules over $\mathbb{Z}G$. For example, suppose $J \subset \mathbb{Z}G$ is a left ideal of finite index m, where m is relatively prime to $|G|$. Then $|G|$ is invertible in $M = \mathbb{Z}G/J$, so M is cohomologically trivial. It now follows as in the proof of (i) $\Rightarrow$ (ii) above that J is projective. (See also exercise 7 of §VIII.2 for an alternative proof that proj dim $M \leq 1$ when $|G|$ is invertible in M.)

EXERCISES

1. Give an example where $\hat{H}^*(G, M) = 0$ but M is not cohomologically trivial. [Hint: Take G to be cyclic.]

2. Prove the following improvement of 8.11: If M is cohomologically trivial and $\mathbb{Z}$-free, then $\mathrm{Hom}(M, K)$ is cohomologically trivial for any G-module K. [Use 8.10. Alternatively, choose an exact sequence $0 \to L \to F \to K \to 0$ with F free, and apply 8.11 to $\mathrm{Hom}(M, L)$ and $\mathrm{Hom}(M, F)$.]

3. (a) If M and P are $\mathbb{Z}G$-modules such that M is $\mathbb{Z}$-free and P is $\mathbb{Z}G$-projective, show that any exact sequence $0 \to P \to E \to M \to 0$ splits. [Hint: Argue as in the proof of 8.10, or, more simply, use the fact that P is relatively injective.]

   (b) Using (a), give a direct proof (without using the results of this section) that (iii) $\Rightarrow$ (ii) in 8.12.

4. Let G be a group such that there exists a free, finite G-CW-complex X with $H_*(X) \approx H_*(S^{2k-1})$. Prove that G admits a complete resolution which is periodic of period $2k$. [Hint: If B is the module of $(2k-1)$-boundaries of $C_*(X)$, then $C_{2k-1}(X)/B$ is $\mathbb{Z}$-free and has finite projective dimension.]

9 Groups with Periodic Cohomology

A finite group G is said to have *periodic cohomology* if for some $d \neq 0$ there is an element $u \in \hat{H}^d(G, \mathbb{Z})$ which is invertible in the ring $\hat{H}^*(G, \mathbb{Z})$. Cup product with u then gives a *periodicity isomorphism*

$$u \cup - : \hat{H}^n(G, M) \xrightarrow{\approx} \hat{H}^{n+d}(G, M)$$

for all $n \in \mathbb{Z}$ and all G-modules M. In particular, taking $n = 0$ and $M = \mathbb{Z}$, we see that $\hat{H}^d(G, \mathbb{Z}) \approx \mathbb{Z}/|G| \cdot \mathbb{Z}$ and that u generates $\hat{H}^d(G, \mathbb{Z})$.

If we know that a group G has periodic cohomology, then the task of computing $\hat{H}^*(G)$ is obviously enormously simplified. It is therefore of interest to find criteria for deciding whether G has periodic cohomology, and that is what we will do in this section.

(9.1) Theorem. *The following conditions are equivalent:*

(i) *G has periodic cohomology.*
(ii) *There exist integers n and d, with $d \neq 0$, such that $\hat{H}^n(G, M) \approx \hat{H}^{n+d}(G, M)$ for all G-modules M.*
(iii) *For some $d \neq 0$, $\hat{H}^d(G, \mathbb{Z}) \approx \mathbb{Z}/|G| \cdot \mathbb{Z}$.*
(iv) *For some $d \neq 0$, $\hat{H}^d(G, \mathbb{Z})$ contains an element u of order $|G|$.*

PROOF. (i) $\Rightarrow$ (ii) trivially. If (ii) holds then we have $\hat{H}^n(G, M) \approx \hat{H}^{n+d}(G, M)$ for *all* $n \in \mathbb{Z}$ by dimension-shifting; taking $n = 0$ and $M = \mathbb{Z}$, we obtain (iii). (iii) $\Rightarrow$ (iv) trivially. Finally, suppose (iv) holds. Then there is a map $\hat{H}^d(G, \mathbb{Z}) \to \mathbb{Q}/\mathbb{Z}$ such that $u \mapsto 1/|G| \bmod \mathbb{Z}$ (cf. proof of III.4.3). By the duality theorem 7.4, this map is given by

$$\hat{H}^d(G, \mathbb{Z}) \xrightarrow{-\cup v} \hat{H}^0(G, \mathbb{Z}) \approx |G|^{-1}\mathbb{Z}/\mathbb{Z} \hookrightarrow \mathbb{Q}/\mathbb{Z}$$

for some $v \in \hat{H}^{-d}(G, \mathbb{Z})$. Thus $uv = 1$, whence (i). □

For example, if G acts freely on S^{2k-1} as in §I.6, then G obviously admits a periodic complete resolution of period $d = 2k$, so 9.1(ii) holds and G has periodic cohomology. [Alternatively, one can show directly, without using 9.1, that $\hat{H}^d(G, \mathbb{Z})$ contains an invertible element; cf. §V.3, exercise 3.]

One way to obtain examples of this is to start with a linear action of a finite group G on an even-dimensional real vector space V, such that G acts freely on $V - \{0\}$ (i.e., such that 1 is not an eigenvalue of any non-trivial element $g \in G$); such an action is called a *fixed-point-free representation* of G. By choosing a G-invariant sphere $S \subset V - \{0\}$ (e.g., the unit sphere relative to a G-invariant inner product on V), we obtain a free action of G on an odd-dimensional sphere. In order to apply §I.6, of course, we must verify that S can be chosen to have a G-equivariant CW-structure. This follows from a general triangulation theorem, but in the present case there is a much more elementary argument, due to Illman [1978]: Let $(e_i)_{1 \le i \le 2k}$ be a basis for V and let C be the convex hull of the finite set $\{\pm ge_i : g \in G, 1 \le i \le 2k\}$. Then C is a convex cell containing 0 as an interior point, hence its boundary S is a topological sphere in $V - \{0\}$, with a CW-structure given by the natural (rectilinear) faces of C (see, for example, Hudson [1969], Chapter I). Since G acts linearly on V and maps C into itself, it is clear that S is a G-complex.

Remark. The question of the existence of an equivariant CW-structure is actually a red herring. For one can show, without assuming an equivariant

CW-structure, that if G acts freely on S^{2k-1} then $\hat{H}^*(G)$ is periodic of period $2k$. A proof will be outlined in exercise 4c of §VII.7.

We can now give some specific examples of groups with periodic cohomology:

(9.2) EXAMPLES

1. Any finite cyclic group G admits a 2-dimensional fixed-point-free representation as a group of rotations, hence $\hat{H}^*(G)$ is periodic of period 2.

2. Let $\mathbb{H}$ be the quaternion algebra $\mathbb{R} \oplus \mathbb{R}i \oplus \mathbb{R}j \oplus \mathbb{R}k$. If G is a finite subgroup of the multiplicative group $\mathbb{H}^*$, then the multiplication action of G on $\mathbb{H}$ yields a 4-dimensional fixed-point-free representation of G, since $\mathbb{H}$ is a division algebra. Thus $\hat{H}^*(G)$ is periodic of period 4. The most obvious examples of finite subgroups of $\mathbb{H}^*$ are the generalized quaternion groups Q_{4m} $(m \geq 2)$ which we studied in §IV.4. In addition to these (and, of course, the cyclic groups), there are precisely three finite subgroups of $\mathbb{H}^*$, up to conjugacy (cf. Wolf [1974], §2.6): the binary tetrahedral group (of order 24), the binary octahedral group (of order 48), and the binary icosahedral group (of order 120). The latter group G is particularly interesting because it is a *perfect* group, i.e., $G_{ab} = 0$. Using Poincaré duality, one can deduce from this that the quotient manifold S^3/G has the same homology as S^3. This was discovered by Poincaré [1904]; it was the first example of a 3-manifold homologically equivalent to S^3 but not homeomorphic to S^3.[2] Another interesting fact about the binary icosahedral group G is given in the following surprising theorem of Zassenhaus (cf. Wolf [1974], 6.2): Up to isomorphism, G is the unique perfect group which admits a fixed-point-free representation.

3. Let m, n, and r be positive integers such that $r^n \equiv 1 \bmod m$. Assume that the following two conditions hold: (a) m and n are relatively prime; (b) if k is the multiplicative order of $r \bmod m$ (so necessarily $k \mid n$), then every prime divisor of n divides n/k. Let G be the semi-direct product $\mathbb{Z}_m \rtimes \mathbb{Z}_n$, where the generator of $\mathbb{Z}_n$ acts on $\mathbb{Z}_m$ by multiplication by r. I claim that G admits a $2k$-dimensional fixed-point-free representation and hence $\hat{H}^*(G)$ is periodic of period $2k$. To see this, note first that the subgroup $A \subset \mathbb{Z}_n$ of index k acts trivially on $\mathbb{Z}_m$, so G contains $C = \mathbb{Z}_m \times A$, which is cyclic by (a). From (b) we see that A contains every subgroup of $\mathbb{Z}_n$ of prime order, and it follows easily that C contains every subgroup of G of prime order. Choose a fixed-point-free representation of C on a 2-dimensional vector space W, and form the induced

[2] Poincaré had asked in an earlier paper whether such a 3-manifold could exist. In view of this example, Poincaré reformulated his question, adding the condition that the manifold be simply-connected. This question, or rather the assertion that there can be no such manifold, has since become known as the Poincaré conjecture.

module $V = \mathbb{Z}G \otimes_{\mathbb{Z}C} W = \mathbb{R}G \otimes_{\mathbb{R}C} W$. Since $C \lhd G$, we have $\operatorname{Res}_C^G V = \bigoplus_{g \in G/C} gW$ by III.5.6b, and each gW is clearly a fixed-point-free representation of C. Consequently, V is a fixed-point-free representation of C. But then V is also fixed-point-free as a representation of G, for a non-trivial isotropy group G_v ($v \in V - \{0\}$) would contain a subgroup of prime order and hence would meet C nontrivially.

See Wolf [1974] for a complete classification of groups which admit a fixed-point-free representation.

Remarks

1. If G has periodic cohomology then so does any subgroup H. This follows from the fact that the restriction map $\hat{H}^*(G, \mathbb{Z}) \to \hat{H}^*(H, \mathbb{Z})$, being a ring homomorphism, must take invertible elements to invertible elements. [Alternatively, use Shapiro's lemma.]

2. If G is abelian but not cyclic, then G does *not* have periodic cohomology. For G must contain a subgroup isomorphic to $\mathbb{Z}_p \times \mathbb{Z}_p$ for some prime p, and direct calculation (using the Künneth formula, for example) shows that $\hat{H}^n(\mathbb{Z}_p \times \mathbb{Z}_p; \mathbb{Z}_p)$ has $\mathbb{Z}_p$-dimension $n + 1$ for $n \geq 0$ and hence is not periodic.

We return now to the general question of finding criteria for a group to have periodic cohomology. In case G is a p-group, the question is completely settled by the following result. Recall that an *elementary abelian p-group* of rank $r \geq 0$ is a group isomorphic to $\mathbb{Z}_p^r = \mathbb{Z}_p \times \cdots \times \mathbb{Z}_p$ (r factors).

(9.3) Proposition. *If G is a p-group for some prime p, then the following conditions are equivalent:*

(i) *G has periodic cohomology.*
(ii) *Every abelian subgroup of G is cyclic.*
(iii) *Every elementary abelian p-subgroup of G has rank ≤ 1.*
(iv) *G has a unique subgroup of order p.*
(v) *G is a cyclic or generalized quaternion group.*

(The generalized quaternion groups, of course, can only occur if $p = 2$.)

PROOF. We have (v) $\Rightarrow$ (i) $\Rightarrow$ (ii) $\Rightarrow$ (iii) by the examples and remarks above. (iii)$\Rightarrow$(iv): Since G is a p-group, one knows that G contains a central subgroup of order p (cf. Hall [1959], Theorem 4.3.1). If there were any other subgroup of order p, the two would generate an elementary abelian p-group of rank 2, contrary to (iii). Finally, the implication (iv) $\Rightarrow$ (v) is Theorem IV.4.3. □

To pass from the p-group case to the general case, we have:

(9.4) Proposition. *A finite group G has periodic cohomology if and only if every Sylow subgroup has periodic cohomology.*

PROOF. The "only if" part follows from Remark 1 above. To prove the "if" part, fix a prime p and let $H \subseteq G$ be a p-Sylow subgroup. By hypothesis, $\hat{H}^*(H, \mathbb{Z})$ contains an invertible element u of degree $d \neq 0$, and I claim that some power of u is G-invariant in the sense of §III.10. In fact, let $e > 0$ be an integer such that $a^e = 1$ for all $a \in (\mathbb{Z}/|H| \cdot \mathbb{Z})^*$, the latter being the group of units in $\mathbb{Z}/|H| \cdot \mathbb{Z}$. For any $g \in G$, the elements $v = \text{res}^H_{H \cap gHg^{-1}} u$ and $w = \text{res}^{gHg^{-1}}_{H \cap gHg^{-1}} gu$ are invertible in $\hat{H}^*(H \cap gHg^{-1}, \mathbb{Z})$ and hence generate the cyclic group $\hat{H}^d(H \cap gHg^{-1}, \mathbb{Z})$. We therefore have $v = aw$ for some $a \in (\mathbb{Z}/|H| \cdot \mathbb{Z})^*$, since the map $(\mathbb{Z}/|H| \cdot \mathbb{Z})^* \to (\mathbb{Z}/|H'| \cdot \mathbb{Z})^*$ is surjective for any $H' \subseteq H$. Thus $v^e = w^e$, and it follows that $u^e \in \hat{H}^{de}(H, \mathbb{Z})$ is invariant. Similarly, u^{-e} is invariant. In view of the Tate cohomology version of Theorem III.10.3, this proves that the ring $\hat{H}^*(G, \mathbb{Z})_{(p)}$ has an invertible element $u(p)$ of non-zero degree $d(p)$. Finally, replacing $u(p)$ by a suitable power $u(p)^{e(p)}$, we may assume that $d(p)$ is the same for all primes $p \,||\, G|$; since $\hat{H}^*(G, \mathbb{Z})$ is the direct product of the rings $\hat{H}^*(G, \mathbb{Z})_{(p)}$ (cf. exercise 2 of §5), it follows that $\hat{H}^*(G, \mathbb{Z})$ contains an invertible element of positive degree. □

Combining 9.3 and 9.4, we have:

(9.5) Theorem. *The following conditions are equivalent for a finite group G:*

(i) *G has periodic cohomology.*
(ii) *Every abelian subgroup of G is cyclic.*
(iii) *Every elementary abelian p-subgroup of G (p prime) has rank ≤ 1.*
(iv) *The Sylow subgroups of G are cyclic or generalized quaternion groups.*

Using the criterion (iv) of this theorem, we can now give some examples of groups with periodic cohomology, beyond those of 9.2.

(9.6) EXAMPLES

1. If m and n are relatively prime integers, then any semi-direct product $\mathbb{Z}_m \rtimes \mathbb{Z}_n$ (arbitrary action) has cyclic Sylow subgroups and hence has periodic cohomology. [Conversely, a theorem of Burnside (cf. Wolf [1974], 5.4) implies that any group with cyclic Sylow subgroups is a semi-direct product of this type.] This generalizes example 3 of 9.2.

2. Let p be a prime and let $G = SL_2(\mathbb{F}_p)$, the group of 2×2 matrices of determinant 1 over the prime field $\mathbb{F}_p$. Then G has periodic cohomology. To see this, note first that $|G| = p(p^2 - 1)$. [There are $p^2 - 1$ possibilities for the first column of a matrix in G; given the first column, the number of possibilities for the second is $(p^2 - p)/(p - 1) = p$.] Thus a p-Sylow subgroup of G is of order p and hence is cyclic. Now let l be a prime $\neq p$. Then the polynomial $x^l - 1$ is separable over $\mathbb{F}_p$, so a matrix of order l is diagonalizable over the finite extension k obtained from $\mathbb{F}_p$ by adjoining the l-th roots of unity. Two

commuting elements of G of order l are then simultaneously diagonalizable and hence generate a group isomorphic to

$$\left\{\begin{pmatrix}\lambda & 0\\ 0 & \lambda^{-1}\end{pmatrix} : \lambda \in k, \lambda^l = 1\right\},$$

which is of order l. Thus G does not contain a subgroup isomorphic to $\mathbb{Z}_l \times \mathbb{Z}_l$ and hence G has periodic cohomology. [See exercise 8 below for a different proof, which consists of explicitly exhibiting cyclic or quaternionic Sylow subgroups.]

Remarks

1. The groups $SL_2(\mathbb{F}_p)$ for $p = 2, 3, 5$ have occurred earlier in our examples. Namely, $SL_2(\mathbb{F}_2) \approx \mathbb{Z}_3 \rtimes \mathbb{Z}_2$, $SL_2(\mathbb{F}_3)$ is isomorphic to the binary tetrahedral group, and $SL_2(\mathbb{F}_5)$ is isomorphic to the binary icosahedral group.

2. A complete classification of the groups with periodic cohomology has been obtained by Suzuki [1955], extending earlier work of Zassenhaus. (See also Wolf [1974], 6.1 and 6.3.) Not all of these groups can act freely on a sphere. In fact, Milnor [1957a] showed that a group which acts freely on a sphere has at most one element of order 2. (Thus, for example, the dihedral group D_{2n} for n odd cannot act freely on a sphere, although all of its Sylow subgroups are cyclic.) Conversely, Madsen, Thomas, and Wall [1976] have proven that if G has periodic cohomology and at most one element of order 2 then G acts freely on a sphere. This applies, for example, to the groups $SL_2(\mathbb{F}_p)$ (p odd) since

$$\begin{pmatrix}-1 & 0\\ 0 & -1\end{pmatrix}$$

is the only element of order 2. These groups do not, however, admit a free *orthogonal* action on a sphere (i.e., a fixed-point-free representation), except for $p = 3$ and $p = 5$.

The condition that *all* Sylow subgroups of a group G be cyclic or generalized quaternion is obviously very restrictive. What is much more common, however, is for there to be at least one prime p such that the p-Sylow subgroups are cyclic or generalized quaternion. It is therefore of interest to observe that the results of this section can be "localized" at a single prime p. Thus we say that G has *p-periodic cohomology* if the ring $\hat{H}^*(G, \mathbb{Z})_{(p)}$ contains an invertible element of non-zero degree d. Cup product with such an element then gives periodicity isomorphisms $\hat{H}^n(G, M)_{(p)} \approx \hat{H}^{n+d}(G, M)_{(p)}$ for any G-module M. The reader can easily state and prove analogues of 9.1, 9.4, and 9.5 for this situation. In particular, one can prove:

(9.7) Theorem. *For any finite group G and prime p dividing $|G|$, the following conditions are equivalent:*

(i) *G has p-periodic cohomology.*
(ii) *The ring $\hat{H}^*(G, \mathbb{Z}_p)$ contains an invertible element of non-zero degree.*
(iii) *Every elementary abelian p-subgroup of G has rank ≤ 1.*
(iv) *The p-Sylow subgroups of G are cyclic or generalized quaternion.*

We close this section by stating without proof a beautiful generalization of this theorem, which was conjectured by Atiyah and Swan and proved by Quillen [1971]. Fix a prime p and let $H(G)$ denote the commutative $\mathbb{Z}_p$-algebra $\bigoplus_{n\geq 0} H^{2n}(G, \mathbb{Z}_p)$. If the maximal rank of an elementary abelian p-subgroup of G is 1, then 9.7 implies that $H(G)$ is a finitely generated module over a polynomial subalgebra $\mathbb{Z}_p[u]$. Consequently, $H(G)$ has Krull dimension 1. (See, for example, Serre [1965] for a treatment of dimension theory of commutative rings.) Quillen's generalization is:

(9.8) Theorem. *For any finite group G, the Krull dimension of $H(G)$ is equal to the maximal rank of an elementary abelian p-subgroup of G.*

Roughly speaking, then, the complexity of the ring $H^*(G, \mathbb{Z}_p)$ is determined by the complexity of the p-subgroups of G.

EXERCISES

1. If G is non-trivial and has periodic cohomology of period d, prove that d is even. [Hint: Use anti-commutativity.]

2. Prove that G has periodic cohomology of period 2 if and only if G is cyclic. [Hint: What can you say about $H_1 G$ if $\hat{H}^*(G)$ has period 2?]

3. Compute $H^n(G, \mathbb{Z})$ and $H_n(G, \mathbb{Z})$ for all n if $\hat{H}^*(G)$ is periodic of period 4. Note, in particular, that $H_2(G) = \hat{H}^{-3}(G) = \hat{H}^1(G) = \operatorname{Hom}(G, \mathbb{Z}) = 0$, so we obtain a new proof of the result of exercise 7a of §II.5.

4. If G has periodic cohomology, show that $\hat{H}^i(G, \mathbb{Z}) = 0$ for i odd. [Hint: It suffices to do this when G is a p-group.]

5. Give a direct proof that (iv) $\Rightarrow$ (i) in 9.3. [Hint: Use induced representations as in example 3 of 9.2.]

6. Let $G = \mathbb{Z}_m \rtimes \mathbb{Z}_n$, where m and n are relatively prime and $\mathbb{Z}_n$ acts on $\mathbb{Z}_m$ via a homomorphism $\mathbb{Z}_n \to \mathbb{Z}_m^*$ whose image has order k. Prove that the minimal period of $\hat{H}^*(G)$ is $2k$. [Hint: If $p \mid n$ then $\hat{H}^*(G, \mathbb{Z})_{(p)} \approx \hat{H}^*(\mathbb{Z}_n, \mathbb{Z})_{(p)}$. If $p \mid m$ then $\hat{H}^*(G)_{(p)} \approx \hat{H}^*(\mathbb{Z}_m)_{(p)}^{\mathbb{Z}_n}$.]

7. Let $\mathbb{F}_q$ be a field with q elements, where q is a prime power. Show that $SL_n(\mathbb{F}_q)$ does not have periodic cohomology if $n \geq 3$ or if q is not prime.

8. Let $k = \mathbb{F}_q$, where q is an odd prime power.

   (a) Let M be the subgroup of $SL_2(k)$ consisting of the monomial matrices

   $$\begin{pmatrix} \lambda & 0 \\ 0 & \lambda^{-1} \end{pmatrix} \quad \text{and} \quad \begin{pmatrix} 0 & -\lambda^{-1} \\ \lambda & 0 \end{pmatrix}, \qquad \lambda \in k^*.$$

   Show that $M \approx Q_{2(q-1)}$ if $q \neq 3$.

   (b) Let $K \supset k$ be a quadratic extension (so $K = \mathbb{F}_{q^2}$). We regard K as a 2-dimensional vector space over k and denote by $GL(K)$ (resp. $SL(K)$) the group of vector space automorphisms of K (resp. the group of automorphisms of determinant 1). Thus $GL(K) \approx GL_2(k)$ and $SL(K) \approx SL_2(k)$. We have $\mathrm{Gal}(K/k) \subset GL(K)$ and $K^* \hookrightarrow GL(K)$, where the latter embedding corresponds to the multiplication action of K^* on K. Show that $\mathrm{Gal}(K/k)$ and K^* generate a subgroup of $GL(K)$ of order $2(q^2 - 1)$, whose intersection with $SL(K)$ is isomorphic to $Q_{2(q+1)}$. [Hint: The composite $K^* \hookrightarrow GL(K) \overset{\det}{\to} k^*$ is simply the norm map of Galois theory, which in this case is given by $\lambda \mapsto \lambda^{q+1}$. Also, if $\sigma \in \mathrm{Gal}(K/k)$ is the non-trivial element, then $\det \sigma = -1$.]

   (c) Using (a) and (b) and the fact that $|SL_2(k)| = q(q^2 - 1)$, show that $SL_2(k)$ has a quaternionic 2-Sylow subgroup and a cyclic l-Sylow subgroup for any odd prime $l \neq \mathrm{char}\, k$.

9. Suppose that G has p-periodic cohomology. Let $P \subseteq G$ be a subgroup of order p, let $N(P)$ (resp. $C(P)$) be the normalizer (resp. centralizer) of P in G, and let $W = N(P)/C(P)$. Prove that $\hat{H}^*(G, M)_{(p)} \approx \hat{H}^*(N(P), M)_{(p)} \approx \hat{H}^*(C(P), M)_{(p)}^W$. [Hint: For the first isomorphism, choose a p-Sylow subgroup $H \supseteq P$ and show that every $N(P)$-invariant in $\hat{H}^*(H)$ is G-invariant. For the second isomorphism, note that $p \nmid |W|$.]

10. For any finite group G, show that the augmentation ideal $I \subset \mathbb{Z}G$ is a cyclic G-module iff G is a cyclic group. [Hint: If I is cyclic, then G admits a periodic resolution of period 2.]

CHAPTER VII

Equivariant Homology and Spectral Sequences

1 Introduction

If a group G operates on a topological space X, then one can define *equivariant homology* and *cohomology* groups, which can be thought of heuristically as a "mixture" of $H(G)$ and $H(X)$. This equivariant theory provides a powerful tool for extracting homological information about G from the action of G on X. It is in this way, for example, that Quillen proved his theorem about the Krull dimension of $H^*(G, \mathbb{Z}_p)$ for G finite (VI.9.8).

The main purpose of this chapter is to construct the equivariant homology theory and give its basic properties, including two spectral sequences. This theory will play a crucial role in the remaining chapters of this book. For simplicity, we will confine ourselves to the case where X is a *G-complex* in the sense of §I.4. See Grothendieck [1957] or Quillen [1971] for a more general point of view.

We begin with a brief treatment of the theory of spectral sequences.

2 The Spectral Sequence of a Filtered Complex

If C is a chain complex and C' is a subcomplex, then there is a long exact sequence which gives information about $H_*(C)$ in terms of $H_*(C')$ and $H_*(C/C')$. Now suppose we are given, instead of a single subcomplex C', a sequence of subcomplexes $\{F_pC\}_{p\in\mathbb{Z}}$, with $F_{p-1}C \subseteq F_pC$. It is reasonable, then, to try to get information about $H_*(C)$ in terms of the groups $H_*(F_pC/F_{p-1}C)$. In this section, we will describe a method for doing this.

What one obtains, roughly speaking, is a sequence of successive approximations E^r ($r \geq 0$) to $H_*(C)$, such that E^1 consists of the groups $H_*(F_pC/F_{p-1}C)$.

We will omit most of the proofs in this section. The reader can either supply the missing proofs (which are routine) or consult any text which treats spectral sequences, e.g., Spanier [1966] or MacLane [1963].

Let R be an arbitrary ring. (In our applications, we will usually take $R = \mathbb{Z}$.) By an *increasing filtration* on an R-module M we mean a family of submodules F_pM ($p \in \mathbb{Z}$) such that $F_pM \subseteq F_{p+1}M$. The filtration is said to be *finite* if $F_pM = 0$ for p sufficiently small and $F_pM = M$ for p sufficiently large. Given a filtration on M, the *associated graded module* Gr M is defined by $\mathrm{Gr}_pM = F_pM/F_{p-1}M$. One thinks of M, then, as being built up from the "pieces" $\mathrm{Gr}_p M$. The following elementary lemma shows that, in some sense, we do not lose too much information by passing from M to Gr M:

(2.1) Lemma. *Let $f: M \to M'$ be a filtration-preserving map, where M and M' are modules with finite filtrations. If* Gr f: Gr $M \to$ Gr M' *is an isomorphism, then f is an isomorphism.*

Another elementary (but important) observation is that if we have a notion of "rank" for R-modules (e.g., if R is a field or $R = \mathbb{Z}$), such that $\mathrm{rk}\, M = \mathrm{rk}\, M' + \mathrm{rk}\, M''$ for every short exact sequence $0 \to M' \to M \to M'' \to 0$, then the rank of a finitely-filtered module can be computed from the associated graded module:

(2.2) $$\mathrm{rk}\, M = \sum_{p \in \mathbb{Z}} \mathrm{rk}\, \mathrm{Gr}_p\, M.$$

If the filtered module M is itself graded (and each F_pM is a graded submodule), then we have for each $n \in \mathbb{Z}$ a filtration $\{F_pM_n\}$ on M_n, and hence there is an obvious way of associating a *bigraded* module to M. The usual notational convention in this case is to set $\mathrm{Gr}_{pq} M = F_pM_{p+q}/F_{p-1}M_{p+q}$. An element of $\mathrm{Gr}_{pq} M$ is said to have *filtration degree* p, *complementary degree* q, and *total degree* $p + q$. To simplify the notation, we will sometimes suppress the second subscript and simply write $\mathrm{Gr}_p M = F_pM/F_{p-1}M$.

Now let $C = (C_n)_{n \in \mathbb{Z}}$ be a filtered chain complex (with each F_pC a subcomplex). For simplicity, we will always assume that the filtration is *dimension-wise finite*, i.e., that $\{F_pC_n\}_{p \in \mathbb{Z}}$ is a finite filtration of C_n for each n. There is an induced filtration on the homology $H(C)$, defined by $F_pH(C) = \mathrm{Im}\{H(F_pC) \to H(C)\}$. We can identify $F_pH(C)$ with $(F_pC \cap Z)/(F_pC \cap B)$, where Z (resp. B) is the module of cycles (resp. boundaries) of C. The associated bigraded module Gr $H(C)$ is given by

(2.3) $$\mathrm{Gr}_p\, H(C) = (F_pC \cap Z)/((F_pC \cap B) + (F_{p-1}C \cap Z)).$$

We now describe the *spectral sequence* associated to the filtered complex

C. This is a sequence $\{E^r\}_{r \geq 0}$ of "successive approximations" to Gr $H(C)$. Let $Z_p^r = F_pC \cap \partial^{-1}F_{p-r}C$. (More precisely,

$$Z_{pq}^r = F_pC_{p+q} \cap \partial^{-1}F_{p-r}C_{p+q-1}.)$$

Let $Z_p^\infty = F_pC \cap Z$. We have $F_pC = Z_p^0 \supseteq Z_p^1 \supseteq \cdots \supseteq Z_p^\infty$. Since the filtration $\{F_pC\}$ is dimension-wise finite, this sequence of inclusions stabilizes dimension-wise to a sequence of equalities, i.e., for fixed (p, q) we have $Z_{pq}^r = Z_{pq}^{r+1} = \cdots = Z_{pq}^\infty$ for r sufficiently large. Let $B_p^r = F_pC \cap \partial F_{p+r-1}C = \partial Z_{p+r-1}^{r-1}$, and let $B_p^\infty = F_pC \cap B$. Then $B_p^0 \subseteq B_p^1 \subseteq \cdots \subseteq B_p^\infty$, and again this stabilizes dimension-wise to a sequence of equalities. We now have

$$B_p^0 \subseteq B_p^1 \subseteq \cdots \subseteq B_p^\infty \subseteq Z_p^\infty \subseteq \cdots \subseteq Z_p^1 \subseteq Z_p^0 = F_pC,$$

and we set

$$E_p^r = Z_p^r/(B_p^r + Z_{p-1}^{r-1}) = Z_p^r/(B_p^r + (F_{p-1}C \cap Z_p^r))$$

and

$$E_p^\infty = Z_p^\infty/(B_p^\infty + Z_{p-1}^\infty) = \mathrm{Gr}_p\, H(C)$$

(cf. 2.3). For fixed (p, q), we have

$$E_{pq}^r = E_{pq}^{r+1} = \cdots = E_{pq}^\infty$$

for r sufficiently large, so the sequence $\{E^r\}$ "converges" to Gr $H(C)$ as $r \to \infty$. (One often suppresses the "Gr" here and simply says that the spectral sequence *converges to* $H(C)$ or that $H(C)$ is the *abutment* of the spectral sequence.)

The modules E^r are particularly easy to describe for $r = 0$ and 1:

(2.4) $$E_p^0 = F_pC/F_{p-1}C = \mathrm{Gr}_p C$$

(2.5) $$E_p^1 = (F_pC \cap \partial^{-1}F_{p-1}C)/(\partial F_pC + F_{p-1}C) \approx H(F_pC/F_{p-1}C).$$

(More precisely, $E_{pq}^1 \approx H_{p+q}(F_pC/F_{p-1}C)$.) Thus E^1 is the homology of E^0, relative to the differential induced on E^0 by ∂. More generally, one can show that ∂ induces a differential d^r on E^r of bidegree $(-r, r-1)$ (i.e., $d^r: E_{pq}^r \to E_{p-r,q+r-1}^r$), and that $E^{r+1} \approx H(E^r)$. An important consequence of this is:

(2.6) Proposition. *Let $\tau: C \to C'$ be a filtration-preserving chain map, where C and C' have dimension-wise finite filtrations. If the induced map $E^r(\tau): E^r(C) \to E^r(C')$ of spectral sequences is an isomorphism for some r, then $H(\tau): H(C) \to H(C')$ is an isomorphism.*

PROOF. If $E^r(\tau)$ is an isomorphism then so is $E^s(\tau)$ for $s > r$, since $E^s = H(E^{s-1})$. Hence $E^\infty(\tau) = \mathrm{Gr}\, H(\tau)$ is an isomorphism, and the proposition now follows from 2.1. □

Another important fact is that spectral sequences can be used to compute Euler characteristics. Suppose, for example, that $R = \mathbb{Z}$ or R is a field. For any graded R-module $A = (A_n)$ (resp. bigraded module $A = (A_{pq})$), let $\chi(A) = \sum_n (-1)^n \operatorname{rk} A_n$ (resp. $\chi(A) = \sum_{p,q} (-1)^{p+q} \operatorname{rk} A_{pq}$), provided all the ranks are finite and almost all of them are zero. In case A has a differential, it is well-known that $\chi(A) = \chi(H(A))$ (cf. Spanier [1966], 4.3.14). In particular, if C is a filtered complex as above and if $\chi(E^r)$ is defined for some r, then it follows that

$$\chi(E^r) = \chi(E^{r+1}) = \cdots = \chi(E^\infty) = \chi(H(C)), \tag{2.7}$$

the last equality being a consequence of 2.2.

Finally, we briefly describe the notational conventions in case C is indexed as a *cochain* complex $(C^n)_{n \in \mathbb{Z}}$, with differential of degree $+1$. In this case the filtration is denoted $\{F^pC\}$ and is assumed to be *decreasing* (i.e., $F^pC \supseteq F^{p+1}C$). The terms of the resulting *cohomology spectral sequence* are denoted E_r^{pq}, and one has $E_\infty^{pq} = \operatorname{Gr}^{pq} H(C) = F^pH^{p+q}(C)/F^{p+1}H^{p+q}(C)$. The differential d_r is of bidegree $(r, -r+1)$, so that $d_r\colon E_r^{pq} \to E_r^{p+r,q-r+1}$. If we think of E_r^{pq} as sitting on the lattice point (p, q) of the plane, then d_r can be visualized as an arrow pointing to the right and downward for $r \geq 2$. By contrast, the differentials d^r in a homology spectral sequence point to the left and upward.

The canonical example of a filtered cochain complex is $\mathscr{Hom}(C, M)$, where C is a filtered chain complex. One sets

$$F^p\mathscr{Hom}(C, M) = \mathscr{Hom}(C/F_{p-1}C, M).$$

EXERCISE

Let $0 \to C' \to C \to C'' \to 0$ be a short exact sequence of chain complexes. Let $\{F_pC\}$ be the filtration such that $F_0C = 0$, $F_1C = C'$, and $F_2C = C$. Describe the modules E_{pq}^1 and deduce from the spectral sequence the familiar long exact homology sequence. [Thus the spectral sequence of a filtered complex can be regarded as a generalization of the long exact sequence associated to a chain complex and a subcomplex, as we implied in the introductory paragraph of this section.]

3 Double Complexes

By a *double complex* we mean a bigraded module $C = (C_{pq})_{p,q \in \mathbb{Z}}$ with a "horizontal" differential ∂' of bidegree $(-1, 0)$ and a "vertical" differential ∂'' of bidegree $(0, -1)$, such that $\partial'\partial'' = \partial''\partial'$:

$$\begin{array}{ccc} C_{p-1,q} & \xleftarrow{\partial'} & C_{pq} \\ \Big\downarrow{\scriptstyle \partial''} & & \Big\downarrow{\scriptstyle \partial''} \\ C_{p-1,q-1} & \xleftarrow{\partial'} & C_{p,q-1}. \end{array}$$

Note that a double complex can be regarded, in two different ways, as a "chain complex in the category of chain complexes." Thus for each q we have a horizontal chain complex $C_{*,q}$ with differential ∂', and we are given chain maps $\partial'': C_{*,q} \to C_{*,q-1}$ such that $\partial''\partial'' = 0$. Similarly, for each p we have a vertical chain complex $C_{p,*}$ with differential ∂'', and we are given chain maps $\partial': C_{p,*} \to C_{p-1,*}$ with $\partial'\partial' = 0$.

A double complex C gives rise to an ordinary chain complex TC, called the *total complex*, as follows: $(TC)_n = \bigoplus_{p+q=n} C_{pq}$, with differential ∂ given by $\partial | C_{pq} = \partial' + (-1)^p\partial''$. The tensor product of two chain complexes C' and C'' provides a familiar example of this construction. Indeed, we have a double complex C with $C_{pq} = C'_p \otimes C''_q$, and TC is simply the usual tensor product $C' \otimes C''$ of chain complexes.

We now filter TC by setting $F_p(TC)_n = \bigoplus_{i \le p} C_{i,n-i}$. This is a dimension-wise finite filtration provided C has only finitely many non-zero modules C_{pq} in any given total degree $p + q$. (This holds automatically, for example, if C is a *first quadrant* double complex, i.e., if $C_{pq} = 0$ when $p < 0$ or $q < 0$.) Thus we have a spectral sequence $\{E^r\}$ converging to $H_*(TC)$. It is immediate from the definitions that $E^0_{pq} = C_{pq}$ and that $d^0 = \pm\, \partial''$. Consequently, E^1 is the vertical homology of C, i.e., $E^1_{pq} = H_q(C_{p,*})$. The differential $d^1: E^1_{pq} \to E^1_{p-1,q}$ is easily seen to be the map induced by the chain map $\partial': C_{p,*} \to C_{p-1,*}$; for an element of E^1_{pq} is represented by an element $c \in C_{pq}$ such that $\partial''c = 0$, and for such a c one has $\partial c = \partial'c$. Thus E^2 can be described as the horizontal homology of the vertical homology of C.

One could equally well filter TC by $F_p(TC)_n = \bigoplus_{j \le p} C_{n-j,j}$. We obtain, then, a second spectral sequence converging to $H(TC)$, this time with $E^0_{pq} = C_{qp}$, $E^1_{pq} = H_q(C_{*,p})$, and $d^1: E^1_{pq} \to E^1_{p-1,q}$ equal (up to sign) to the map induced by $\partial'': C_{*,p} \to C_{*,p-1}$.

[Warning: Even though the two spectral sequences have the same abutment $H_*(TC)$, they do not in general have the same E^∞-term; for we have two different filtrations on $H(TC)$ and hence two different E^∞-terms $\operatorname{Gr} H(TC)$.]

A similar discussion applies to double cochain complexes $C = C^{pq}$. One has a total cochain complex TC with two decreasing filtrations, and hence two spectral sequences of cohomological type converging to $H^*(TC)$ (provided that C has only finitely many non-zero modules in any given total degree). Details are left to the reader.

Exercise

Let C be a first-quadrant double complex such that one of the associated spectral sequences (the first one, say) has $E^1_{pq} = 0$ for $q \neq 0$. Let D be the chain complex $E^1_{*,0}$ with differential d^1.

(a) Show that there is an isomorphism $\varphi: H_*(TC) \xrightarrow{\approx} H_*(D)$.

(b) Make this result more precise by showing that there is a weak equivalence $\tau\colon TC \to D$. [Hint: Take τ to be the canonical surjection which comes from the fact that D is the vertical 0-dimensional homology of C. Show that $\tau_* = \varphi$ by directly examining the definitions in the construction of the spectral sequence. Alternatively, you can avoid getting your hands dirty by observing that τ can be viewed as a map of double complexes $C \to D$ (where D is regarded as a double complex concentrated on the line $q = 0$); the induced map of spectral sequences is an isomorphism at the E^1-level, so τ induces an isomorphism $H_*(TC) \to H_*(TD) = H_*(D)$.]

4 Example: The Homology of a Union

Let X be a CW-complex which is the union of a family of non-empty subcomplexes X_α, where α ranges over some totally ordered index set J. In case $J = \{1, 2\}$, one has a short exact sequence of chain complexes

$$0 \to C(X_1 \cap X_2) \to C(X_1) \oplus C(X_2) \to C(X) \to 0$$

(where $C(\ )$ denotes the *cellular* chain complex), from which one obtains the familiar Mayer–Vietoris sequence. In the general case, we will show that the exact sequence above is replaced by an exact sequence

$$\cdots \to \bigoplus_{\alpha<\beta<\gamma} C(X_\alpha \cap X_\beta \cap X_\gamma) \to \bigoplus_{\alpha<\beta} C(X_\alpha \cap X_\beta) \to \bigoplus_{\alpha} C(X_\alpha) \to C(X) \to 0$$

and that the Mayer–Vietoris sequence is replaced by a spectral sequence. The details are as follows.

Let K be the abstract simplicial complex whose vertex set is J and whose simplices are the non-empty finite subsets σ of J such that the intersection $X_\sigma = \bigcap_{\alpha\in\sigma} X_\alpha$ is non-empty; K is called the *nerve* of the covering $\{X_\alpha\}$. For $p \geq 0$ let C_p be the chain complex $\bigoplus_{\sigma\in K^{(p)}} C(X_\sigma)$, where $K^{(p)}$ is the set of p-simplices of K. If σ has vertices $\alpha_0 < \cdots < \alpha_p$, we denote by $\partial_i\sigma$ $(0 \leq i \leq p)$ the $(p-1)$-simplex $\{\alpha_0, \ldots, \hat{\alpha}_i, \ldots, \alpha_p\}$. The inclusions $C(X_\sigma) \hookrightarrow C(X_{\partial_i\sigma})$ induce a chain map $\partial_i\colon C_p \to C_{p-1}$ for $p \geq 1$, and we set $\partial = \sum_{i=0}^{p}(-1)^i\partial_i$. Similarly, the inclusions $C(X_\alpha) \hookrightarrow C(X)$ induce a chain map $\varepsilon\colon C_0 \to C(X)$. We have, then, an augmented chain complex

$$\cdots \to C_p \xrightarrow{\partial} C_{p-1} \to \cdots \to C_0 \xrightarrow{\varepsilon} C(X) \to 0 \tag{4.1}$$

in the category of chain complexes, and hence a double complex C with C_{pq} equal to the group of q-chains of C_p, i.e., $C_{pq} = \bigoplus_{\sigma\in K^{(p)}} C_q(X_\sigma)$.

I claim now that 4.1 is exact. To see this, we give an alternative description of C_{pq}. For any cell e of X, let K_e be the subcomplex of K consisting of the simplices σ such that $e \subseteq X_\sigma$. Then C_{pq} has one basis element for every pair (σ, e) such that e is a q-cell of X and σ is a p-simplex of K_e. In other words, $C_{pq} \approx \bigoplus_{e\in X^{(q)}} C_p(K_e)$, where $X^{(q)}$ is the set of q-cells of X. Moreover, an examination of the definitions of the maps ∂ and ε in 4.1 shows that $\partial\colon C_{pq} \to$

$C_{p-1,q}$ is equal to $\bigoplus_{e \in X^{(q)}} \{\partial\colon C_p(K_e) \to C_{p-1}(K_e)\}$, and similarly for ε. Thus the q-dimensional part of 4.1,

$$\cdots \to C_{pq} \to \cdots \to C_{0q} \to C_q(X) \to 0,$$

is isomorphic to $\bigoplus_{e \in X^{(q)}} \tilde{C}(K_e)$, where $\tilde{C}(K_e)$ is the augmented simplicial chain complex of K_e. But K_e is acyclic. Indeed, its simplices are all of the finite subsets of the non-empty set $J_e = \{\alpha \in J : e \subseteq X_\alpha\}$, so K_e is the "simplex" spanned by the (possibly infinite) set J_e. Thus $\bigoplus_e \tilde{C}(K_e)$ is acyclic, which proves the claim.

Consider, now, the two spectral sequences of the double complex (C_{pq}). In view of the exactness of 4.1, the second spectral sequence has

$$E^1_{pq} = H_q(C_{*,p}) = \begin{cases} 0 & \text{if } q \neq 0 \\ C_p(X) & \text{if } q = 0. \end{cases}$$

Taking the homology with respect to p, we obtain

$$E^2_{pq} = \begin{cases} 0 & \text{if } q \neq 0 \\ H_p(X) & \text{if } q = 0. \end{cases}$$

The spectral sequence therefore "collapses" to yield:

$$H_*(TC) \approx H_*(X). \tag{4.2}$$

The first spectral sequence, on the other hand, has

$$E^1_{pq} = H_q(C_p) = \bigoplus_{\sigma \in K^{(p)}} H_q(X_\sigma). \tag{4.3}$$

Moreover, in view of the isomorphism 4.2, we can regard $H_*(X)$ as the abutment of the spectral sequence, i.e., $E^\infty \approx \operatorname{Gr} H_*(X)$ relative to some filtration on $H_*(X)$. This first spectral sequence, then, is our desired generalization of the Mayer–Vietoris sequence. It approximates $H_* X$ in terms of the homology of the X_α and their intersections.

To describe the E^2-term, we need the notion of "coefficient system" on a simplicial complex K. By this we will mean a family $\mathscr{A} = \{A_\sigma\}$ of abelian groups, where σ ranges over the simplices of K, together with a map $f_{\sigma\tau}\colon A_\sigma \to A_\tau$ whenever τ is a face of σ (written $\tau \subset \sigma$), such that $f_{\tau\rho} f_{\sigma\tau} = f_{\sigma\rho}$ if $\rho \subset \tau \subset \sigma$. There is an obvious way to construct a chain complex $C(K, \mathscr{A})$, with $C_p(K, \mathscr{A}) = \bigoplus_{\sigma \in K^{(p)}} A_\sigma$, and one therefore has homology groups $H_*(K, \mathscr{A})$. (See Godement [1958], I.3.3, for a detailed treatment of the cohomological version of this.)

Returning now to the situation where K is the nerve of the covering $\{X_\alpha\}$, we have for each $q \geq 0$ a coefficient system $\mathscr{H}_q = \{H_q(X_\sigma)\}$ on K, where $f_{\sigma\tau}\colon H_q(X_\sigma) \to H_q(X_\tau)$ is induced by the inclusion $X_\sigma \hookrightarrow X_\tau$. It is immediate from the definition that the E^1-term in 4.3 is equal to $C_p(K, \mathscr{H}_q)$, hence $E^2_{pq} = H_p(K, \mathscr{H}_q)$. We summarize this discussion by writing

$$E^2_{pq} = H_p(K, \mathscr{H}_q) \Rightarrow H_{p+q}(X),$$

where the symbol "$\Rightarrow$" indicates that the spectral sequence converges to what appears on the right-hand side.

In case each X_σ is acyclic, we have $\mathscr{H}_0 = \mathbb{Z}$ and $\mathscr{H}_q = 0$ for $q \neq 0$. The spectral sequence therefore collapses at E^2 to yield the following result, which seems to be essentially due to Leray:

(4.4) Theorem. *Suppose X is the union of subcomplexes X_α such that every non-empty intersection $X_{\alpha_0} \cap \cdots \cap X_{\alpha_p}$ $(p \geq 0)$ is acyclic. Then $H_*(X) \approx H_*(K)$, where K is the nerve of the cover.*

EXERCISES

1. Make 4.4 more precise by showing that, under the given hypotheses, there is a chain complex T which admits weak equivalences $T \to C(X)$ and $T \to C(K)$. [Take $T = TC$, where C is the double complex used in this section, and use part (b) of the exercise of §3.]

*2. Make 4.4 still more precise by showing that there is a CW-complex Y which maps to both K and X by maps inducing homology isomorphisms. [Hint: Let Y be the union of the cells $\sigma \times e$ in $K \times X$ such that $e \subseteq X_\sigma$ (or, equivalently, such that σ is in K_e); note that $C(Y)$ can be identified with the complex T of exercise 1.]

Remark. The map $Y \to X$ is in fact a homotopy equivalence. This can be seen, for example, by an inductive argument over the skeleta of X, using the fact that each complex K_e is contractible. Consequently, one obtains a canonical homotopy class of maps $X \to K$ inducing a homology isomorphism. In case each X_σ is contractible (and not just acyclic), one can similarly show that the map $Y \to K$ is a homotopy equivalence, so that X and K are homotopy equivalent. This sort of result was first proved by Weil [1952], §5; see also Borel–Serre [1974], §8, and Quillen [1978], 1.6–1.8.

5 Homology of a Group with Coefficients in a Chain Complex

Recall that $H_*(G, M)$ is defined to be $H_*(F \otimes_G M)$, where F is a projective resolution of $\mathbb{Z}$ over $\mathbb{Z}G$. It is useful to generalize this by allowing a non-negative chain complex $C = (C_n)_{n \geq 0}$ of coefficients. Thus we set

$$H_*(G, C) = H_*(F \otimes_G C);$$

as usual, this is well-defined up to canonical isomorphism. If C consists of a single module M concentrated in dimension 0, then $H_*(G, C)$ reduces to $H_*(G, M)$.

Since $F \otimes_G C$ is the total complex of the double complex of abelian groups $(F_p \otimes_G C_q)$, we have by §3 two spectral sequences converging to $H_*(G, C)$. The first of these has $E^1_{pq} = H_q(F_p \otimes_G C_*) = F_p \otimes_G H_q(C)$ since $F_p \otimes_G -$ is an exact functor. Taking now the homology with respect to p, we obtain $E^2_{pq} = H_p(G, H_q C)$. Thus the spectral sequence has the form

$$E^2_{pq} = H_p(G, H_q C) \Rightarrow H_{p+q}(G, C). \tag{5.1}$$

An important consequence of this is that $H_*(G, C)$ is an invariant of the "weak homotopy type" of C:

(5.2) Proposition. *If $\tau: C \to C'$ is a weak equivalence of G-chain complexes, then τ induces an isomorphism $H_*(G, C) \xrightarrow{\approx} H_*(G, C')$.*

PROOF. τ induces a map of spectral sequences which is an isomorphism at the E^2-level, hence τ induces an isomorphism on the abutments by 2.6. □

[This result also follows, of course, from I.8.6, but it still provides a nice illustration of spectral sequence techniques.]

The second spectral sequence has $E^1_{pq} = H_q(F_* \otimes_G C_p) = H_q(G, C_p)$. Thus we have:

$$E^1_{pq} = H_q(G, C_p) \Rightarrow H_{p+q}(G, C). \tag{5.3}$$

The group E^2_{pq} can therefore be described as the p-th homology group of the complex obtained from C by applying the functor $H_q(G, —)$ dimension-wise. Both spectral sequences, then, can be thought of as giving approximations to $H_*(G, C)$ in terms of *ordinary* homology groups $H_*(G, M)$.

To get a better feeling for $H_*(G, C)$, let's look at some special cases. Suppose, first, that G acts trivially on C. Then $F \otimes_G C \approx F_G \otimes C$, so there is a *Künneth formula*

$$0 \to \bigoplus_{p+q=n} H_p G \otimes H_q C \to H_n(G, C) \to \bigoplus_{p+q=n-1} \operatorname{Tor}(H_p G, H_q C) \to 0$$

expressing $H_*(G, C)$ in terms of $H_* G$ and $H_* C$. Similarly, if k is a field and C is a complex of k-vector spaces (with trivial G-action), then

$$H_*(G,C) \approx H_*(G, k) \otimes_k H_* C. \tag{5.4}$$

At the other extreme, suppose that each C_p is a *free* $\mathbb{Z}G$-module, or, more generally, an H_*-acyclic G-module (e.g., projective or induced). The E^1-term in 5.3 is then concentrated on the line $q = 0$, and $E^1_{p,0} = (C_p)_G$. The spectral sequence therefore collapses at E^2 to yield:

$$H_*(G, C) \approx H_*(C_G). \tag{5.5}$$

The abutment of the spectral sequence (5.1) can therefore be identified with $H_*(C_G)$; this proves:

(5.6) Proposition. *Let C be a non-negative chain complex of G-modules such that each C_n is H_*-acyclic. Then there is a spectral sequence of the form*

$$E^2_{pq} = H_p(G, H_q C) \Rightarrow H_{p+q}(C_G).$$

Remark. This spectral sequence is a special case of the "universal coefficient spectral sequence," cf. Godement [1958], I.5.5.1.

We can also define *cohomology groups* $H^*(G, C)$, where $C = (C^n)_{n \geq 0}$ is a non-negative cochain complex; namely, we set

$$H^*(G, C) = H^*(\mathcal{H}om_G(F, C)),$$

where F is a projective resolution of $\mathbb{Z}$ over $\mathbb{Z}G$. Now $\mathcal{H}om_G(F, C)$ is in fact the total complex associated to a double complex with $C^{pq} = \operatorname{Hom}_G(F_q, C^p)$. Indeed, we have $\mathcal{H}om_G(F, C)^n = \mathcal{H}om_G(F, C)_{-n} = \prod_{p+q=n} \operatorname{Hom}_G(F_q, C_{-p}) = \bigoplus_{p+q=n} \operatorname{Hom}_G(F_q, C^p)$. And if we take δ' to be the coboundary operator in $\mathcal{H}om_G(F_q, C)$ [where q is fixed and the module F_q is regarded as a chain complex concentrated in dimension 0] and δ'' to be the coboundary operator in $\mathcal{H}om_G(F, C^p)$, then the resulting total coboundary operator is the same as that in $\mathcal{H}om_G(F, C)$. One can now easily deduce cohomology analogues of 5.1–5.6.

Similarly, if G is finite, one can define *Tate cohomology groups* $\hat{H}^*(G, C)$ by using a *complete* resolution F. We assume, here, that $C^n = 0$ for $n \gg 0$, so that the double complex $\operatorname{Hom}_G(F_p, C^q)$ has only finitely many non-zero groups in each total degree.

Finally, we remark that all of the formal properties of homology and cohomology given in Chapters III and V extend without difficulty to homology and cohomology with coefficients in a (co-) chain complex. In particular, one has long exact sequences, Shapiro's lemma, restriction and corestriction maps, cup products, etc. Moreover, the cohomology spectral sequences analogous to 5.1 and 5.3 are compatible with the cup products. More precisely, suppose C and D are G-cochain complexes, and consider the cochain cup product

$$\mathcal{H}om_G(F, C) \otimes \mathcal{H}om_G(F, D) \to \mathcal{H}om_G(F, C \otimes D)$$

defined via some diagonal approximation $F \to F \otimes F$. One checks easily that this cup product is compatible with the filtrations defining the two spectral sequences, i.e., that it induces a product

$$F^p\mathcal{H}om_G(F, C) \times F^{p'}\mathcal{H}om_G(F, D) \to F^{p+p'}\mathcal{H}om_G(F, C \otimes D).$$

It follows that there is an induced cup product

$$E_r^{pq}(C) \otimes E_r^{p'q'}(D) \to E_r^{p+p', q+q'}(C \otimes D),$$

where $\{E_r(\ \ )\}$ denotes the cohomology version of either 5.1 or 5.3. We will see applications of this product structure in Chapter X.

6 Example: The Hochschild–Serre Spectral Sequence

Let $1 \to H \to G \to Q \to 1$ be a group extension. If F is a projective resolution of $\mathbb{Z}$ over $\mathbb{Z}G$ and M is a G-module, then $F \otimes_G M = (F \otimes M)_G$ can be computed in two steps (cf. §II.2, exercise 3), by first dividing out by the action of H on $F \otimes M$ and then dividing out by the action of Q:

$$F \otimes_G M = ((F \otimes M)_H)_Q = (F \otimes_H M)_Q.$$

Thus

$$H_*(G, M) = H_*(C_Q), \tag{6.1}$$

where $C = F \otimes_H M$. Note also that we have a Q-module isomorphism

$$H_*(H, M) \approx H_*(C), \tag{6.2}$$

where the Q-action on $H_*(H, M)$ is that of III.8.2. Finally, I claim that the Q-modules $C_p = (F_p \otimes M)_H$ are H_*-acyclic. In fact, it suffices to show that $(\mathbb{Z}G \otimes M)_H$ is H_*-acyclic; and for this one need only observe (by using III.5.7, for example) that $(\mathbb{Z}G \otimes M)_H$ is an induced Q-module $\mathbb{Z}Q \otimes A$.

We are now in a position to apply 5.6 to the Q-complex C. In view of 6.1 and 6.2 we obtain:

(6.3) Theorem (Hochschild–Serre [1953]). *For any group extension $1 \to H \to G \to Q \to 1$ and any G-module M, there is a spectral sequence of the form*

$$E^2_{pq} = H_p(Q, H_q(H, M)) \Rightarrow H_{p+q}(G, M).$$

(There is, of course, an analogous spectral sequence in cohomology, derived from the observation that $\mathcal{H}om_G(F, M) = (\mathcal{H}om_H(F, M))^Q$.)

An important consequence of 6.3 is the following *5-term exact sequence* of low-dimensional homology groups, generalizing the 5-term exact sequence of exercise 6 of §II.5:

(6.4) Corollary. *Under the hypotheses of* 6.3 *there is an exact sequence*

$$H_2(G, M) \to H_2(Q, M_H) \to H_1(H, M)_Q \to H_1(G, M) \to H_1(Q, M_H) \to 0.$$

Proof. Since $E^\infty = \operatorname{Gr} H(G, M)$, we have a short exact sequence

$$0 \to E^\infty_{0,1} \to H_1(G, M) \to E^\infty_{1,0} \to 0.$$

Now there are no non-zero differentials involving $E^r_{1,0}$ $(r \geq 2)$, so

$$E^\infty_{1,0} = E^2_{1,0}.$$

And the only possible non-zero differential involving $E^r_{0,1}$ or $E^r_{2,0}$ is

$$d^2\colon E^2_{2,0} \to E^2_{0,1},$$

so there is an exact sequence

$$0 \to E^\infty_{2,0} \to E^2_{2,0} \xrightarrow{d^2} E^2_{0,1} \to E^\infty_{0,1} \to 0.$$

The short-exact sequence above now yields an exact sequence

$$0 \to E^\infty_{2,0} \to E^2_{2,0} \to E^2_{0,1} \to H_1(G, M) \to E^2_{1,0} \to 0.$$

Since $E^2_{pq} = H_p(Q, H_q(H, M))$ and $E^\infty_{2,0}$ is a quotient of $H_2(G, M)$, this yields the desired 5-term exact sequence. □

7 Equivariant Homology

We now specialize the theory of §5 to the case where the chain complex C is the cellular chain complex $C(X)$ of a G-complex X (cf. §I.4). The resulting homology groups $H_*(G, C(X))$ are denoted $H^G_*(X)$ and called the *equivariant homology groups* of (G, X). More generally, if M is a G-module, then there is a diagonal G-action on $C(X, M) = C(X) \otimes M$, and we set

$$H^G_*(X, M) = H_*(G, C(X, M)).$$

Similarly, the *equivariant cohomology groups* are defined by

$$H^*_G(X, M) = H^*(G, C^*(X, M)).$$

In case X is finite dimensional and G is finite, we can also define *equivariant Tate cohomology groups* by

$$\hat{H}^*_G(X, M) = \hat{H}^*(G, C^*(X, M)).$$

Finally, if $Y \subseteq X$ is a G-invariant subcomplex, then there are relative equivariant homology and cohomology groups, defined by means of the relative complex $C(X, Y) = C(X)/C(Y)$.

For simplicity, we will confine ourselves in this section to a discussion of the (absolute) homology groups $H^G_*(X, M)$. There are, of course, relative versions, as well as cohomology analogues, of all the results.

Note first that $H^G_*(\text{pt.}, M) = H_*(G, M)$. Since any G-complex X admits a (unique) G-map to a point, we deduce that there is a *canonical map*

(7.1) $$H^G_*(X, M) \to H_*(G, M).$$

Next we record the spectral sequence 5.1 in the present context:

(7.2) $$E^2_{pq} = H_p(G, H_q(X, M)) \Rightarrow H^G_{p+q}(X, M).$$

(Note that the E^2-term here involves the *diagonal action* of G on $H_*(X, M)$, induced by the action of G on X and M.)

As a consequence of 7.2, we have (cf. 5.2):

(7.3) Proposition. *If $f: X \to Y$ is a cellular map of G-complexes such that $f_*: H_*X \to H_*Y$ is an isomorphism, then f induces an isomorphism $H_*^G(X, M) \xrightarrow{\approx} H_*^G(Y, M)$ for any G-module M. In particular, if X is acyclic then the canonical map 7.1 is an isomorphism $H_*^G(X, M) \xrightarrow{\approx} H_*(G, M)$.*

To analyze the second spectral sequence (5.3), we decompose $C_p(X) \otimes M$ as in III.5.5b. For each p-cell σ of X we have a G_σ-module $\mathbb{Z}_\sigma$ which is additively isomorphic to $\mathbb{Z}$, on which G_σ operates by the "orientation character" $\chi_\sigma: G_\sigma \to \{\pm 1\}$. Let

$$M_\sigma = \mathbb{Z}_\sigma \otimes M;$$

thus M_σ is a G_σ-module which is additively isomorphic to M, with the G_σ-action "twisted" by χ_σ. Let X_p be the set of p-cells of X and let Σ_p be a set of representatives for X_p/G. We then have an additive decomposition

$$C_p(X, M) = C_p(X) \otimes M = \bigoplus_{\sigma \in X_p} M_\sigma, \tag{7.4}$$

from which we obtain a G-module decomposition

$$C_p(X, M) \approx \bigoplus_{\sigma \in \Sigma_p} \operatorname{Ind}_{G_\sigma}^G M_\sigma. \tag{7.5}$$

Shapiro's lemma now yields

$$H_q(G, C_p(X, M)) \approx \bigoplus_{\sigma \in \Sigma_p} H_q(G_\sigma, M_\sigma), \tag{7.6}$$

so that 5.3 takes the form:

$$E_{pq}^1 = \bigoplus_{\sigma \in \Sigma_p} H_q(G_\sigma, M_\sigma) \Rightarrow H_{p+q}^G(X, M). \tag{7.7}$$

Suppose, for example, that the G-action is free, so that each $G_\sigma = \{1\}$, and assume for simplicity that $M = \mathbb{Z}$. The spectral sequence then collapses at E^2 to yield (cf. 5.5):

$$H_*^G(X) \approx H_*(C(X)_G) = H_*(X/G). \tag{7.8}$$

Combining 7.8 and 7.2, we conclude (cf. 5.6):

(7.9) Theorem. *If X is a free G-complex, then there is a spectral sequence of the form*

$$E_{pq}^2 = H_p(G, H_q X) \Rightarrow H_{p+q}(X/G).$$

Remark. This spectral sequence is called the *Cartan–Leray spectral sequence* associated to the regular covering map $X \to X/G$. Theorem 7.9 remains true if we introduce an arbitrary G-module M of coefficients, but the resulting homology groups $H_*(X/G, M)$ have to be interpreted as homology groups with local coefficients if G acts non-trivially on M.

Consider now the situation where X is acyclic (but the action of G is not necessarily free), so that

$$H_*^G(X, M) \approx H_*(G, M)$$

by 7.3. The spectral sequence 7.7 can then be rewritten:

$$E_{pq}^1 = \bigoplus_{\sigma \in \Sigma_p} H_q(G_\sigma, M_\sigma) \Rightarrow H_{p+q}(G, M). \tag{7.10}$$

This is an important computational tool in the homology theory of groups. To use it, of course, one must find an interesting acyclic space X on which G acts. If G is an amalgamated free product, for example, we will see in §9 that a suitable choice of X leads to the Mayer–Vietoris sequence which we derived earlier from a different point of view (II.7.7 and the exercise of §III.6).

Exercises

1. Consider the "left-hand edge" $E_{0,*}^r$ of the spectral sequence 7.7. Since d^r is zero on this edge for $r \geq 1$, the spectral sequence gives us maps

$$\bigoplus_{v \in \Sigma_0} H_*(G_v, M) = E_{0,*}^1 \twoheadrightarrow E_{0,*}^\infty = \mathrm{Gr}_0 H_*^G(X, M) \hookrightarrow H_*^G(X, M).$$

[We have written $H_*(G_v, M)$ here instead of $H_*(G_v, M_v)$ because a zero-cell has a unique orientation, so that M_v is canonically isomorphic to M as a G_v-module.] The composite

$$\bigoplus_{v \in \Sigma_0} H_*(G_v, M) \to H_*^G(X, M)$$

is one of the two "edge homomorphisms" associated to the spectral sequence. Show that its restriction to $H_*(G_v, M)$ is the map

$$H_*(G_v, M) = H_*^{G_v}(v, M) \to H_*^G(X, M)$$

induced by the inclusion $(G_v, v) \hookrightarrow (G, X)$. [Hint: This can be done by straightforward (but tedious) definition-checking. It is easier, however, to use a naturality argument to reduce to the case where X consists of the single vertex v, in which case the definition-checking is trivial.]

2. Let X be a G-complex such that for each cell σ of X, the isotropy group G_σ fixes σ pointwise. In this case it is easy to see that the orbit space X/G inherits a CW-structure. If, in addition, each G_σ is finite, show that

$$H_*^G(X, \mathbb{Q}) \approx H_*(X/G, \mathbb{Q}).$$

Deduce, in particular, that

$$H_*(G, \mathbb{Q}) \approx H_*(X/G, \mathbb{Q})$$

if X is contractible. [Remark: The hypothesis that G_σ fixes σ pointwise is not very restrictive in practice. In the case of a simplicial action, for example, it can always be achieved by passage to the barycentric subdivision.]

3. Show that $H_*^G(X) \approx H_*(X \times^G E)$, where E is a free, contractible G-complex (i.e., the universal cover of a $K(G, 1)$) and $X \times^G E$ denotes $X \times E$ modulo the diagonal G-action. [Hint: There is a G-map $X \times E \to X$ which is a homotopy equivalence, and G acts freely on $X \times E$.]

Remark. The isomorphism of exercise 3 is often taken as a definition. From this point of view one obtains a spectral sequence of the form 7.2 by noting that there is a fiber bundle $X \times^G E \to K(G, 1)$ with fiber X.

4. (a) Let X be a G-space (not necessarily a G-complex). Use singular chains to define equivariant singular homology $H_*^G(X)$ and construct a spectral sequence of the form 7.2.

(b) Suppose that G acts freely on X and that the projection $X \to X/G$ is a covering map. (It is then a regular G-cover in the sense of the appendix to Chapter I.) Prove that $H_*^G(X) \approx H_*(X/G)$. [Hint: 5.5.] Deduce that there is a Cartan–Leray spectral sequence (as in 7.9) in singular homology.

*(c) Suppose $X = S^{2k-1}$ in (b). Prove that G has periodic cohomology of period $2k$. [Method 1: X/G is a finite dimensional manifold and hence has vanishing homology in high dimensions, with arbitrary local coefficients. The differential in the spectral sequence therefore yields $H_i(G, M) \approx H_{i+2k}(G, M)$ for $i \gg 0$. Method 2: We have a sphere bundle $S^{2k-1} \times^G E \to K(G, 1)$ (with E as in exercise 3) whose total space is homotopy equivalent to S^{2k-1}/G. Up to homotopy, then, the inclusion of the fiber in the total space is a map of degree $|G|$ of orientable $(2k-1)$-manifolds. The Gysin sequence now shows that the Euler class of the sphere bundle is an element of order $|G|$ in $H^{2k}(G, \mathbb{Z})$, so the result follows from VI.9.1(iv). See Jackowski [1978] for further results obtainable from the study of sphere bundles over $K(G, 1)$.]

5. Let X be a G-complex. If N is a normal subgroup of G which acts freely on X, show that $H_*^G(X) \approx H_*^{G/N}(X/N)$ with any G/N-module of coefficients. [Hint: Apply 7.7 to (G, X) and $(G/N, X/N)$.]

6. Let Y be a connected CW-complex with $\pi_i Y = 0$ for $1 < i < n$ as in II.5.2. Deduce from the Cartan–Leray spectral sequence of the universal cover of Y that there is a 5-term exact sequence

$$H_{n+1}Y \to H_{n+1}\pi \to (\pi_n Y)_\pi \to H_n Y \to H_n\pi \to 0,$$

where $\pi = \pi_1 Y$.

8 Computation of d^1

Let X be a G-complex and M a G-module. In this section we will compute the differential d^1 in the spectral sequence 7.7.

If σ is a p-cell of X and τ is a $(p-1)$-cell, we denote by $\partial_{\sigma\tau}: M_\sigma \to M_\tau$ the (σ, τ)-component of the boundary operator $\partial: C_p(X, M) \to C_{p-1}(X, M)$

(cf. 7.4). Let $\mathscr{F}_\sigma = \{\tau : \partial_{\sigma\tau} \neq 0\}$. This is a finite set of $(p-1)$-cells and is G_σ-invariant. Writing $G_{\sigma\tau} = G_\sigma \cap G_\tau$, we conclude that $(G_\sigma : G_{\sigma\tau}) < \infty$ for $\tau \in \mathscr{F}_\sigma$. There is therefore a transfer map

$$t_{\sigma\tau}: H_*(G_\sigma, M_\sigma) \to H_*(G_{\sigma\tau}, M_\sigma).$$

Next, note that $\partial_{\sigma\tau}: M_\sigma \to M_\tau$ is a $G_{\sigma\tau}$-map, since ∂ is a G-map; so $\partial_{\sigma\tau}$ and the inclusion $G_{\sigma\tau} \hookrightarrow G_\tau$ induce a map

$$u_{\sigma\tau}: H_*(G_{\sigma\tau}, M_\sigma) \to H_*(G_\tau, M_\tau).$$

Finally, let $\tau_0 \in \Sigma_{p-1}$ be the representative of the G-orbit of τ, and choose $g(\tau) \in G$ such that $g(\tau)\tau = \tau_0$. Then the action of $g(\tau)$ on $C_{p-1}(X, M)$ gives an isomorphism $M_\tau \to M_{\tau_0}$ compatible with the conjugation isomorphism $G_\tau \to G_{\tau_0}$ given by $g \mapsto g(\tau)gg(\tau)^{-1}$; thus there is an induced isomorphism

$$v_\tau: H_*(G_\tau, M_\tau) \to H_*(G_{\tau_0}, M_{\tau_0}).$$

[As the notation suggests, one can show that v_τ depends only on τ, and not on the choice of $g(\tau)$; but we will not need to know this].

We can now define a map

$$\varphi: \bigoplus_{\sigma \in \Sigma_p} H_*(G_\sigma, M_\sigma) \to \bigoplus_{\tau \in \Sigma_{p-1}} H_*(G_\tau, M_\tau)$$

by

$$\varphi \mid H_*(G_\sigma, M_\sigma) = \sum_{\tau \in \mathscr{F}'_\sigma} v_\tau u_{\sigma\tau} t_{\sigma\tau},$$

where $\mathscr{F}'_\sigma$ is a set of representatives for $\mathscr{F}_\sigma / G_\sigma$.

(8.1) Proposition. *Up to sign, the map φ is the differential $d^1: E^1_{p,*} \to E^1_{p-1,*}$ in 7.7.*

Proof. What we must prove is that the diagram

$$\begin{array}{ccc} \bigoplus_{\sigma \in \Sigma_p} H_*(G_\sigma, M_\sigma) & \xrightarrow{\varphi} & \bigoplus_{\tau \in \Sigma_{p-1}} H_*(G_\tau, M_\tau) \\ \downarrow \approx & & \downarrow \approx \\ H_*(G, C_p(X, M)) & \xrightarrow{H_*(G, \partial)} & H_*(G, C_{p-1}(X, M)) \end{array}$$

commutes, where the vertical isomorphisms come from 7.6. We will in fact prove commutativity of the corresponding diagram of chain complexes. More precisely, if F is a projective resolution of $\mathbb{Z}$ over $\mathbb{Z}G$, we will compute φ on the chain level as a map (still denoted φ)

$$\bigoplus_{\sigma \in \Sigma_p} F \otimes_{G_\sigma} M_\sigma \to \bigoplus_{\tau \in \Sigma_{p-1}} F \otimes_{G_\tau} M_\tau,$$

and we will prove that the diagram

$$\begin{array}{ccc}
\bigoplus_{\sigma \in \Sigma_p} F \otimes_{G_\sigma} M_\sigma & \xrightarrow{\varphi} & \bigoplus_{\tau \in \Sigma_{p-1}} F \otimes_{G_\tau} M_\tau \\
\Big\downarrow \theta & & \Big\downarrow \psi \\
F \otimes_G C_p(X, M) & \xrightarrow{F \otimes \partial} & F \otimes_G C_{p-1}(X, M)
\end{array}$$

commutes. Here θ and ψ are the chain isomorphisms underlying 7.6; explicitly, a glance at the proof of Shapiro's lemma (III.6.2) shows that θ and ψ are simply the maps induced by the inclusions $M_\sigma \hookrightarrow C_p(X) \otimes M$ and $M_\tau \hookrightarrow C_{p-1}(X) \otimes M$.

We begin by computing the transfer map $t_{\sigma\tau}$ on the chain level by means of Definition (C) of §III.9. For $x \in F$ and $m \in M_\sigma$, we find

$$t_{\sigma\tau}(x \otimes m) = \sum_{g \in G_\sigma/G_{\sigma\tau}} g^{-1}x \otimes g^{-1}m \in F \otimes_{G_{\sigma\tau}} M_\sigma.$$

Next, the map $u_{\sigma\tau}$ is given on the chain level by $F \otimes \partial_{\sigma\tau}: F \otimes_{G_{\sigma\tau}} M_\sigma \to F \otimes_{G_\tau} M_\tau$. Thus

$$\begin{aligned}
u_{\sigma\tau} t_{\sigma\tau}(x \otimes m) &= \sum_{g \in G_\sigma/G_{\sigma\tau}} g^{-1}x \otimes \partial_{\sigma\tau}(g^{-1}m) \\
&= \sum_{g \in G_\sigma/G_{\sigma\tau}} g^{-1}x \otimes g^{-1}\partial_{\sigma, g\tau}(m),
\end{aligned}$$

where the second equality comes from the fact that ∂ is a G-map. We apply now the map v_τ, which can be computed via the "diagonal action" of $g(\tau)$ (cf. §III.8, paragraph preceding III.8.1), and we sum over $\tau \in \mathscr{F}'_\sigma$; this gives

$$\varphi(x \otimes m) = \sum_{\tau \in \mathscr{F}'_\sigma} \sum_{g \in G_\sigma/G_{\sigma\tau}} g(\tau)g^{-1}x \otimes g(\tau)g^{-1}\partial_{\sigma, g\tau}(m).$$

We can now establish the desired commutative diagram of chain complexes. Using the tensor product relations $gu \otimes gv = u \otimes v$ in $- \otimes_G -$, we deduce from the formula for φ above that

$$\psi\varphi(x \otimes m) = \sum_{\tau \in \mathscr{F}'_\sigma} \sum_{g \in G_\sigma/G_{\sigma\tau}} x \otimes \partial_{\sigma, g\tau}(m).$$

But it is clear from the definition of $\mathscr{F}'_\sigma$ that the right-hand side of this equation is simply

$$\sum_{\tau \in \mathscr{F}_\sigma} x \otimes \partial_{\sigma\tau}(m) = x \otimes \partial m.$$

Thus $\psi\varphi = (F \otimes \partial)\theta$, as required. □

The description of E^1 and d^1 simplifies greatly if X has the property that G_σ fixes σ pointwise for every cell σ, as in exercise 2 of §7. Indeed, one can then orient each cell of X in such a way that the G-action preserves orientations, and such a choice of orientations determines for each σ an isomorphism

$M_\sigma \approx M$ of G_σ-modules. We can therefore write $H_q(G_\sigma, M)$ instead of $H_q(G_\sigma, M_\sigma)$. More significantly, we have $G_{\sigma\tau} = G_\sigma$ for $\tau \in \mathscr{F}_\sigma$, since such a τ is necessarily in the closure of σ and hence is fixed by G_σ. Thus the transfer map $t_{\sigma\tau}$ is the identity map, and also $\mathscr{F}'_\sigma = \mathscr{F}_\sigma$.

Roughly speaking, then, we have a "coefficient system"

$$\mathscr{H}_q = \{H_q(G_\sigma, M)\}$$

on the orbit complex X/G, and (E^1, d^1) is the chain complex of X/G with coefficients in this system. The E^2-term of the spectral sequence can therefore be thought of as $H_p(X/G, \mathscr{H}_q)$. We will not attempt to make these remarks precise.

9 Example: Amalgamations

Consider an amalgamation $G = H *_A K$, where $A \hookrightarrow H$ and $A \hookrightarrow K$. Let X be the tree associated to G (cf. appendix to Chapter II), and recall that G acts on X with no "inversions," i.e., no element of G interchanges the vertices of a 1-simplex of X. [Thus we are in the situation described at the end of §8.] There is a single 1-simplex e which maps isomorphically onto the quotient graph X/G, and the isotropy groups of e and its vertices v and w are given by

$$G_v = H, \quad G_w = K, \quad \text{and} \quad G_e = A.$$

We therefore have, for any G-module M, a spectral sequence converging to $H_*(G, M)$ (cf. 7.10), with

$$E^1_{p,*} = \begin{cases} H_*(H, M) \oplus H_*(K, M) & \text{if } p = 0 \\ H_*(A, M) & \text{if } p = 1 \\ 0 & \text{if } p > 1. \end{cases}$$

(We have taken, here, $\Sigma_0 = \{v, w\}$ and $\Sigma_1 = \{e\}$.) The spectral sequence therefore collapses at E^2 to yield a Mayer–Vietoris sequence

(9.1) $$\cdots \to H_n(A, M) \xrightarrow{\alpha} H_n(H, M) \oplus H_n(K, M) \xrightarrow{\beta} H_n(G, M) \to H_{n-1}(A, M) \to \cdots.$$

The map β here is the edge homomorphism discussed in exercise 1 of §7, and hence it is the map induced by the inclusions $H \hookrightarrow G$ and $K \hookrightarrow G$. The map α is the differential d^1 given by 8.1. The latter is particularly easy to apply here because the maps $t_{\sigma\tau}$ and v_τ of §8 are identity maps. Since $\partial_{e,w}\colon M \to M$ is id_M and $\partial_{e,v}\colon M \to M$ is $-\mathrm{id}_M$, we conclude that α is simply the difference of the maps $H_*(A) \to H_*(H)$ and $H_*(A) \to H_*(K)$ induced by the inclusions $A \hookrightarrow H$ and $A \hookrightarrow K$.

Next we wish to briefly describe a generalization of the notion of amalgamation, for which one can also construct a Mayer–Vietoris sequence. By a *graph of groups* we mean a connected graph Y together with (a) groups G_v and G_e, where v (resp. e) ranges over the vertices (resp. edges) of Y, and (b) monomorphisms $\theta_0: G_e \hookrightarrow G_v$ and $\theta_1: G_e \hookrightarrow G_w$ for every edge e, where v and w are the vertices of e. [Note: We allow the possibility that $v = w$, but θ_0 and θ_1 may still differ.] One can then define the *fundamental group* G of the graph of groups by using a suitable notion of "path": one regards an element of G_v as a path from v to v, and one imposes the relations

$$\theta_0(a)e = e\theta_1(a)$$

for every $a \in G_e$, i.e., one identifies the composite paths

$\theta_0(a)$ ○ v —e→ w and v —e→ w ○ $\theta_1(a)$

from v to w. (See Serre [1977a] for more details.)

For example, the amalgam $H *_A K$ is the fundamental group of the graph

H —A— K

where we have labelled the vertices and edge with the associated groups. (In this case θ_0 and θ_1 are the inclusions $A \hookrightarrow H, A \hookrightarrow K$.)

Another important group theoretic construction which fits into this framework is the *HNN extension* $H *_A$. Here we are given a group H, a subgroup A, and a monomorphism $\theta: A \to H$, and $H *_A$ is obtained by adjoining an element t to H subject to the relations $t^{-1}at = \theta(a)$ $(a \in A)$. This group can be realized as the fundamental group of the graph

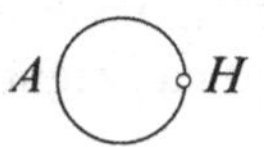

with one vertex and one edge, where θ_0 and θ_1 are the inclusion $i: A \hookrightarrow H$ and the given map $\theta: A \to H$.

Just as in the case of amalgamation, there is a tree X associated to a graph of groups, and the fundamental group G acts on X without inversion. The original graph Y is the orbit graph X/G, and the groups G_v and G_e are the isotropy subgroups of G at suitable liftings to X of the vertices and edges of Y. Consequently, one obtains from 7.10 a Mayer–Vietoris sequence

$$\cdots \to \bigoplus_{e \in Y_1} H_n(G_e, M) \to \bigoplus_{v \in Y_0} H_n(G_v, M) \to H_n(G, M) \to \bigoplus_{e \in Y_1} H_{n-1}(G_e, M) \to \cdots, \tag{9.2}$$

generalizing 9.1 (Here Y_i is the set of i-cells of Y.) This exact sequence was first written down explicitly by Chiswell [1976a].

[This exact sequence can, of course, be derived without the use of spectral sequences. Namely, one argues as in the exercise of §III.6, using the exact sequence of G-modules

$$(9.3) \qquad 0 \to \bigoplus_{e \in Y_1} \mathbb{Z}[G/G_e] \to \bigoplus_{v \in Y_0} \mathbb{Z}[G/G_v] \to \mathbb{Z} \to 0$$

given by the augmented chain complex of the tree X.]

Consider, for example, the HNN extension $G = H *_A$ described above. The sequence 9.2 then has the form

$$(9.4) \qquad \cdots \to H_n(A, M) \xrightarrow{\alpha} H_n(H, M) \xrightarrow{\beta} H_n(G, M) \to H_{n-1}(A, M) \to \cdots.$$

As in 9.1, β is induced by the inclusion $H \to G$. To evaluate $\alpha = d^1$, we need to know the following fact about the tree X: the unique edge of Y lifts to an edge e of X whose vertices have the form v, tv. Taking $\Sigma_0 = \{v\}$ and $\Sigma_1 = \{e\}$ in the notation of §8, we find that v_τ $(\tau = tv)$ is the conjugation isomorphism $c(t^{-1})_*: H_*(tHt^{-1}, M) \to H_*(H, M)$. Since the composite

$$(A, M) \to (tHt^{-1}, M) \xrightarrow{c(t^{-1})} (H, M)$$

is equal to (θ, t^{-1}), it follows that $\alpha = (\theta, t^{-1})_* - i_*$.

10 Equivariant Tate Cohomology

Recall from §7 that $\hat{H}_G^*(X, M)$ is defined if G is finite and the G-complex X is finite dimensional. This theory was introduced by Swan [1960a], who in fact constructed a *topological* version of the theory (i.e., X is not required to be a CW-complex).

The usefulness of equivariant Tate cohomology comes from the fact that it provides a machine for systematically ignoring free actions:

(10.1) Proposition. *Let Y be a G-invariant subcomplex of X such that the isotropy group G_σ is trivial for every cell σ of X that is not in Y. Then the inclusion $Y \hookrightarrow X$ induces an isomorphism $\hat{H}_G^*(X, M) \xrightarrow{\approx} \hat{H}_G^*(Y, M)$.*

Proof. Consider the Tate cohomology version of the spectral sequence 7.7 for (G, X) and (G, Y). By hypothesis, $\hat{H}^*(G_\sigma, M) = 0$ if σ is cell of $X - Y$. Thus the inclusion $Y \hookrightarrow X$ induces an isomorphism of spectral sequences and hence an isomorphism of the abutments $\hat{H}_G^*(X, M)$ and $\hat{H}_G^*(Y, M)$. □

As a simple illustration of 10.1, we prove the following result (cf. §VI.8, exercise 4):

(10.2) Proposition. *Let G be a finite group which admits a finite dimensional free G-complex X such that $H_*(X) \approx H_*(S^{2k-1})$. Then G has periodic cohomology of period $2k$.*

PROOF. By 10.1 (with $Y = \varnothing$), $\hat{H}_G^*(X, M) = 0$. On the other hand, we have a spectral sequence (cf. 7.2)

$$E_2^{pq} = \hat{H}^p(G, H^q(X, M)) \Rightarrow \hat{H}_G^{p+q}(X, M). \tag{10.3}$$

Since the spectral sequence is concentrated on the horizontal lines $q = 0$ and $q = 2k - 1$, it follows that the differential d_{2k} is an isomorphism

$$\hat{H}^p(G, H^{2k-1}(X, M)) \xrightarrow{\approx} \hat{H}^{p+2k}(G, M). \tag{10.4}$$

In view of the universal coefficient isomorphism

$$H^{2k-1}(X, M) \approx \operatorname{Hom}(H_{2k-1}X, M),$$

which is natural and hence an isomorphism of G-modules, the proof will be complete if we show that G acts trivially on the infinite cyclic group $H_{2k-1}X$. This can be deduced from a general Lefschetz fixed-point theorem (cf. §IX.5, exercise 1), but in the present situation there is a much easier proof. It suffices to prove that every cyclic subgroup of G acts trivially on $H_{2k-1}X$, so we are immediately reduced to the case where G is cyclic (and non-trivial). In this case we apply 10.4 with $M = \mathbb{Z}$ and $p = 0$ to conclude that

$$\hat{H}^0(G, H^{2k-1}(X, \mathbb{Z})) \approx \hat{H}^{2k}(G, \mathbb{Z}) \approx \mathbb{Z}/|G| \cdot \mathbb{Z}.$$

But this implies that $H^{2k-1}(X, \mathbb{Z})^G \neq 0$, which can only happen if G acts trivially on $H_{2k-1}X$. □

Remark. The advantage of this spectral sequence proof of 10.2 is that it applies verbatim to the topological situation, where X is not assumed to be a CW-complex, cf. Swan [1960a].

As another application of equivariant Tate cohomology theory (also due to Swan), we will prove some classical results of fixed-point theory. (See Bredon [1972] or Borel et al [1960] for the standard proofs of these results, based on "Smith theory.")

(10.5) Theorem. *Let X be a finite-dimensional G-complex, where G has prime order p. Assume that the fixed-point set X^G is a subcomplex.*

(a) *If $H^*(X, \mathbb{Z}_p)$ is finitely generated, then so is $H^*(X^G, \mathbb{Z}_p)$.*
(b) *If X is acyclic* mod p (i.e., $H^*(X, \mathbb{Z}_p) \approx H^*(\text{pt.}, \mathbb{Z}_p)$), *then so is X^G.*
(c) *If X is a cohomology sphere* mod p (i.e., $H^*(X, \mathbb{Z}_p) \approx H^*(S^n, \mathbb{Z}_p)$ *for some $n \geq 0$), then so is X^G, provided $X^G \neq \varnothing$.*

(The hypothesis that X^G is a subcomplex holds, for example, if X satisfies the hypothesis of exercise 2 of §7.)

PROOF. We have

$$\hat{H}_G^*(X, \mathbb{Z}_p) \approx \hat{H}_G^*(X^G, \mathbb{Z}_p)$$

by 10.1. On the other hand, since G acts trivially on X^G, we have a Künneth isomorphism

$$\hat{H}_G^*(X^G, \mathbb{Z}_p) \approx \hat{H}^*(G, \mathbb{Z}_p) \otimes_{\mathbb{Z}_p} H^*(X^G, \mathbb{Z}_p),$$

cf. 5.4. Since $\dim_{\mathbb{Z}_p} \hat{H}^n(G, \mathbb{Z}_p) = 1$ for all $n \in \mathbb{Z}$, it follows that

$$\dim_{\mathbb{Z}_p} \hat{H}_G^n(X, \mathbb{Z}_p) = \sum_{i \geq 0} \dim_{\mathbb{Z}_p} H^i(X^G, \mathbb{Z}_p) \tag{10.6}$$

for all n. Assuming now that $H^*(X, \mathbb{Z}_p)$ is finitely generated, the spectral sequence 10.3 (with $M = \mathbb{Z}_p$) implies that $\hat{H}_G^n(X, \mathbb{Z}_p)$ is finitely generated, so (a) follows from 10.6. Similarly, if X is acyclic mod p then the left side of 10.6 is equal to 1, hence so is the right side; this proves (b). The proof of (c) is identical to that of (b), except that one uses relative cohomology groups of the pair (X, v), where v is some vertex of X^G. □

EXERCISES

1. Where did the proof of 10.5 use the assumption that $|G| = p$? [There are two places.]

2. Extend 10.5 to the case where G is an arbitrary p-group and X^H is a subcomplex of X for every $H \subseteq G$. [Hint: Use induction on $|G|$, noting that $X^G = (X^N)^{G/N}$ if $N \lhd G$.]

3. Using the method of proof of 10.2, show that if X is a finite-dimensional free G-complex (G finite) with $H_*(X) \approx H_*(S^{2k})$, then every non-trivial element of G acts non-trivially on $H_{2k}(X)$, and hence $|G| \leq 2$.

CHAPTER VIII

Finiteness Conditions

1 Introduction

Recall that the definition of $H_*(G, M)$ and $H^*(G, M)$ allows us to choose an arbitrary projective resolution $P = (P_i)_{i \geq 0}$ of $\mathbb{Z}$ over $\mathbb{Z}G$. Similarly, if we wish to take the topological point of view, then we can compute $H_*(G, M)$ and $H^*(G, M)$ in terms of an arbitrary $K(G, 1)$-complex Y. Since we have this freedom of choice, it is reasonable to try to choose P (or Y) to be as "small" as possible, and this leads to various finiteness conditions on G.

For example, if we interpret "small" in terms of the length of P (or the dimension of Y), then we are led to the notion of *cohomological dimension*. Or if we interpret small to mean that each P_i should be finitely generated (or that Y should have only finitely many cells), then we are led to the so-called "*FP*" and "*FL*" conditions.

Our goal in this chapter is to introduce these and related finiteness conditions and to give some examples. Our treatment will by no means be complete; for the most part we will give complete proofs only for those results which will be needed in Chapters IX and X. See Bieri [1976] and Serre [1971, 1979] for a much more thorough treatment of the subject and a guide to the literature. See also Wall [1979] for a list of open questions concerning finiteness conditions.

Finally, a word about notation: In the theory of discrete subgroups of Lie groups, which is the source of many of our examples, it is customary to denote the Lie group by G and the discrete subgroup by Γ. In order to be consistent with this, we will use the letter "Γ" from now on (instead of "G") to denote a typical abstract group.

2 Cohomological Dimension

We begin with a basic lemma which characterizes projective dimension in terms of the vanishing of cohomology functors. Recall first (cf. §VI.8) that if R is a ring, M is an R-module, and n is a non-negative integer, then $\operatorname{proj\,dim}_R M \leq n$ if and only if M admits a projective resolution

$$0 \to P_n \to \cdots \to P_0 \to M \to 0$$

of length n. Recall also (cf. §III.2) that the Ext functors are defined by

$$\operatorname{Ext}^i_R(M, -) = H^i(\mathscr{H}om_R(P, -)),$$

where P is a projective resolution of M. In particular,

$$\operatorname{Ext}^i_{\mathbb{Z}\Gamma}(\mathbb{Z}, -) = H^i(\Gamma, -).$$

(2.1) Lemma. *The following conditions are equivalent:*

(i) $\operatorname{proj\,dim}_R M \leq n$.
(ii) $\operatorname{Ext}^i_R(M, -) = 0$ *for* $i > n$.
(iii) $\operatorname{Ext}^{n+1}_R(M, -) = 0$.
(iv) *If* $0 \to K \to P_{n-1} \to \cdots \to P_0 \to M \to 0$ *is any exact sequence of* R-*modules with each* P_i *projective, then* K *is projective.*

PROOF. It is obvious that (iv) $\Rightarrow$ (i) $\Rightarrow$ (ii) $\Rightarrow$ (iii), so we need only prove (iii) $\Rightarrow$ (iv). Given a partial resolution as in (iv), complete it arbitrarily to a projective resolution

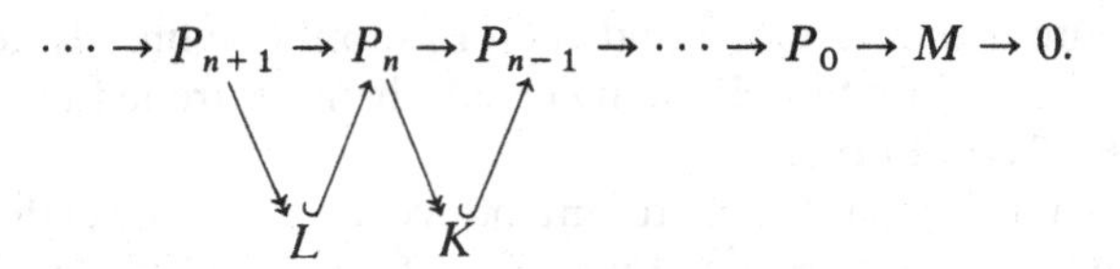

For any R-module N, an $(n + 1)$-cocycle in $\mathscr{H}om_R(P, N)$ is a map $P_{n+1} \to N$ whose composition with $P_{n+2} \to P_{n+1}$ is zero. Such a cocycle, therefore, can be regarded as a map $\varphi: L \to N$. The cocycle is a coboundary if and only if φ extends to a map $P_n \to N$. Thus (iii) implies that every map on L extends to P_n. In particular, the identity map on L extends to P_n, so $P_n \approx L \oplus K$ and hence K is projective. □

The implication (i) $\Rightarrow$ (iv) of 2.1 is very useful. It shows that if there exist projective resolutions of length n, then we don't have to be clever to find one—*any* partial resolution of length $n - 1$ can be completed to a projective resolution of length n. We will give another proof of this important fact later (remark 3 following Lemma 4.4).

We now specialize to the case $R = \mathbb{Z}\Gamma$, $M = \mathbb{Z}$. The *cohomological dimension* of Γ, denoted $\operatorname{cd} \Gamma$, is defined to be the smallest integer n such that the conditions of the lemma hold, provided there exist such integers n; otherwise we set cd $\Gamma = \infty$. Thus

$$\begin{aligned} \operatorname{cd}\Gamma &= \operatorname{proj\,dim}_{\mathbb{Z}\Gamma}\mathbb{Z} \\ &= \inf\{n: \mathbb{Z} \text{ admits a projective resolution of length } n\} \\ &= \inf\{n: H^i(\Gamma, -) = 0 \text{ for } i > n \\ &= \sup\{n: H^n(\Gamma, M) \neq 0 \text{ for some } \Gamma\text{-module } M\}. \end{aligned}$$

There is an obvious topological analogue of cd Γ: The *geometric dimension* of Γ, denoted geom dim Γ, is defined to be the minimal dimension of a $K(\Gamma, 1)$-complex. Since the cellular chain complex of the universal cover of a $K(\Gamma, 1)$ complex Y yields a free resolution of $\mathbb{Z}$ over $\mathbb{Z}\Gamma$ (of length equal to the dimension of Y), we clearly have:

(2.2) Proposition. cd $\Gamma \leq$ geom dim Γ.

We will return to this in §7, where we will prove that equality usually holds.

EXAMPLES

1. cd $\Gamma = 0$ if and only if Γ is the trivial group (cf. exercise 1 below).

2. If Γ is free and non-trivial then cd $\Gamma = 1$ (cf. I.4.3). Conversely, a deep theorem of Stallings [1968] and Swan [1969] says that every group of cohomological dimension 1 is free. In view of Chapter IV, this result can be restated as follows: If Γ is a group which admits no non-split extension with abelian kernel, then Γ admits no non-split extension at all.

3. Let Γ be the fundamental group of a connected closed surface Y other than S^2 or P^2. Then Y is a 2-dimensional $K(\Gamma, 1)$ (cf. §II.4, example 2), so cd $\Gamma \leq 2$. And $H^2(\Gamma, \mathbb{Z}_2) \approx H^2(Y, \mathbb{Z}_2) \neq 0$, so cd $\Gamma = 2$.

4. More generally, suppose Γ is a one-relator group whose relator is not a proper power. Then cd $\Gamma \leq 2$ by Lyndon's theorem which we quoted in §II.4, example 3.

5. If $\Gamma = \mathbb{Z}^n$ then the n-dimensional torus $Y = S^1 \times \cdots \times S^1$ is a $K(\Gamma, 1)$ with $H^n(Y, \mathbb{Z}) \approx \mathbb{Z} \neq 0$, hence cd $\Gamma = n$.

6. Let Γ be the group of 3×3 strictly upper triangular matrices with integral entries:

$$\Gamma = \begin{pmatrix} 1 & * & * \\ 0 & 1 & * \\ 0 & 0 & 1 \end{pmatrix}$$

We will show that cd $\Gamma = 3$. Let Γ' be the central subgroup

$$\begin{pmatrix} 1 & 0 & * \\ 0 & 1 & 0 \\ 0 & 0 & 1 \end{pmatrix}$$

of Γ, and note that the quotient $\Gamma'' = \Gamma/\Gamma'$ is free abelian of rank 2. Consider the Hochschild–Serre spectral sequence

$$E_2^{pq} = H^p(\Gamma'', H^q(\Gamma', M)) \Rightarrow H^{p+q}(\Gamma, M),$$

where M is an arbitrary Γ-module. Example 5 shows that this spectral sequence is concentrated in the rectangle $0 \leq p \leq 2$, $0 \leq q \leq 1$. Consequently, $H^i(\Gamma, M) = 0$ for $i > 3$, so cd $\Gamma \leq 3$. Moreover, since the differential d_r maps E_r^{pq} to $E_r^{p+r, q-r+1}$, there can be no non-zero differentials involving the upper right-hand corner $E_r^{2,1}$. This being the only non-trivial term of total degree 3, it follows that

$$H^3(\Gamma, M) = E_\infty^{2,1} = E_2^{2,1} = H^2(\Gamma'', H^1(\Gamma', M)).$$

In particular, taking $M = \mathbb{Z}$ and recalling that Γ' is central in Γ, we see that $H^3(\Gamma, \mathbb{Z}) \approx \mathbb{Z}$ (cf. exercise 1 of §III.8); hence cd $\Gamma = 3$.

7. More generally, let Γ be an arbitrary finitely generated, torsion-free, nilpotent group. One can show that Γ admits a central series

$$\Gamma = \Gamma_0 \supset \Gamma_1 \supset \cdots \supset \Gamma_n = \{1\}$$

with free abelian quotients Γ_i/Γ_{i+1}. The sum of the ranks of these quotients is independent of the choice of central series and is called the *rank* (or *Hirsch number*) of Γ. Arguing as in example 6, one finds

$$\text{cd}\,\Gamma = \text{rank}\,\Gamma.$$

Details are omitted. See Bieri [1976], §7.3, and Gruenberg [1970], §8.8, for more details and for further results of this type. See also example 2 of §9 below for an indication of a different proof of this result.

The rest of this section will be devoted to some elementary properties of cohomological dimension.

(2.3) Proposition. *If* cd $\Gamma < \infty$ *then*

$$\text{cd}\,\Gamma = \sup\{n\colon H^n(\Gamma, F) \neq 0 \textit{ for some free } \mathbb{Z}\Gamma\textit{-module } F\}.$$

Proof. Let $n = \text{cd}\,\Gamma$. In view of the long exact cohomology sequence (III.6.1(ii')), the functor $H^n(\Gamma, -)$ is right exact. Since $H^n(\Gamma, M) \neq 0$ for some M, it follows that $H^n(\Gamma, F) \neq 0$ for any free module F which maps onto M. □

(2.4) Proposition.

(a) *If* $\Gamma' \subset \Gamma$ *then*

$$\operatorname{cd} \Gamma' \leq \operatorname{cd} \Gamma;$$

equality holds if $\operatorname{cd} \Gamma < \infty$ *and* $(\Gamma : \Gamma') < \infty$.

(b) *If* $1 \to \Gamma' \to \Gamma \to \Gamma'' \to 1$ *is a short exact sequence of groups, then*

$$\operatorname{cd} \Gamma \leq \operatorname{cd} \Gamma' + \operatorname{cd} \Gamma''.$$

(c) *If* $\Gamma = \Gamma_1 *_A \Gamma_2$ *(where* $A \hookrightarrow \Gamma_i$*), then*

$$\operatorname{cd} \Gamma \leq \max\{\operatorname{cd} \Gamma_1, \operatorname{cd} \Gamma_2, 1 + \operatorname{cd} A\}.$$

PROOF. The inequality in (a) follows immediately from Shapiro's lemma or, alternatively, from the fact that a projective resolution of $\mathbb{Z}$ over $\mathbb{Z}\Gamma$ can also be regarded as a projective resolution of $\mathbb{Z}$ over $\mathbb{Z}\Gamma'$. To prove the second part of (a), suppose $\operatorname{cd} \Gamma = n < \infty$. By 2.3 there is a free $\mathbb{Z}\Gamma$-module F with $H^n(\Gamma, F) \neq 0$. If F' is a free $\mathbb{Z}\Gamma'$-module of the same rank, then $F \approx \operatorname{Ind}_{\Gamma'}^{\Gamma} F'$, so Shapiro's lemma yields $H^n(\Gamma', F') \approx H^n(\Gamma, F) \neq 0$. Thus $\operatorname{cd} \Gamma' \geq n$, whence (*a*). [Exercise: Where did we use the hypothesis that $(\Gamma : \Gamma') < \infty$?] (b) is an immediate consequence of the Hochschild–Serre spectral sequence, as in Example 6 above. Finally, (c) follows from the cohomology version of the Mayer–Vietoris sequence (VII.9.1 or exercise of §III.6). □

(2.5) Corollary. *If* $\operatorname{cd} \Gamma < \infty$ *then* Γ *is torsion-free.*

PROOF. If Γ is not torsion-free then Γ contains a nontrivial finite cyclic subgroup Γ'. Such a Γ' has $\operatorname{cd} \Gamma' = \infty$, since $H^{2k}(\Gamma', \mathbb{Z}) \neq 0$ for all k (cf. §III.1, Example 2) so 2.4a implies that $\operatorname{cd} \Gamma = \infty$. □

If $\Gamma' \subset \Gamma$ and $(\Gamma : \Gamma') < \infty$, then 2.4a shows that $\operatorname{cd} \Gamma' = \operatorname{cd} \Gamma$ unless the following occurs:

$$(*) \qquad \operatorname{cd} \Gamma = \infty \quad \text{and} \quad \operatorname{cd} \Gamma' < \infty.$$

Easy examples show that (*) can in fact occur; e.g., take Γ of order 2 and $\Gamma' = \{1\}$. More generally, if Γ is any group with torsion, then $\operatorname{cd} \Gamma = \infty$ by 2.5, but Γ may very well have torsion-free subgroups Γ' of finite index with $\operatorname{cd} \Gamma' < \infty$. We will prove in the next section a theorem of Serre which says that *all* examples of (*) are of this type, i.e., that (*) cannot occur if Γ is torsion-free.

Our last result shows that, as far as cohomological dimension is concerned, we never need to use projective resolutions which are not free:

(2.6) Proposition. *For any group* Γ *there is a free resolution of* $\mathbb{Z}$ *over* $\mathbb{Z}\Gamma$ *of length equal to* $\operatorname{cd} \Gamma$.

The proof requires the following "Eilenberg trick":

(2.7) Lemma. *If P is a projective module over an arbitrary ring R, then there is a free module F such that $P \oplus F \approx F$.*

(Warning: F will be of infinite rank, in general, even if P is finitely generated.)

PROOF. Since P is projective there is a module Q such that $P \oplus Q$ is free. Let F be the countable direct sum

$$(P \oplus Q) \oplus (P \oplus Q) \oplus \cdots.$$

Then F is free, being a direct sum of free modules. But F can also be described as the sum of a countable number of copies of P and a countable number of copies of Q. Adding one more copy of P does not change this, so $P \oplus F \approx F$. □

PROOF OF 2.6. Let $n = \operatorname{cd} \Gamma$. We may assume $0 < n < \infty$. Choose a partial free resolution

$$F_{n-1} \xrightarrow{\partial} \cdots \to F_0 \to \mathbb{Z} \to 0$$

of length $n - 1$ and let $P = \ker\{\partial : F_{n-1} \to F_{n-2}\}$. (Here we set $F_{-1} = \mathbb{Z}$ if $n = 1$.) Then P is projective by 2.1, so 2.7 gives us a free module F such that $P \oplus F$ is free. Thus if we replace F_{n-1} by $F_{n-1} \oplus F$ and set $\partial | F = 0$, we obtain a partial free resolution of length $n - 1$ with $\ker \partial$ free, whence the proposition. □

EXERCISES

1. Prove that the trivial group is the only group of cohomological dimension zero. [Hint: $\operatorname{cd} \Gamma = 0 \Leftrightarrow \mathbb{Z}$ is a projective $\mathbb{Z}\Gamma$-module $\Leftrightarrow$ the canonical surjection $\varepsilon : \mathbb{Z}\Gamma \to \mathbb{Z}$ splits.]

2. Give an example to show that the hypothesis $\operatorname{cd} \Gamma < \infty$ is necessary in 2.3.

3. (a) Show that geom dim $\Gamma_1 *_A \Gamma_2 \le \max\{\text{geom dim } \Gamma_1, \text{geom dim } \Gamma_2, 1 + \text{geom dim } A\}$.

   *(b) If $1 \to \Gamma' \to \Gamma \to \Gamma'' \to 1$ is exact, show that geom dim $\Gamma \le$ geom dim Γ' + geom dim Γ''.

4. Prove the following generalization of 2.4c: If Γ is an arbitrary group and X is an acyclic Γ-complex, then

$$\operatorname{cd} \Gamma \le \sup_{\sigma}\{\operatorname{cd} \Gamma_\sigma + \dim \sigma\},$$

where σ ranges over a set of representatives for the cells of X mod Γ. (Taking X to be the tree associated to an amalgam, we recover 2.4c.) [Hint: Use the cohomology version of the spectral sequence VII.7.10. Or see Serre [1971] for an alternative proof.]

5. The purpose of this exercise is to give a cohomological criterion for finite-dimensionality (up to homotopy) of an arbitrary chain complex. In case the chain complex is a projective resolution, this result ((b) below) reduces to the equivalence (i)$\Leftrightarrow$(ii)$\Leftrightarrow$(iv) of Lemma 2.1. Let C be a chain complex over an arbitrary ring R.

   (a) For any integer n, prove that the following two conditions are equivalent:

   (i) $H^{n+1}(\mathcal{H}om_R(C, M)) = 0$ for all R-modules M.
   (ii) $H_{n+1}C = 0$ and the module B_n of n-boundaries is a direct summand of C_n.

   [Hint: Examine cocycles and coboundaries in $\mathcal{H}om_R(C, M)$ as in the proof of 2.1.]

   (b) For any integer n, prove that the following three conditions are equivalent:

   (i) C is homotopy equivalent to a complex C' such that $C'_i = 0$ for $i > n$.
   (ii) $H^i(\mathcal{H}om_R(C, M)) = 0$ for all $i > n$ and all R-modules M.
   (iii) If C' is the quotient of C defined by

   $$C'_i = \begin{cases} C_i & \text{if } i < n \\ C_n/B_n & \text{if } i = n \\ 0 & \text{if } i > n, \end{cases}$$

   then the quotient map $C \to C'$ is a homotopy equivalence.

   If these conditions hold and C is a complex of *projective* R-modules, prove that the complex C' in (iii) is also a complex of projectives.

   [Hint for (ii) $\Rightarrow$ (iii): If (ii) holds, deduce from (a) that $C \approx C' \oplus C''$, where C'' is contractible.]

*6. Suppose $\operatorname{cd} \Gamma = n < \infty$, and let $\Gamma' \subseteq \Gamma$ be a subgroup of finite index.

   (a) For any Γ-module M, show that the transfer map $\operatorname{tr}: H^n(\Gamma', M) \to H^n(\Gamma, M)$ is surjective. [Hint: Compute tr on the chain level, using a projective resolution of $\mathbb{Z}$ over $\mathbb{Z}\Gamma$ of length n; note that the chain map is surjective. Alternatively, see Serre [1971], 1.3, Lemme 2.]

   (b) Suppose that Γ' is normal in Γ. If M is a Γ-module and d is an integer such that $H^i(\Gamma', M) = 0$ for $i > d$, show that $H^i(\Gamma, M) = 0$ for $i > d$ and that tr induces an isomorphism

   $$H^d(\Gamma', M)_{\Gamma/\Gamma'} \xrightarrow{\approx} H^d(\Gamma, M).$$

   In particular, this holds with $d = n$ for any M. [See Brown–Kahn [1977], Prop. 1.2.]

7. (a) Show that induced Γ-modules $\mathbb{Z}\Gamma \otimes A$ have projective dimension ≤ 1.

   (b) If $\operatorname{proj\,dim}_R M \leq n$, show that $\operatorname{proj\,dim}_R M' \leq n$ for any direct summand M' of M. [Hint: 2.1.]

   (c) Deduce from (a) and (b) the following result (which is also an immediate consequence of VI.8.12): If Γ is finite and M is a Γ-module in which $|\Gamma|$ is invertible, then $\operatorname{proj\,dim}_{\mathbb{Z}\Gamma} M \leq 1$. [Hint: M is a direct summand of an induced module.]

3 Serre's Theorem

(3.1) Theorem (Serre [1971]). *If* Γ *is a torsion-free group and* Γ' *is a subgroup of finite index, then* $\operatorname{cd}\Gamma' = \operatorname{cd}\Gamma$.

PROOF. In view of 2.4a, we need only show that if $\operatorname{cd}\Gamma' < \infty$ then $\operatorname{cd}\Gamma < \infty$. It is possible, as we will indicate below, to give a purely algebraic proof of this result; but we will give instead a topological proof, when will yield important information that will be needed in Chapters IX and X. This topological proof requires the following result of Eilenberg and Ganea [1957], which we will prove later as part of Theorem 7.1: If Γ is a group such that $\operatorname{cd}\Gamma < \infty$, then there exists a finite dimensional $K(\Gamma, 1)$-complex.

Returning now to the proof of 3.1, we are given that $\operatorname{cd}\Gamma' < \infty$, so there is a finite-dimensional $K(\Gamma', 1)$. Its universal cover X' is then a finite dimensional, contractible, free Γ'-complex. To prove $\operatorname{cd}\Gamma < \infty$, we will construct from X' a finite dimensional, contractible, free Γ-complex X. The construction, which is a straightforward analogue of the co-induction construction for modules, goes as follows:

The underlying set of X is defined by $X = \operatorname{Hom}_{\Gamma'}(\Gamma, X')$, where Γ' acts on Γ by left translation and $\operatorname{Hom}_{\Gamma'}(\ ,\ )$ denotes maps in the category of left Γ'-sets. Since the right action of Γ on itself commutes with the left action of Γ' on Γ, there is an induced (left) action of Γ on X, given by $(\gamma_0 f)(\gamma) = f(\gamma\gamma_0)$ for $f \in X$, $\gamma, \gamma_0 \in \Gamma$.

If we choose a set of coset representatives $\gamma_1, \ldots, \gamma_n$ for $\Gamma' \backslash \Gamma$, then we obtain a bijection

$$\varphi : X \xrightarrow{\approx} \prod_{i=1}^{n} X',$$

given by evaluation at $\gamma_1, \ldots, \gamma_n$. Since the product on the right has a natural CW-structure (with the cells being the products of the cells of the factors), we can use φ to give X a topology and a CW-structure. [Note: The product is to be given the CW-topology, i.e., the "weak topology" with respect to the cells; this agrees with the usual product topology if X' is countable or locally compact, cf. Lundell–Weingram [1969], §II.5. In any case, the two topologies agree on all compact subsets.]

This structure is independent of the choice of coset representatives; for if we replace $\gamma_1, \ldots, \gamma_n$ by $\gamma'_1\gamma_1, \ldots, \gamma'_n\gamma_n$ ($\gamma'_i \in \Gamma'$), then the new φ is obtained from the old one by composition with the CW-isomorphism

$$\prod_{i=1}^{n} \gamma'_i : \prod_{i=1}^{n} X' \to \prod_{i=1}^{n} X'.$$

The structure is also independent of the ordering of the cosets.

It now follows that the Γ-action on X preserves the CW-structure.

Indeed, for any $\gamma \in \Gamma$ we have a commutative diagram

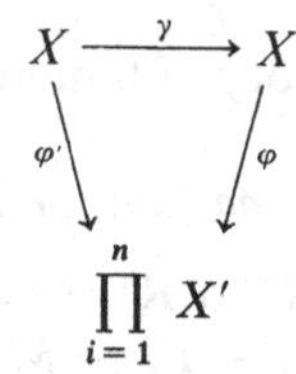

where φ is defined via coset representatives $(\gamma_i)_{1 \le i \le n}$ and φ' is defined via $(\gamma_i \gamma)_{1 \le i \le n}$. Thus X is a well-defined Γ-complex, which is clearly contractible and finite dimensional.

To complete the proof that $\operatorname{cd} \Gamma < \infty$, we will show that Γ acts freely on X. There is a canonical map $X \to X'$, given by evaluation at $1 \in \Gamma$. This map is Γ'-equivariant and takes cells to cells. Since Γ' acts freely on X', it follows that Γ' acts freely on X. For any cell σ of X, then, we have $\Gamma_\sigma \cap \Gamma' = \{1\}$, hence Γ_σ is finite. But Γ is torsion-free, so these finite isotropy groups Γ_σ are trivial. □

Remark. The interested reader can translate the proof above into a purely algebraic proof. One works directly with projective resolutions, and one does the "co-induction" construction using tensor products of chain complexes rather than cartesian products of *CW*-complexes. More details can be found in Swan [1969], Theorem 9.2.

Exercise

Note that the proof of 3.1 is valid even if Γ has torsion, except for the last sentence of the proof. Consequently, if Γ is a group which contains a subgroup Γ' of finite index such that $\operatorname{cd} \Gamma' < \infty$, then we can construct a contractible, finite dimensional Γ-complex X, with finite isotropy groups Γ_σ. Show that this complex has the following additional property: For every finite subgroup $H \subseteq \Gamma$, the fixed-point set X^H is contractible. [Hint: Show that X, as an H-complex, is isomorphic to the product of $(\Gamma : \Gamma')$ copies of the complex X', with H acting by permuting the factors according to the (free) right action of H on $\Gamma' \backslash \Gamma$. One can see this either by using the double coset formula for co-induction or simply by direct inspection, using coset representatives of the form $\{\gamma_i h\}_{1 \le i \le d, h \in H}$, where $d = (\Gamma : \Gamma')/|H|$. Hence X^H is homeomorphic to the product of d copies of X'.]

4 Resolutions of Finite Type

Our next goal is to study a different sort of finiteness condition, where we require that there be a projective resolution P with each P_i finitely generated. In this section we collect some general facts about such resolutions over an arbitrary ring R. Then in the next two sections we specialize to resolutions of $\mathbb{Z}$ over $\mathbb{Z}\Gamma$.

We begin by reviewing the theory of finitely presented modules:

(4.1) Proposition. *The following conditions on an R-module M are equivalent:*

(i) *There is an exact sequence $R^m \to R^n \to M \to 0$ for some integers m, n.*
(ii) *There is an exact sequence $P_1 \to P_0 \to M \to 0$ for some finitely generated projectives P_0, P_1.*
(iii) *M is finitely generated, and for every surjection $\varepsilon: P \twoheadrightarrow M$ with P finitely generated and projective,* ker ε *is finitely generated.*

The proof is based on "Schanuel's lemma":

(4.2) Lemma. Let $0 \to K \to P \to M \to 0$ and $0 \to K' \to P' \to M \to 0$ *be exact sequences with P and P' projective. Then $P \oplus K' \approx P' \oplus K$.*

Proof. Let Q be the pullback of the given maps

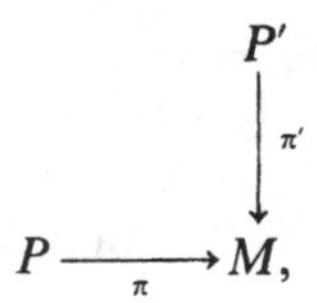

i.e., Q is the submodule of $P \times P'$ consisting of those pairs (x, x') such that $\pi(x) = \pi'(x')$. One then verifies easily that there is a commutative diagram

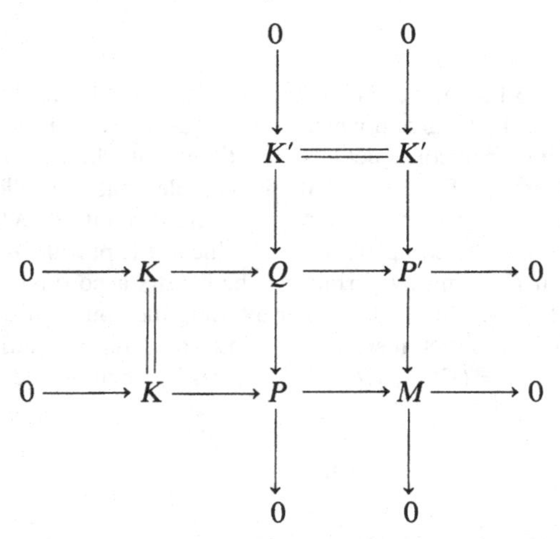

with exact rows and columns. Since P and P' are projective, the two exact sequences involving Q must split, yielding $P \oplus K' \approx Q \approx P' \oplus K$. □

Proof of 4.1. Clearly (iii)$\Rightarrow$(i)$\Rightarrow$(ii). To prove (ii)$\Rightarrow$(iii), note first that M is certainly finitely generated if (ii) holds, since P_0 is finitely generated. Now apply Schanuel to get $P \oplus \ker\{P_0 \to M\} \approx P_0 \oplus \ker \varepsilon$. The lefthand side being finitely generated by hypothesis, it follows that $P_0 \oplus \ker \varepsilon$ is finitely generated, hence so is $\ker \varepsilon$. $\square$

M is said to be *finitely presented* if the conditions of 4.1 hold. An exact sequence as in (i) is said to give a *finite presentation* of M with n generators and m relations. Note that the implication (i) $\Rightarrow$ (iii), applied when P is free, reduces to the following well-known fact: If M admits *some* finite presentation, then *every* finite set $x_1, \ldots, x_k$ of generators of M has the property that the relations among them (i.e., the k-tuples $(r_1, \ldots, r_k)$ such that $\sum r_i x_i = 0$) form a finitely generated submodule of R^k.

It is natural to generalize finite presentation as follows: A resolution or partial resolution (P_i) is said to be of *finite type* if each P_i is finitely generated. A module M is said to be of *type* FP_n $(n \geq 0)$ if there is a partial projective resolution $P_n \to \cdots \to P_0 \to M \to 0$ of finite type. Thus the FP_0 condition is simply finite generation, FP_1 is finite presentation, and the conditions $FP_2, FP_3, \ldots$ are successive strengthenings of finite presentation.

Generalizing 4.1, we have:

(4.3) Proposition. *For any module M and integer $n \geq 0$ the following conditions are equivalent*:

(i) *There is a partial resolution $F_n \to \cdots \to F_0 \to M \to 0$ with each F_i free of finite rank.*
(ii) *M is of type FP_n.*
(iii) *M is finitely generated, and for every partial projective resolution $P_k \to \cdots \to P_0 \to M \to 0$ of finite type with $k < n$, $\ker\{P_k \to P_{k-1}\}$ is finitely generated.*

For the proof we will need the following generalization of Schanuel's lemma:

(4.4) Lemma. *Let*

$$0 \to P_n \to P_{n-1} \to \cdots \to P_0 \to M \to 0$$

and

$$0 \to P'_n \to P'_{n-1} \to \cdots \to P'_0 \to M \to 0$$

be exact sequences with P_i and P'_i projective for $i \leq n - 1$. Then

$$P_0 \oplus P'_1 \oplus P_2 \oplus P'_3 \oplus \cdots \approx P'_0 \oplus P_1 \oplus P'_2 \oplus P_3 \oplus \cdots.$$

Consequently, if P_i and P'_i are finitely generated for $i \leq n - 1$, then P_n is finitely generated if and only if P'_n is finitely generated.

Proof. We argue by induction on n. Let K (resp. K') be the kernel of $P_{n-2} \to P_{n-3}$ (resp. $P'_{n-2} \to P'_{n-3}$). By the induction hypothesis we have

$$K \oplus Q \approx K' \oplus Q',$$

where

$$Q = P'_{n-2} \oplus P_{n-3} \oplus \cdots$$

and

$$Q' = P_{n-2} \oplus P'_{n-3} \oplus \cdots.$$

On the other hand, we have exact sequences

$$\begin{array}{c} 0 \to P_n \to P_{n-1} \oplus Q \to K \oplus Q \to 0 \\ \wr\wr \\ 0 \to P'_n \to P'_{n-1} \oplus Q' \to K' \oplus Q' \to 0. \end{array}$$

Since $P_{n-1} \oplus Q$ and $P'_{n-1} \oplus Q'$ are projective, 4.2 implies that

$$P_n \oplus P'_{n-1} \oplus Q' \approx P'_n \oplus P_{n-1} \oplus Q,$$

which is the desired isomorphism. □

Proof of 4.3. (i)$\Rightarrow$(ii) trivially. (ii)$\Rightarrow$(iii): If M is of type FP_n, then for any $k < n$ there is a partial projective resolution $P_k \to \cdots \to P_0 \to M \to 0$ of finite type with $\ker\{P_k \to P_{k-1}\}$ finitely generated. It follows from 4.4 that any other partial projective resolution $P'_k \to \cdots \to P'_0 \to M \to 0$ of finite type (and the same length k) has $\ker\{P'_k \to P'_{k-1}\}$ finitely generated, so (iii) holds. (iii)$\Rightarrow$(i): If (iii) holds then we can construct the desired $F_n \to \cdots \to F_0 \to M \to 0$ step by step. □

Remarks

1. One can remember the isomorphism of 4.4 by formally "transposing" terms to obtain an "Euler characteristic" equality

$$P_0 - P_1 + P_2 - \cdots = P'_0 - P'_1 + P'_2 - \cdots.$$

2. If we regard 4.4 as a comparison theorem for resolutions, it is natural to ask if it can be deduced from the results of §I.7. This can in fact be done; here is an outline: By I.7.4 we can choose an augmentation-preserving chain map $P' \to P$; let C be its mapping cone. Then C is acyclic and hence breaks up into short exact sequences. Using the fact that C is projective except in the top two dimensions, one can prove inductively that these short exact sequences split (cf. exercise 3a of §I.8). Thus $C_1 \approx C_0 \oplus Z_1$, $C_2 \approx Z_1 \oplus Z_2$, $C_3 \approx Z_2 \oplus Z_3$, etc., whence

$$C_0 \oplus C_2 \oplus \cdots \approx C_1 \oplus C_3 \oplus \cdots.$$

Recalling that $C_i = P_i \oplus P'_{i-1}$, we obtain the desired isomorphism.

3. It follows from the isomorphism of 4.4 that P_n is projective if and only if P'_n is projective. Thus 4.4 yields an alternative proof of the implication (i) ⇒ (iv) of 2.1.

We will be primarily interested in the case where the conditions of 4.3 hold for all integers $n \geq 0$. This situation is characterized as follows:

(4.5) Proposition. *The following conditions on a module M are equivalent:*

(i) *M admits a free resolution of finite type.*
(ii) *M admits a projective resolution of finite type.*
(iii) *M is of type FP_n for all integers $n \geq 0$.*

PROOF. (i) ⇒ (ii) ⇒ (iii) trivially. If (iii) holds then we can use 4.3(iii) to construct a free resolution of finite type step by step, so (iii) ⇒ (i). □

We say that M is of *type FP_∞* if these conditions hold.

We close this section by mentioning some useful formal properties which the homology and cohomology functors $\mathrm{Tor}_*^R(M, —)$ and $\mathrm{Ext}_R^*(M, —)$ satisfy when M is of type FP_n. (The reader should keep in mind, during this discussion, the main case of interest: $R = \mathbb{Z}\Gamma$, $M = \mathbb{Z}$. In this case the Tor and Ext functors are simply $H_*(\Gamma, —)$ and $H^*(\Gamma, —)$.)

Note first that the homology functors $\mathrm{Tor}_*^R(M, —)$ always commute with direct limits, in the following sense: Let $\{N_i\}_{i \in I}$ be a direct system of R-modules, where I is a directed set, and let $N = \varinjlim_{i \in I} N_i$. Thus N is the universal target of a compatible family of maps $N_i \to N$ $(i \in I)$. These maps induce a compatible family of abelian group homomorphisms $\mathrm{Tor}_*^R(M, N_i) \to \mathrm{Tor}_*^R(M, N)$, from which we obtain a map

$$\varphi : \varinjlim_{i \in I} \mathrm{Tor}_*(M, N_i) \to \mathrm{Tor}_*(M, N);$$

the assertion, then, is that φ is an isomorphism. This follows directly from the definition of $\mathrm{Tor}_*^R(M, —)$ as $H_*(P \otimes_R —)$, where P is a projective resolution of M, together with the following two facts:

(a) $P \otimes_R —$ commutes with direct limits (cf. Spanier [1966], 5.1.9);
(b) $H_*(—)$ commutes with direct limits (cf. Spanier [1966], 4.1.7).

Functors of the form $\mathscr{H}om_R(P, —)$, however, do not in general commute with direct limits, so we cannot expect $\mathrm{Ext}_R^*(M, —)$ to commute with direct limits. But we can prove that this does hold under suitable FP_n hypotheses. For simplicity, we will confine ourselves to the case $n = \infty$.

(4.6) Proposition. *If M is of type FP_∞ then* $\mathrm{Ext}_R^*(M, —)$ *commutes with direct limits.*

This follows immediately from:

(4.7) Lemma. *If P is a finitely generated projective R-module then* $\operatorname{Hom}_R(P, -)$ *commutes with direct limits.*

Proof. The duality isomorphism I.8.3b shows that the functor $\operatorname{Hom}_R(P, -)$ is equivalent to the functor $P^* \otimes_R -$, and we know that the latter commutes with direct limits. □

(See Exercise 4 below for a generalization of 4.7 and an indication of an alternative proof.)

Similarly, one can show that $\operatorname{Tor}_*^R(M, -)$ commutes with direct products if M is of type FP_∞ (whereas $\operatorname{Ext}_R^*(M, -)$ commutes with products for *any* M).

Surprisingly, these formal properties of $\operatorname{Ext}_R^*(M, -)$ and $\operatorname{Tor}_*^R(M, -)$ characterize the FP_∞ property. In fact, one can prove:

(4.8) Theorem. *The following conditions are equivalent:*

(i) *M is of type FP_∞.*
(ii) $\operatorname{Ext}_R^*(M, -)$ *commutes with direct limits.*
(iii) $\operatorname{Tor}_*^R(M, -)$ *commutes with direct products.*

We omit the proof, since we will not be making serious use of this result. The equivalence of (i) and (iii) was first proved by Bieri and Eckmann [1974]. See Brown [1975a] for the equivalence of (i) and (ii), as well as for a generalization of 4.8. See also Strebel [1976] for further results of this type.

Exercises

1. Let Γ be a group. Show that $\mathbb{Z}$ is finitely presented as a $\mathbb{Z}\Gamma$-module if and only if Γ is a finitely generated group. [Hint: Use Exercise 1d of §I.2.]

2. Let M' and M'' be modules and let $M = M' \oplus M''$. Show that M is of type FP_n if and only if M' and M'' are of type FP_n. [Hint: Suppose P' and P'' are finite type partial resolutions of M' and M'' of length k, and let $P = P' \oplus P''$. Note that P has finitely generated kernel if and only if P' and P'' do.]

3. Let C and C' be non-negative chain complexes of projective modules. If C and C' are homotopy equivalent, show that
$$C_0 \oplus C_1' \oplus C_2 \oplus \cdots \approx C_0' \oplus C_1 \oplus C_2' \oplus \cdots.$$
[Hint: Imitate the alternative proof of 4.4 outlined in Remark 2 following 4.4.]

4. If M is a finitely presented R-module, show that $\operatorname{Hom}_R(M, -)$ commutes with direct limits. [Hint: This can be deduced from 4.7. Alternatively, give a direct proof based on the fact that a map from M to any other module can be specified by giving a finite set of elements of the target module which satisfy a certain finite set of relations.]

5 Groups of Type FP_n

We now specialize the theory of §4 to the case $R = \mathbb{Z}\Gamma$, $M = \mathbb{Z}$. We will say that Γ is of *type* FP_n $(0 \le n \le \infty)$ if $\mathbb{Z}$ is of type FP_n as a $\mathbb{Z}\Gamma$-module. Thus every group is of type FP_0, and it is easy to see that Γ is of type FP_1 if and only if it is finitely generated (cf. Exercise 1 of §4). The FP_2 condition is less well-understood, however. One knows that finitely presented groups Γ are of type FP_2, for if Y is a finite 2-complex with $\pi_1 Y = \Gamma$ then the cellular chain complex of the universal cover of Y is a partial free resolution of $\mathbb{Z}$ over $\mathbb{Z}\Gamma$ of length 2 and of finite type. [See also exercise 4c of §IV.2 for an algebraic proof.] But it is not known whether the converse is true. See Exercise 3 below for a reformulation of the problem.[3]

We will not pause now to discuss examples, since there will be plenty of examples of groups of type FP_∞ in §§6, 9, and 11. The reader who wants to see examples for $n < \infty$ of groups of type FP_n but not of type FP_{n+1} should consult Bieri [1976], §2.6, and Stuhler [1980]; see also Exercise 2 below for the case $n = 1$ and Stallings [1963] for the case $n = 2$.

The FP_n conditions behave nicely with respect to subgroups of finite index:

(5.1) Proposition. *Let $\Gamma' \subseteq \Gamma$ be a subgroup of finite index. Then Γ is of type FP_n $(0 \le n \le \infty)$ if and only if Γ' is of type FP_n.*

Proof. Any (partial) projective resolution of $\mathbb{Z}$ over $\mathbb{Z}\Gamma$ of finite type can also be regarded as a (partial) projective resolution of $\mathbb{Z}$ over $\mathbb{Z}\Gamma'$, and as such it is still of finite type since $(\Gamma : \Gamma') < \infty$. This proves the "only if" part. Conversely, suppose Γ' is of type FP_n. We will show that the $\mathbb{Z}\Gamma$-module $\mathbb{Z}$ satisfies condition 4.3(iii). Let P be a partial projective resolution of $\mathbb{Z}$ over $\mathbb{Z}\Gamma$ of finite type and of length $k < n$. Regarding P as a partial resolution of $\mathbb{Z}$ over $\mathbb{Z}\Gamma'$, we can apply 4.3(iii) to conclude that $\ker\{P_k \to P_{k-1}\}$ is finitely generated over $\mathbb{Z}\Gamma'$. But then $\ker\{P_k \to P_{k-1}\}$ is certainly finitely generated over $\mathbb{Z}\Gamma$, so 4.3(iii) holds for $\mathbb{Z}$ over $\mathbb{Z}\Gamma$. □

Taking $n = 1$, for example, we recover the well-known (but not completely obvious) fact that Γ is finitely generated if and only if Γ' is finitely generated.

Finally, we record for future reference an important consequence of 4.6:

(5.2) Proposition. *Let Γ be a group of type FP_∞ and let n be an integer such that $H^n(\Gamma, \mathbb{Z}\Gamma) = 0$. Then $H^n(\Gamma, F) = 0$ for all free $\mathbb{Z}\Gamma$-modules F.*

Proof. If F is of finite rank then the result follows from the additivity of $H^n(\Gamma, -)$. In the general case, choose a basis $(e_i)_{i \in I}$ for F and note that $F = \varinjlim F_J$, where J ranges over the finite subsets of I and F_J is the

[3] For solvable Γ, substantial progress on this question has been made by Bieri and Strebel. For references, see the appendix to the forthcoming second edition of Bieri [1976].

submodule of F generated by the e_i for $i \in J$. Then F_J is free of finite rank; using 4.6 we conclude that $H^n(\Gamma, F) = \varinjlim H^n(\Gamma, F_J) = 0$. □

Exercises

1. Let Γ be a group of type FP_n and let M be a Γ-module which is finitely generated as an abelian group. Show that $H_i(\Gamma, M)$ and $H^i(\Gamma, M)$ are finitely generated abelian groups for $i \leq n$.

2. Let Γ be an amalgamation $\Gamma_1 *_A \Gamma_2$ where Γ_1 and Γ_2 are free of finite rank and A is free of infinite rank. Show that Γ is finitely generated but not of type FP_2. In particular, Γ is not finitely presented. [Hint: Use the Mayer-Vietoris sequence to show that $H_2\Gamma$ is not finitely generated.]

3. (a) If Γ is of type FP_2 and N is a perfect normal subgroup (i.e., a normal subgroup such that $N = [N, N]$), prove that Γ/N is of type FP_2. [Hint: Apply the functor $(\ )_N$ to a partial resolution of $\mathbb{Z}$ over $\mathbb{Z}\Gamma$ of length 2 and finite type; the resulting complex will still be acyclic in dimension one because its one-dimensional homology is $H_1 N = 0$.]

   (b) Let $\Gamma = F/R$ where F is a finitely generated free group. Prove that the following conditions are equivalent:

   (i) Γ is of type FP_2.
   (ii) The relation module R_{ab} is a finitely generated Γ-module.
   (iii) $\Gamma \approx \tilde{\Gamma}/N$, where $\tilde{\Gamma}$ is finitely presented and N is a perfect normal subgroup.

   [Hint: For (i) ⇒ (ii) use II.5.4.]

   Remark. In view of the equivalence of (i) and (iii), we can reformulate as follows the question as to whether every group of type FP_2 is finitely presented: Is a perfect normal subgroup of a finitely presented group necessarily finitely generated as a normal subgroup?

4. For any group Γ, note that the cohomology groups $H^*(\Gamma, \mathbb{Z}\Gamma)$ have a canonical structure of *right* Γ-module. Indeed, the coefficient module $\mathbb{Z}\Gamma$, which is thought of as a left module for the purpose of defining $H^*(\Gamma, \mathbb{Z}\Gamma)$, also admits a right Γ-action (by right translation) which commutes with the left action; this right action induces a right action of Γ on the cohomology groups $H^*(\Gamma, \mathbb{Z}\Gamma)$. In this exercise you will prove two "universal coefficient" formulas involving these right Γ-modules.

   (a) If Γ is a group of type FP_∞, prove that

   $$H^*(\Gamma, F) \approx H^*(\Gamma, \mathbb{Z}\Gamma) \otimes_{\mathbb{Z}\Gamma} F$$

   for any flat $\mathbb{Z}\Gamma$-module F. [Hint: Use I.8.3b to rewrite $\mathcal{H}om_\Gamma(P, F)$ as a tensor product, where P is a projective resolution of finite type.]

   (b) If Γ is of type FP_∞ and Q is an injective $\mathbb{Z}\Gamma$-module, prove that

   $$H_*(\Gamma, Q) \approx \mathrm{Hom}_\Gamma(H^*(\Gamma, \mathbb{Z}\Gamma), Q).$$

   [Here we should take Q to be a *right* Γ-module so that the Hom makes sense.]

5. Let Γ be a free product $\Gamma_1 * \Gamma_2$, where Γ_1 and Γ_2 are infinite and of type FP_∞. For any integer i show that there is a map of (right) Γ-modules
$$H^i(\Gamma, \mathbb{Z}\Gamma) \to \mathrm{Ind}_{\Gamma_1}^{\Gamma} H^i(\Gamma_1, \mathbb{Z}\Gamma_1) \oplus \mathrm{Ind}_{\Gamma_2}^{\Gamma} H^i(\Gamma_2, \mathbb{Z}\Gamma_2)$$
which is an isomorphism if $i > 1$ and an epimorphism if $i = 1$ with kernel a free $\mathbb{Z}\Gamma$-module of rank 1. [Hint: Use the Mayer–Vietoris sequence, and apply exercise 4a to compute $H^*(\Gamma_j, \mathbb{Z}\Gamma)$, $j = 1, 2$.]

6 Groups of Type *FP* and *FL*

We now combine the two types of finiteness conditions which we have considered in this chapter. A resolution is said to be *finite* if it is both of finite type and of finite length. A group Γ is said to be of *type FP* if $\mathbb{Z}$ admits a finite projective resolution over $\mathbb{Z}\Gamma$.

(6.1) Proposition. *Γ is of type FP if and only if* (i) $\mathrm{cd}\,\Gamma < \infty$ *and* (ii) *Γ is of type FP_∞.*

Proof. The "only if" part is obvious. Conversely, if $\mathrm{cd}\,\Gamma < \infty$ and Γ is of type FP_∞, then we can construct a finite resolution as follows: Take a partial resolution $P_{n-1} \to \cdots \to P_0 \to \mathbb{Z} \to 0$ of finite type, where $n = \mathrm{cd}\,\Gamma$, and let $P_n = \ker\{P_{n-1} \to P_{n-2}\}$. Then P_n is projective (2.1(iv)) and finitely generated (4.3(iii)), so we have a finite projective resolution
$$0 \to P_n \to \cdots \to P_0 \to \mathbb{Z} \to 0. \qquad \square$$

Note that it would have been enough in the proof above to assume that Γ was of type FP_n instead of FP_∞, where $n = \mathrm{cd}\,\Gamma$. It follows, for instance, that a finitely presented group Γ with $\mathrm{cd}\,\Gamma = 2$ is of type FP. Note also that the partial resolution $(P_i)_{i \le n-1}$ above could have been taken free. Hence if Γ is of type FP then there is a finite projective resolution

$$0 \to P \to F_{n-1} \to \cdots \to F_0 \to \mathbb{Z} \to 0 \tag{6.2}$$

with each F_i free. (See also exercise 2 below.) But there is no reason to expect to be able to take P free. Thus, for the first time in this book there really seems to be a difference between what can be done with projective resolutions and what can be done using only free resolutions.

We are therefore led to introduce a still stronger finiteness condition: Γ is of *type FL* if $\mathbb{Z}$ admits a finite free resolution over $\mathbb{Z}\Gamma$. [*Warning*: It is a common mistake to assume that "*FL*" stands for "finite length," and thereby to confuse the *FL* property with the much weaker property of having finite cohomological dimension. In fact, the "*L*" in "*FL*" stands for "libre," not for "length." One also finds "*FF*" ("*f*inite *f*ree resolution") in the literature instead of "*FL*."]

It is obvious how to use topology to obtain examples of groups of type FL:

(6.3) Proposition. *If there exists a* $K(\Gamma, 1)$ *which is a finite complex, then* Γ *is of type* FL.

(We will see in §7 that the converse of this is also true, at least if we assume that Γ is finitely presented.)

Looking at the examples in §2, we immediately deduce the following examples of groups of type FL: free groups of finite rank; surface groups; finitely generated one-relator groups whose relator is not a power; and free abelian groups of finite rank. With a little more work (cf. exercise 8 below), one can also show that torsion-free, finitely generated, nilpotent groups are of type FL. We will reprove this result and give many additional examples of groups of type FL in §9.

The FP property also admits a topological interpretation, for which we need the following notion: A space Y is *finitely dominated* if there is a finite complex K such that Y is a retract of K in the homotopy category (i.e., we require maps $i: Y \to K$ and $r: K \to Y$ with $ri \simeq \mathrm{id}_Y$).

(6.4) Proposition. *If there exists a finitely dominated* $K(\Gamma, 1)$, *then* Γ *is of type* FP.

(Again the converse is also true, and will be proved in §7, provided Γ is finitely presented.)

We will not be making any use of this result, so we confine ourselves to a brief sketch of the proof:

Let Y be a $K(\Gamma, 1)$-complex dominated by a finite complex K. One can choose K so that the maps $Y \leftrightarrows K$ induce π_1-isomorphisms. Letting $\tilde{Y}$ and $\tilde{K}$ be the universal covers, one deduces that the cellular chain complex $C(\tilde{Y})$ is a retract of $C(\tilde{K})$ in the homotopy category of chain complexes over $\mathbb{Z}\Gamma$. Since $C(\tilde{Y})$ is a free resolution of $\mathbb{Z}$ and $C(\tilde{K})$ is a finite free complex, it follows that $H^*(\Gamma, -)$ commutes with direct limits and that $H^i(\Gamma, -) = 0$ for $i > \dim K$. In view of 4.8, this implies that Γ is of type FP. □

Remark. Propositions 6.3 and 6.4 and their converses are special cases of the following result due to Wall [1965, 1966]: Let Y be a connected CW-complex whose fundamental group π is finitely presented, and let C be the chain complex of the universal cover of Y, regarded as a complex of π-modules. Then (a) Y is homotopy equivalent to a finite CW-complex if and only if C is homotopy equivalent to a finite free chain complex; and (b) Y is finitely dominated if and only if C is homotopy equivalent to a finite projective chain complex.

Having carefully explained the difference between the FP condition and the FL condition, we are now forced to admit that there are no known examples of groups of type FP which are not of type FL. [From the topological point of view, we have the following situation: Although there are plenty of known examples of finitely dominated spaces which do not have the homotopy type of a finite complex, none of these known examples are $K(\Gamma, 1)$'s.]

In order to better appreciate the problem, let's see what the obstruction is to proving that a group of type FP is of type FL. Suppose Γ is of type FP, and choose a finite projective resolution as in 6.2 which is free except in the top dimension. The projective P which occurs in the top dimension might have the property that $P \oplus F$ is free for some free module F of finite rank. In this case P is said to be *stably free*, and we can modify 6.2 exactly as in the proof of 2.6 to obtain a finite free resolution. Conversely, if there exists a finite free resolution, then we can compare it to 6.2 via 4.4 to deduce that P is stably free. [Note: The two resolutions might not have the same length, but 4.4 is still applicable; for we can extend the shorter resolution by zeroes.] Thus we have:

(6.5) Proposition. *Let Γ be a group of type FP and let $0 \to P \to F_{n-1} \to \cdots \to F_0 \to \mathbb{Z} \to 0$ be a finite projective resolution of $\mathbb{Z}$ over $\mathbb{Z}\Gamma$ with each F_i free. Then Γ is of type FL if and only if P is stably free.*

Thus the question as to whether there exist groups of type FP which are not of type FL has led to a more fundamental question: Do there exist finitely generated projectives which are not stably free? Over a general ring the answer is certainly "yes," and there are even known examples over integral group rings $\mathbb{Z}\Gamma$, the simplest example being with $\Gamma = \mathbb{Z}_{23}$ (cf. Milnor [1971], §3). The surprising fact, however, is that there are no known examples with Γ *torsion-free*, and a group of type FP is necessarily torsion-free by 2.5.

We remark, finally, that there do exist resolutions as in 6.2 in which P is not free. The first such example was given by Dunwoody [1972], with Γ equal to the trefoil group. In all of the known examples, however, the group Γ is known to be of type FL (and hence P is stably free).

In spite of this lack of examples, we will see later (cf. IX.6.4 and Remark 1 following its proof) that there are concrete results whose proofs require that we consider the FP property, and not just the FL property. One reason for this is that we can prove the following result about groups of type FP, the analogue of which for groups of type FL is not known:

(6.6) Proposition. *Let Γ be a torsion-free group and Γ' a subgroup of finite index. Then Γ is of type FP if and only if Γ' is of type FP.*

PROOF. This follows from 3.1 and 5.1. □

In particular, if Γ is a torsion-free group which contains a subgroup of finite index which is of type FL, then we know from 6.6 that Γ is of type FP, even though we don't know that Γ is of type FL.

We close this section by discussing some special features of the top-dimensional cohomology of a group of type FP. First we note the following improvement of 2.3:

(6.7) Proposition. *If Γ is of type FP then* $\operatorname{cd}\Gamma = \max\{n: H^n(\Gamma, \mathbb{Z}\Gamma) \neq 0\}$.

PROOF. This follows from 2.3 and 5.2. □

Recall that the cohomology groups $H^i(\Gamma, \mathbb{Z}\Gamma)$ admit a canonical *right Γ-module structure* (cf. exercise 4 of §5). For any (left) Γ-module M, we can therefore form the tensor product $H^i(\Gamma, \mathbb{Z}\Gamma) \otimes_{\mathbb{Z}\Gamma} M$, and there is a canonical map

$$\varphi: H^*(\Gamma, \mathbb{Z}\Gamma) \otimes_{\mathbb{Z}\Gamma} M \to H^*(\Gamma, M),$$

defined as follows on the cochain level: Let P be a projective resolution of $\mathbb{Z}$ over $\mathbb{Z}\Gamma$; given a cochain $u \in \mathcal{H}om_\Gamma(P, \mathbb{Z}\Gamma)$ and an element $m \in M$, we send $u \otimes m$ to the cochain $x \mapsto u(x)m$ $(x \in P)$ in $\mathcal{H}om_\Gamma(P, M)$.

We can now prove the following "universal coefficient theorem" for the top-dimensional cohomology of a group of type FP:

(6.8) Proposition. *If Γ is of type FP and $n = \operatorname{cd}\Gamma$, then*

$$\varphi: H^n(\Gamma, \mathbb{Z}\Gamma) \otimes_{\mathbb{Z}\Gamma} M \to H^n(\Gamma, M)$$

is an isomorphism for all Γ-modules M.

PROOF. We will give two proofs, both of which are instructive.

Proof 1: Regard φ as a natural transformation between functors of M. Since both functors are right exact, it suffices to prove that φ is an isomorphism when M is free. [The general case is then obtained by considering an exact sequence $F' \to F \to M \to 0$ with F and F' free.] Since both functors are additive and commute with direct limits, it suffices to consider the case where M is free of rank 1, i.e., we may assume $M = \mathbb{Z}\Gamma$. But in this case $\varphi: H^n(\Gamma, \mathbb{Z}\Gamma) \otimes_{\mathbb{Z}\Gamma} \mathbb{Z}\Gamma \to H^n(\Gamma, \mathbb{Z}\Gamma)$ is simply the well-known isomorphism $u \otimes r \mapsto ur$ (with inverse $u \mapsto u \otimes 1$).

Proof 2: Let P be a finite projective resolution of $\mathbb{Z}$ over $\mathbb{Z}\Gamma$ of length n and let $\bar{P}$ be the dual complex $\mathcal{H}om_\Gamma(P, \mathbb{Z}\Gamma)$ of right Γ-modules. Then $H^*(\Gamma, \mathbb{Z}\Gamma) \approx H^*(\bar{P})$. In particular, we have an exact sequence

$$\bar{P}^{n-1} \to \bar{P}^n \to H^n(\Gamma, \mathbb{Z}\Gamma) \to 0$$

for the top-dimensional cohomology module. Tensoring with M and applying the duality isomorphism I.8.3b, we obtain the diagram

$$\begin{array}{ccccccc} \bar{P}^{n-1} \otimes_{\mathbb{Z}\Gamma} M & \longrightarrow & \bar{P}^{n} \otimes_{\mathbb{Z}\Gamma} M & \longrightarrow & H^n(\Gamma, \mathbb{Z}\Gamma) \otimes_{\mathbb{Z}\Gamma} M & \longrightarrow & 0 \\ \downarrow \approx & & \downarrow \approx & & \downarrow \varphi & & \\ \mathrm{Hom}_\Gamma(P_{n-1}, M) & \longrightarrow & \mathrm{Hom}_\Gamma(P_n, M) & \longrightarrow & H^n(\Gamma, M) & \longrightarrow & 0. \end{array}$$

which commutes and has exact rows. It follows that φ is an isomorphism. □

The isomorphism of 6.8 can be written in the following suggestive way: Let D be the Γ-module $H^n(\Gamma, \mathbb{Z}\Gamma)$, so that $H^n(\Gamma, \mathbb{Z}\Gamma) \otimes_{\mathbb{Z}\Gamma} M = D \otimes_{\mathbb{Z}\Gamma} M = (D \otimes M)_\Gamma = H_0(\Gamma, D \otimes M)$, where $D \otimes M = D \otimes_{\mathbb{Z}} M$ with the diagonal Γ-action. [Note: Since D is a right module and M is a left module, the diagonal action is defined by $\gamma \cdot (d \otimes m) = d\gamma^{-1} \otimes \gamma m$ for $\gamma \in \Gamma$, $d \in D$, $m \in M$.] Consequently, 6.8 can be viewed as an isomorphism

$$H^n(\Gamma, M) \approx H_0(\Gamma, D \otimes M). \tag{6.9}$$

We will return to this point of view in §10.

Exercises

1. If Γ is of type FL and $\mathrm{cd}\,\Gamma = n$, show that $\mathbb{Z}$ admits a finite free resolution over $\mathbb{Z}\Gamma$ of length n.

2. Let Γ be of type FP and let m be an arbitrary integer ≥ 0. Prove that $\mathbb{Z}$ admits a finite projective resolution P over $\mathbb{Z}\Gamma$ such that P_i is free for $i \neq m$. Moreover, P can be taken to have length equal to $\max\{m, \mathrm{cd}\,\Gamma\}$. [Hint: Given a resolution, you can modify it by taking the direct sum with a complex of the form $0 \to Q \xrightarrow{\mathrm{id}} Q \to 0$.]

3. If Γ is of type FP and $n = \mathrm{cd}\,\Gamma$, show that $H^n(\Gamma, \mathbb{Z}\Gamma)$ is a finitely generated Γ-module.

*4. Let Γ be of type FP, let $n = \mathrm{cd}\,\Gamma$, and suppose $H^n(\Gamma, \mathbb{Z}\Gamma)$ is finitely generated *as an abelian group*. If $\Gamma' \subset \Gamma$ is a subgroup of infinite index, show that $\mathrm{cd}\,\Gamma' < n$. [Hint: For any Γ'-module M', $H^n(\Gamma', M') \approx H^n(\Gamma, \mathrm{Coind}_{\Gamma'}^{\Gamma} M')$. Now apply 6.9 and exercise 4b of §III.5.] Give an example to show that the hypothesis on $H^n(\Gamma, \mathbb{Z}\Gamma)$ cannot be dropped.

5. In this exercise it will be convenient to think of coefficient modules for homology as being *right* Γ-modules and coefficient modules for cohomology as *left* modules. Recall from §V.3 that there is an evaluation pairing

$$H_i(\Gamma, M) \otimes H^i(\Gamma, N) \to M \otimes_{\mathbb{Z}\Gamma} N.$$

In particular, taking $N = \mathbb{Z}\Gamma$, we obtain a map

$$H_i(\Gamma, M) \otimes H^i(\Gamma, \mathbb{Z}\Gamma) \to M \otimes_{\mathbb{Z}\Gamma} \mathbb{Z}\Gamma \approx M. \tag{$*$}$$

(a) Show that this map is a homomorphism of right Γ-modules, where Γ acts on the domain via its action on $H^i(\Gamma, \mathbb{Z}\Gamma)$. Deduce a map

$$\psi: H_i(\Gamma, M) \to \mathrm{Hom}_\Gamma(H^i(\Gamma, \mathbb{Z}\Gamma), M).$$

[Hint: Use the naturality of the evaluation pairing with respect to module homomorphisms. Alternatively, simply check definitions on the chain level.]

(b) If Γ is of type FP and $n = \mathrm{cd}\,\Gamma$, show that ψ is an isomorphism in dimension n. Thus

$$H_n(\Gamma, M) \approx \mathrm{Hom}_\Gamma(D, M) = \mathrm{H}^0(\Gamma, \mathrm{Hom}(D, M)), \tag{6.10}$$

where $D = H^n(\Gamma, \mathbb{Z}\Gamma)$ as in 6.9. [Hint: Consider the natural map $M \otimes_{\mathbb{Z}\Gamma} P \to \mathcal{H}om_\Gamma(\bar{P}, M)$, where $\bar{P} = \mathcal{H}om_\Gamma(P, \mathbb{Z}\Gamma)$.]

(c) Under the hypotheses of (b), let $z \in H_n(\Gamma, D)$ correspond under 6.10 to $\mathrm{id}_D \in \mathrm{Hom}_\Gamma(D, D)$. We call z the *fundamental class* of Γ. By definition, it is characterized by the equation

$$\langle z, u \rangle = u$$

for every $u \in H^n(\Gamma, \mathbb{Z}\Gamma) = D$, where $\langle\ ,\ \rangle$ denotes the evaluation pairing given by (*) above. Show for any module M that

$$\psi^{-1}: \mathrm{Hom}_\Gamma(D, M) \xrightarrow{\approx} H_n(\Gamma, M)$$

is given by $\psi^{-1}(f) = f_* z$ for $f \in \mathrm{Hom}_\Gamma(D, M)$, where $f_* = H_n(\Gamma, f): H_n(\Gamma, D) \to H_n(\Gamma, M)$. [Hint: This is true by the definition of z if $M = D$ and $f = \mathrm{id}$; the general case follows from the naturality of ψ^{-1}, cf. exercise 3a of §I.7.] Deduce that the isomorphism

$$H^0(\Gamma, \mathrm{Hom}(D, M)) \xrightarrow{\approx} H_n(\Gamma, M)$$

of 6.10 is given by cap product with z; more precisely, it is the composite

$$H^0(\Gamma, \mathrm{Hom}(D, M)) \xrightarrow{\cap z} H_n(\Gamma, \mathrm{Hom}(D, M) \otimes D) \to H_n(\Gamma, M),$$

where the second map is induced by the obvious coefficient homomorphism $\mathrm{Hom}(D, M) \otimes D \to M$. [Hint: Use the description of the cap product $H^0 \otimes H_n \to H_n$ given in exercise 1 of §V.3.]

(d) Show that the isomorphism

$$\varphi^{-1}: H^n(\Gamma, M) \xrightarrow{\approx} D \otimes_{\mathbb{Z}\Gamma} M$$

of 6.8 is given by evaluation on the fundamental class z. [Hint: This is true by the definition of z if $M = \mathbb{Z}\Gamma$; the general case follows by general nonsense, as in the first proof of 6.8.] Deduce that the isomorphism

$$H^n(\Gamma, M) \xrightarrow{\approx} H_0(\Gamma, D \otimes M)$$

of 6.9 is given by cap product with z. [Hint: Use V.3.10.]

6. (a) Let Γ be of type FP and let D and n be as above. Show that $H_n(\Gamma, D) \neq 0$ and deduce that $\mathrm{cd}\,\Gamma = \mathrm{hd}\,\Gamma$, where $\mathrm{hd}\,\Gamma$, the *homological dimension* of Γ is defined by

$$\mathrm{hd}\,\Gamma = \sup\{n: H_n(\Gamma, M) \neq 0 \text{ for some } \Gamma\text{-module } M.\}$$

*(b) Give an example of a group Γ (necessarily *not* of type FP) such that $\mathrm{hd}\,\Gamma < \mathrm{cd}\,\Gamma$.

*7. If Γ is of type FP and cd $\Gamma = n$, show that there are spectral sequences

$$E_{pq}^2 = \operatorname{Tor}_p^\Gamma(H^{n-q}(\Gamma, \mathbb{Z}\Gamma), M) \Rightarrow H^{n-(p+q)}(\Gamma, M)$$

and

$$E_2^{pq} = \operatorname{Ext}_\Gamma^p(H^{n-q}(\Gamma, \mathbb{Z}\Gamma), M) \Rightarrow H_{n-(p+q)}(\Gamma, M).$$

By considering the corner $p = q = 0$, recover the isomorphisms $H_n(\Gamma, M) \approx \operatorname{Hom}_\Gamma(D, M)$ and $H^n(\Gamma, M) \approx D \otimes_{\mathbb{Z}\Gamma} M$. [Hint: Use the universal coefficient spectral sequences, cf. Godement [1958], I.5.4.1 and I.5.5.1.]

*8. Prove that the FP and FL properties behave well with respect to extensions, amalgamations, etc. More precisely:

(a) Suppose $1 \to \Gamma' \to \Gamma \to \Gamma'' \to 1$ is exact. If Γ' and Γ'' are of type FP (resp. FL), then so is Γ.

(b) Let X be an acyclic Γ-complex such that X has only finitely many cells mod Γ. If each isotropy group Γ_σ is of type FP (resp. FL), then so is Γ. In particular, an amalgamation $\Gamma_1 *_A \Gamma_2$ is of type FP (or FL) if Γ_1, Γ_2, and A are of type FP (or FL).
[Hint: See Serre [1971] or Bieri [1976].]

7 Topological Interpretation

We have seen (cf. 2.2) that the existence of a finite dimensional $K(\Gamma, 1)$ implies that Γ has finite cohomological dimension. Similarly, 6.3 and 6.4 show that the existence of a finite (resp. finitely dominated) $K(\Gamma, 1)$ implies that Γ is of type FL (resp. FP). The purpose of this section is to consider the converse implications. We also want to give a topological interpretation of the Γ-modules $H^*(\Gamma, \mathbb{Z}\Gamma)$ which arose in §6. We will require a tiny bit of homotopy theory, namely, the Hurewicz theorem (which we quoted in §II.5).

The following theorem is due to Eilenberg–Ganea [1957] and Wall [1965, 1966]:

(7.1) Theorem. *Let Γ be an arbitrary group and let $n = \max\{\text{cd}\,\Gamma, 3\}$. Then there exists an n-dimensional $K(\Gamma, 1)$-complex Y. If Γ is finitely presented and of type FL (resp. FP) then Y can be taken to be finite (resp. finitely dominated).*

(We allow here the possibility that cd $\Gamma = \infty$, in which case the theorem simply asserts the existence of a $K(\Gamma, 1)$-complex.)

As an immediate consequence of 7.1 we have:

(7.2) Corollary (Eilenberg–Ganea). *If* cd $\Gamma \geq 3$ *then* cd Γ = geom dim Γ.

Of course we also have $\text{cd}\,\Gamma = \text{geom dim}\,\Gamma$ if $\text{cd}\,\Gamma = 0$ (since Γ is then trivial) or if $\text{cd}\,\Gamma = 1$ (by the Stallings–Swan theorem which we quoted in Example 2 of §2). In view of Theorem 7.1, then, we always have $\text{cd}\,\Gamma = \text{geom dim}\,\Gamma$ except possibly if $\text{cd}\,\Gamma = 2$ and $\text{geom dim}\,\Gamma = 3$. It is not known whether this possibility can actually occur.

Proof of 7.1. We will construct the skeleta Y^k of the desired Y inductively. To start the induction, let Y^2 be the 2-complex associated to some presentation of Γ (cf. §II.5, Exercise 2); thus $\pi_1 Y^2 \approx \Gamma$. If Γ is finitely presented, Y^2 can be taken to be finite. Note that its universal cover X^2 has $H_i = 0$ for $0 < i < 2$. Now assume inductively that Y^{k-1} has been constructed and that its universal cover X^{k-1} has $H_i = 0$ for $0 < i < k-1$. If Γ is finitely presented and of type FP, assume further that Y^{k-1} is finite. Choose a set of generators (z_α) for the Γ-module $H_{k-1}X^{k-1}$. By the Hurewicz theorem we can find for each α a map $f_\alpha: S^{k-1} \to X^{k-1}$ which represents z_α, in the sense that $H_{k-1}(f_\alpha): H_{k-1}S^{k-1} \to H_{k-1}X^{k-1}$ sends a generator of $H_{k-1}S^{k-1}$ to z_α. We now set

$$Y^k = Y^{k-1} \cup \bigcup_\alpha e_\alpha^k,$$

where the k-cell e_α^k is attached to Y^{k-1} via the composite

$$S^{k-1} \xrightarrow{f_\alpha} X^{k-1} \to Y^{k-1}.$$

Letting X^k be the universal cover of Y^k, we must verify that $H_i X^k = 0$ for $0 < i < k$. Note first that we can view X^{k-1} as the $(k-1)$-skeleton of X^k; indeed, X^k is obtained from X^{k-1} by attaching k-cells via the maps f_α and their transforms under the action of Γ on X^{k-1}. It is clear, then, that $H_i X^k = H_i X^{k-1} = 0$ for $0 < i < k-1$ and that we have an exact sequence

$$H_k(X^k, X^{k-1}) \xrightarrow{\partial} H_{k-1}X^{k-1} \to H_{k-1}X^k \to 0.$$

It will therefore suffice to show that ∂ is surjective.

Recall that $H_k(X^k, X^{k-1})$ (which is simply $C_k(X^k)$, the k-th cellular chain group) is a free $\mathbb{Z}\Gamma$-module with one basis element for each k-cell of Y^k, i.e., for each index α. Explicitly, there is a basis (v_α) obtained as follows: if $\chi_\alpha: (E^k, S^{k-1}) \to (X^k, X^{k-1})$ is a characteristic map for the cell attached via f_α, then $v_\alpha \in H_k(X^k, X^{k-1})$ is defined to be the image under

$$H_k(\chi_\alpha): H_k(E^k, S^{k-1}) \to H_k(X^k, X^{k-1})$$

of a generator of $H_k(E^k, S^{k-1}) = \mathbb{Z}$. In view of the diagram

$$\begin{array}{ccc} H_k(E^k, S^{k-1}) & \xrightarrow[\approx]{\partial} & H_{k-1}S^{k-1} \\ \downarrow{\scriptstyle H_k(\chi_\alpha)} & & \downarrow{\scriptstyle H_{k-1}(f_\alpha)} \\ H_k(X^k, X^{k-1}) & \xrightarrow{\partial} & H_{k-1}X^{k-1}, \end{array}$$

it follows that $\partial v_\alpha = z_\alpha$ (assuming that the generators of $H_k(E^k, S^{k-1})$ and $H_{k-1}S^{k-1}$ have been chosen compatibly), so that ∂ is indeed surjective.

(Note for future reference that if $H_{k-1}X^{k-1}$ happens to be a *free* $\mathbb{Z}\Gamma$-module with basis (z_α), then $\partial: H_k(X^k, X^{k-1}) \to H_{k-1}X^{k-1}$ is an isomorphism. It then follows from the long exact homology sequence of the pair (X^k, X^{k-1}) that $H_i X^k = 0$ for all $i > 0$, so that Y^k is a $K(\Gamma, 1)$.)

To complete the inductive step, we must show that Y^k can be taken to be finite if Γ is finitely presented and of type FP, i.e., we must show in this case that $H_{k-1}X^{k-1}$ is a finitely generated Γ-module. To see this, we need only note that the cellular chain complex

$$C_{k-1} \to \cdots \to C_0 \to \mathbb{Z} \to 0$$

of X^{k-1} is a partial free resolution of *finite type*, since Y^{k-1} was assumed to be finite. Therefore $H_{k-1} = \ker\{C_{k-1} \to C_{k-2}\}$ is finitely generated by 4.3 since Γ is of type FP.

If $n = \infty$, we now continue this inductive process indefinitely, and $Y = \bigcup_k Y^k$ is the desired $K(\Gamma, 1)$. If $n < \infty$, consider X^{n-1}. (This makes sense because $n - 1 \geq 2$.) Its cellular chain complex

$$C_{n-1} \to \cdots \to C_0 \to \mathbb{Z} \to 0$$

is a partial free resolution of length $n - 1$. Hence 2.1 implies that $H_{n-1}X^{n-1}$ is a projective $\mathbb{Z}\Gamma$-module. By the Eilenberg trick (2.7) there is a free module F such that $H_{n-1}X^{n-1} \oplus F$ is free. We now replace Y^{n-1} by $\bar{Y}^{n-1} = Y^{n-1} \vee S^{n-1} \vee S^{n-1} \vee \cdots$, where there is one copy of S^{n-1} for each basis element of F. The effect of this on $C(X^{n-1})$ is simply to add F to C_{n-1}, with $\partial|F = 0$. The universal cover $\bar{X}^{n-1}$ now has H_{n-1} free. We may therefore attach n-cells e_α^n to $\bar{Y}^{n-1}$ corresponding to basis elements z_α of $H_{n-1}\bar{X}^{n-1}$; as remarked above, the resulting $\bar{Y}^n = \bar{Y}^{n-1} \cup \bigcup e_\alpha^n$ will then be an n-dimensional $K(\Gamma, 1)$.

Suppose now that Γ is finitely presented and of type FL. Then Y^{n-1} is finite and the projective $H_{n-1}X^{n-1}$ is finitely generated. We know from 6.5 that $H_{n-1}X^{n-1}$ is stably free, so that there is a free module F of *finite rank* such that $H_{n-1}X^{n-1} \oplus F$ is free of finite rank. We now proceed as in the previous paragraph, and the resulting $\bar{Y}^n$ will be a finite $K(\Gamma, 1)$.

Finally, suppose that Γ is finitely presented but only of type FP instead of FL. On the one hand, the general inductive step above gives us a finite complex

$$Y^n = Y^{n-1} \cup e^n \cup \cdots \cup e^n$$

whose universal cover has $H_i = 0$ for $0 < i < n$. Hence $\pi_i Y^n \approx \pi_i X^n = 0$ for $1 < i < n$. On the other hand, we know that there is a $K(\Gamma, 1)$ of the form

$$\bar{Y}^n = Y^{n-1} \vee S^{n-1} \vee \cdots \cup e^n \cup \cdots,$$

so that $\pi_i \bar{Y}^n = 0$ for all $i > 1$. I claim that Y^n dominates $\bar{Y}^n$. Indeed, the required maps

$$\bar{Y}^n \xrightarrow{i} Y^n \xrightarrow{r} \bar{Y}^n$$

with $ri \simeq \text{id}$ are easily constructed as follows: i and r are both defined to be the identity on the common subcomplex Y^{n-1}, and they are extended arbitrarily to the cells that were attached. (These extensions exist trivially for each S^{n-1} that was wedged onto Y^{n-1} in forming $\bar{Y}^n$, and they exist for each e^n because $\pi_{n-1}Y^n = 0$ and $\pi_{n-1}\bar{Y}^n = 0$.) Finally, the homotopy $ri \simeq \text{id}$ is defined to be the constant homotopy on Y^{n-1} and is extended to all of $\bar{Y}^n$ by means of the vanishing of $\pi_{n-1}\bar{Y}^n$ and $\pi_n \bar{Y}^n$. □

For future reference we mention the following refinement of 7.1:

(7.3) Addendum. *The complex Y in* 7.1 *can be taken to be a simplicial complex.*

PROOF. Assume inductively that Y^{k-1} is simplicial, where the notation is that of the proof above. By the simplicial approximation theorem, we may then take each attaching map $f_\alpha\colon S^{k-1} \to Y^{k-1}$ to be simplicial relative to some triangulation of S^{k-1}. The resulting space Y^k is then triangulable by Whitehead [1949], §9, Lemma 2. □

Finally, we give a topological interpretation of the right Γ-modules $H^*(\Gamma, \mathbb{Z}\Gamma)$, assuming that Γ is finitely presented and of type FL. We will need the following observation:

(7.4) Lemma. *Let Γ be a group and M a left Γ-module. Let* $\text{Hom}_c(M, \mathbb{Z}) \subseteq \text{Hom}(M, \mathbb{Z})$ *consist of all abelian group homomorphisms $f\colon M \to \mathbb{Z}$ such that, for every $m \in M$, $f(\gamma m) = 0$ for all but finitely many $\gamma \in \Gamma$. Then there is a natural isomorphism*

$$\text{Hom}_\Gamma(M, \mathbb{Z}\Gamma) \approx \text{Hom}_c(M, \mathbb{Z}).$$

Moreover, this is an isomorphism of right Γ-modules, where Γ acts on $\text{Hom}_\Gamma(M, \mathbb{Z}\Gamma)$ *via its right action on $\mathbb{Z}\Gamma$ and Γ acts on* $\text{Hom}_c(M, \mathbb{Z})$ *via its left action on M (i.e., $(f\gamma)(m) = f(\gamma m)$ for $f \in$* $\text{Hom}_c(M, \mathbb{Z})$, *$\gamma \in \Gamma$, $m \in M$).*

PROOF. A $\mathbb{Z}$-module map $F\colon M \to \mathbb{Z}\Gamma$ has the form

$$F(m) = \sum_{\gamma \in \Gamma} f_\gamma(m)\gamma,$$

where $f_\gamma\colon M \to \mathbb{Z}$ and, for each $m \in M$, $f_\gamma(m) = 0$ for almost all $\gamma \in \Gamma$. One checks that such an F is a Γ-module homomorphism if and only if $f_\gamma(m) = f_1(\gamma^{-1}m)$ for all $\gamma \in \Gamma$. We therefore have a map $\text{Hom}_\Gamma(M, \mathbb{Z}\Gamma) \to \text{Hom}_c(M, \mathbb{Z})$ given by $F \mapsto f_1$, and this map is an isomorphism with inverse

$$f \mapsto \left\{m \mapsto \sum_{\gamma \in \Gamma} f(\gamma^{-1}m)\gamma\right\}.$$

The reader can easily verify that this isomorphism is natural and compatible with the right Γ-actions. □

Suppose now that Γ is finitely presented and of type FL. By 7.1 there is a contractible, free Γ-complex X with X/Γ finite. Then $C(X)$ is a finite free resolution of $\mathbb{Z}$ over $\mathbb{Z}\Gamma$, and $H^*(\Gamma, \mathbb{Z}\Gamma)$ is the cohomology of

$$\mathcal{H}om_\Gamma(C(X), \mathbb{Z}\Gamma).$$

In view of the lemma, we have

$$\mathcal{H}om_\Gamma(C(X), \mathbb{Z}\Gamma) \approx \mathcal{H}om_c(C(X), \mathbb{Z}) \subseteq \mathcal{H}om(C(X), \mathbb{Z}),$$

this isomorphism being compatible with the right Γ-action and the coboundary operators (the latter because of the naturality assertion in 7.4).

Recall that $C(X)$ has a $\mathbb{Z}$-basis with one element for each cell σ of X. These basis elements are freely permuted by Γ and fall into finitely many orbits. It follows easily that $\mathcal{H}om_c(C(X), \mathbb{Z})$ consists of those cochains $f \in \mathcal{H}om(C(X), \mathbb{Z})$ such that $f(\sigma) = 0$ for all but finitely many cells σ. The cohomology of this complex is called the cohomology of X with *compact supports*, and is denoted $H_c^*(X; \mathbb{Z})$. We have now established:

(7.5) Proposition. *If X is a contractible, free Γ-complex with compact quotient X/Γ, then there is an isomorphism*

$$H^*(\Gamma, \mathbb{Z}\Gamma) \approx H_c^*(X; \mathbb{Z})$$

of right Γ-modules, where the right action of Γ on $H_c^*(X; \mathbb{Z})$ is induced by the left action of Γ on X.

In view of 6.7, this yields:

(7.6) Corollary. *If X is as in 7.5, then*

$$\operatorname{cd} \Gamma = \max\{n: H_c^n(X; \mathbb{Z}) \neq 0\}.$$

EXERCISES

1. If Γ is a countable group, show that there exists a countable $K(\Gamma, 1)$-complex.

2. Give a topological interpretation of the FP_n conditions ($3 \leq n \leq \infty$) assuming Γ is finitely presented.

3. Suppose $\operatorname{cd} \Gamma = 2$. Show that there exists a 2-dimensional, acyclic, free Γ-complex. [Hint: Write $\Gamma = F/R$, where F is free and $R \lhd F$. Let Y^1 be a 1-complex with $\pi_1 = F$, let X^1 be the covering space corresponding to R, and argue as in the proof of 7.1.] Thus for any Γ we can characterize $\operatorname{cd} \Gamma$ topologically as the minimal dimension of a free, acyclic Γ-complex.

*4. Prove the following generalization of 7.5: Let X be a contractible Γ-complex with finite isotropy groups Γ_σ and with only finitely many cells mod Γ. Then $H^*(\Gamma, \mathbb{Z}\Gamma) \approx H_c^*(X; \mathbb{Z})$. [Hint: First show, by a spectral sequence argument for instance, that $H^*(\Gamma, \mathbb{Z}\Gamma)$ can be computed from $\mathcal{H}om_\Gamma(C(X), \mathbb{Z}\Gamma)$.]

8 Further Topological Results

The purpose of this section is to see what finiteness properties of Γ can be deduced if we are given a $K(\Gamma, 1)$ which is a *manifold*. We will see many examples of this situation in the next section. The proofs to be given in this section require more background in topology than we have assumed elsewhere in this book. The reader without this background, however, can still read and understand the *statements* of all the results.

We begin by noting the analogues of 2.2 and 6.3 for $K(\Gamma, 1)$-manifolds:

(8.1) Proposition. *Suppose Y is a d-dimensional $K(\Gamma, 1)$-manifold (possibly with boundary).*

(a) $\operatorname{cd} \Gamma \leq d$, *with equality if and only if Y is closed (i.e., compact and without boundary).*

(b) *If Y is compact then Γ is of type FL.*

Proof. (a) If Y is smooth, as it will be in all of our examples, then there are several ways to show that Y has the homotopy type of a CW-complex Y' of dimension $\leq d$ (so that $\operatorname{cd} \Gamma \leq d$ by 2.2): one can use Whitehead's triangulation theorem (cf. Munkres [1966], Ch. II), or Morse theory (cf. Milnor [1963]), or the theory of nerves of coverings (cf. Weil [1952]). An alternative, which works even if Y is not smooth, is to deduce from Poincaré duality with local coefficients that $H^i(\Gamma, M) = H^i(Y; M) = 0$ for $i > d$ and all Γ-modules M. This proves the first part of (a). If Y is closed, then we have $H^d(\Gamma, \mathbb{Z}_2) \approx H^d(Y; \mathbb{Z}_2) = \mathbb{Z}_2 \neq 0$, so $\operatorname{cd} \Gamma = d$. If Y is not closed, then one can deduce from Poincaré duality with local coefficients that $H^d(\Gamma, M) = H^d(Y; M) = 0$ for all Γ-modules M, so that $\operatorname{cd} \Gamma < d$; we omit the details since we will not make serious use of this strict inequality. [Alternatively, if Y is assumed to be triangulated, then there is a geometric proof that $\operatorname{cd} \Gamma < d$: one shows that Y admits a deformation retraction onto a subcomplex of dimension $< d$. We will describe such a deformation retraction explicitly in an interesting example in §9, Example 3.]

(b) Suppose Y is compact. In the smooth case, the complex Y' in the proof of (a) can be taken to be finite, so Γ is of type FL by 6.3. In the general case it is still true, but considerably harder to prove, that Y has the homotopy type of a finite complex (cf. Kirby–Siebenmann [1969] or West [1977]), so (b) is proved. [It is worth noting here that if we are content to prove that Γ is of type FP, then there is a quite elementary proof: we need only embed Y in Euclidean space and note that it is a retract of a compact polyhedral neighborhood (cf. Dold [1972], proof of V.4.11); thus Γ is of type FP by 6.4.] □

Next we wish to reinterpret $H^*(\Gamma, \mathbb{Z}\Gamma)$ (cf. 7.5) in case there is a compact $K(\Gamma, 1)$-manifold Y. Let X be the universal cover of Y. Since X is simply-connected, it is certainly orientable, and we denote by Ω its "orientation module." Thus Ω is an infinite cyclic group whose two generators correspond to the two orientations of X. The action of Γ on X induces an action of Γ on Ω, with an element $\gamma \in \Gamma$ acting as ± 1 according as the action of γ on X is orientation-preserving or orientation-reversing. Note that Γ acts trivially on Ω if and only if Y is orientable. Finally, we make the convention that the reduced homology of a space Z, denoted $\tilde{H}_*(Z)$, is the homology of the augmented chain complex of Z. In particular, $\tilde{H}_{-1}(\varnothing) = \mathbb{Z}$. We can now state:

(8.2) Proposition. *Let Y be a compact d-dimensional $K(\Gamma, 1)$-manifold (possibly with boundary). Let X be its universal cover and let Ω be the corresponding orientation module. Then there are Γ-module isomorphisms*

$$H^i(\Gamma, \mathbb{Z}\Gamma) \approx \tilde{H}_{d-i-1}(\partial X) \otimes \Omega$$

for all i. In particular, if Y is a closed manifold, then

$$H^i(\Gamma, \mathbb{Z}\Gamma) \approx \begin{cases} 0 & i \neq d \\ \Omega & i = d. \end{cases}$$

(8.3) Corollary. *Under the hypotheses of* 8.2, *there exists at least one integer k such that $\tilde{H}_k(\partial X) \neq 0$. Moreover, letting*

$$l = 1 + \min\{k : \tilde{H}_k(\partial X) \neq 0\},$$

we have

$$\operatorname{cd} \Gamma = d - l.$$

The corollary is immediate from 8.2 and 6.7. Note that $l > 0$ if $\partial Y \neq \varnothing$. Thus 8.3 makes more precise the inequality $\operatorname{cd} \Gamma < d$ of 8.1a in this case.

PROOF OF 8.2. Since Y is compact and has the homotopy type of a finite complex, we have

$$H^i(\Gamma, \mathbb{Z}\Gamma) \approx H^i_c(X)$$

by 7.5. On the other hand, there is a Poincaré–Lefschetz duality isomorphism

$$H^i_c(X) \approx H_{d-i}(X, \partial X).$$

This, however, is not canonical; it depends on a choice of orientation of X. In particular, it commutes or anti-commutes with the action of an element $\gamma \in \Gamma$ according as γ preserves or reverses the orientation of X. Consequently, we have a Γ-module isomorphism

$$H^i_c(X) \approx H_{d-i}(X, \partial X) \otimes \Omega.$$

Finally, since $\tilde{H}_*(X) = 0$,

$$H_{d-i}(X, \partial X) \approx \tilde{H}_{d-i-1}(\partial X).$$

The proposition follows at once. ☐

Remark. The reader who is uncomfortable with Poincaré–Lefschetz duality for non-compact manifolds with boundary might prefer the following alternative proof, which uses the duality theorem only for the compact manifold Y (but with local coefficients) and which does not use the fact that Y has the homotopy type of a finite complex:

Regarding $\mathbb{Z}\Gamma$ and Ω as local coefficient systems on Y, we have

$$H^i(\Gamma, \mathbb{Z}\Gamma) \approx H^i(Y; \mathbb{Z}\Gamma) \approx H_{d-i}(Y, \partial Y; \mathbb{Z}\Gamma \otimes \Omega).$$

Now $\mathbb{Z}\Gamma \otimes \Omega$, with diagonal Γ-action, is an induced module (cf. III.5.7), and homology with coefficients in an induced module is easily seen to be ordinary homology of the universal cover. One deduces

$$H_{d-i}(Y, \partial Y; \mathbb{Z}\Gamma \otimes \Omega) \approx H_{d-i}(X, \partial X; \Omega) \approx H_{d-i}(X, \partial X) \otimes \Omega,$$

and the result now follows as above.

Exercise

Let Γ be a group such that there exists a closed $K(\Gamma, 1)$-manifold. If $\Gamma' \subset \Gamma$ is a subgroup of infinite index, deduce from 8.1a that $\operatorname{cd} \Gamma' < \operatorname{cd} \Gamma$. [Hint: If $p: \tilde{Y} \to Y$ is a covering map of infinite degree, then $\tilde{Y}$ cannot be compact.] See Exercise 6 of §10 below for a generalization of this result.

9 Further Examples

We saw a few examples in §2 of groups Γ with $\operatorname{cd} \Gamma < \infty$, and we noted in §6 that some of those examples were of type FL. We now want to introduce some additional families of examples, the most interesting of which are the "arithmetic" groups. The study of such examples leads, as we will see in the next chapter, to some remarkable connections between group cohomology theory and number theory. Unfortunately, the assertions which we will make about the examples in this section are considerably less elementary than the corresponding assertions concerning our previous examples, and we will not be able to give the proofs. The reader is therefore advised to casually read through this section, taking on faith a number of deep results which we will have to state without proof.

1. Let Γ be a classical knot group (i.e., the fundamental group of the complement of a non-trivial knot $K \hookrightarrow S^3$). Then Γ is of type FL and $\operatorname{cd} \Gamma = 2$. To see this, let $Y = S^3 - T$, where T is an open tubular neighborhood of K. Then Y is a compact 3-manifold whose boundary is a torus, and a deep theorem of Papakyriakopoulos [1957] says that Y is a $K(\Gamma, 1)$. It now follows from 8.1 that Γ is of type FL. To calculate $\operatorname{cd} \Gamma$, we use another result from knot theory, namely, that $\pi_1(\partial Y)$ injects into $\pi_1 Y$ as a subgroup

of infinite index. If X is the universal cover of Y, it follows that ∂X is the disjoint union of countably many copies of the universal cover $\mathbb{R}^2$ of ∂Y. Thus the integer l in 8.3 is 1, and we have $\operatorname{cd} \Gamma = 3 - 1 = 2$, as claimed. Furthermore, we can use 8.2 to compute $H^*(\Gamma, \mathbb{Z}\Gamma)$. In particular, we see that $H^i(\Gamma, \mathbb{Z}\Gamma) = 0$ for $i \neq 2$ and that $H^2(\Gamma, \mathbb{Z}\Gamma)$ is a free abelian group of countable rank.

2. Let Γ be the $n \times n$ strict upper triangular group over $\mathbb{Z}$. We saw in Example 5 of §II.4 how to construct a $K(\Gamma, 1)$-manifold (without boundary). Namely, take $Y = \Gamma\backslash G$ where G is the $n \times n$ strict upper triangular group over $\mathbb{R}$. It is easy to see that Y is compact (see the exercise below), so 8.1 implies that Γ is of type FL and that

$$\operatorname{cd} \Gamma = \dim Y = n(n-1)/2.$$

(In particular, if $n = 3$, we recover the result of Example 6 of §2.) Moreover, 8.2 shows that

$$H^i(\Gamma, \mathbb{Z}\Gamma) = \begin{cases} 0 & i \neq \dfrac{n(n-1)}{2} \\ \mathbb{Z} & i = \dfrac{n(n-1)}{2}, \end{cases}$$

where Γ acts trivially on $\mathbb{Z}$. More generally, one can prove analogous results for an arbitrary finitely generated, torsion-free, nilpotent group Γ. For Malcev [1949] proved that Γ can be embedded as a discrete subgroup with compact quotient in a nilpotent Lie group G which is homeomorphic to Euclidean space of dimension $d = \operatorname{rank} \Gamma$. [One can view G as a "tensor product" $\mathbb{R} \otimes \Gamma$, cf. Bourbaki, LIE II, pp. 82–83, and LIE III, pp. 283–284.]

3. Consider now the group $SL_n(\mathbb{Z})$, $n \geq 2$. This group has torsion, hence $\operatorname{cd}(SL_n(\mathbb{Z})) = \infty$. We know, however, that it has torsion-free subgroups Γ of finite index, cf. Exercise 3 of §II.4. The intersection of Γ with the strict upper triangular group U has finite index in U and hence has $\operatorname{cd} = n(n-1)/2$ by Example 2 and Prop. 2.4a. Thus

$$\operatorname{cd} \Gamma \geq \frac{n(n-1)}{2}. \tag{9.1}$$

We will now outline a proof, based on the reduction theory of quadratic forms, that Γ is of type FL and that equality holds in 9.1. A different proof of these results will be discussed in Example 5 below.

Let X be the space which we called X_0 in Example 7 of §II.4; thus X is the space of positive definite quadratic forms on $\mathbb{R}^n$, modulo multiplication by positive scalars. Recall that X is a contractible manifold (without boundary) of dimension $n(n+1)/2 - 1$, that $SL_n(\mathbb{R})$ (hence also Γ) acts on X, and that X/Γ is a $K(\Gamma, 1)$. Recall also that if $n = 2$ then we can identify

X with the upper half plane (or, equivalently, with the open unit disk), with $SL_n(\mathbb{R})$ acting by linear fractional transformations.

In view of 8.1a we have

$$\operatorname{cd} \Gamma \leq \frac{n(n+1)}{2} - 1.$$

Moreover, one can show that X/Γ is non-compact, so that 8.1a implies that strict inequality holds above. But we can do substantially better than this by giving a direct geometric construction instead of relying on the generalities given in 8.1. Namely, we will show that X/Γ admits a deformation retraction onto a subspace which is a CW-complex of dimension $n(n-1)/2$. This will show that $\operatorname{cd} \Gamma \leq n(n-1)/2$ and hence will prove our claim that equality holds in 9.1.

We begin by describing how this is done for $n = 2$, using the open unit disk as our model for X. There is a tiling of X by "ideal hyperbolic triangles," which is compatible with the action of $SL_2(\mathbb{Z})$, and which is well-known in the theory of modular forms (see, for instance, Lehner [1964]). It is obtained by starting with a single ideal triangle (i.e., a hyperbolic triangle with vertices on the unit circle) and generating further triangles by successive reflections across the sides. This is illustrated in Fig. 9.2a, where alternate triangles of the tiling are shaded. The vertices of the triangles of the tiling are called cusps, and we denote by X^* the space obtained from X by adjoining the cusps; it can be viewed as a simplicial complex with a simplicial $SL_2(\mathbb{Z})$-action. [Note: We give X^* the usual simplicial topology, rather than the topology it inherits as a subset of the plane. In particular, the set ∂X^* of cusps is a discrete set in

Figure 9.2a

Figure 9.2b

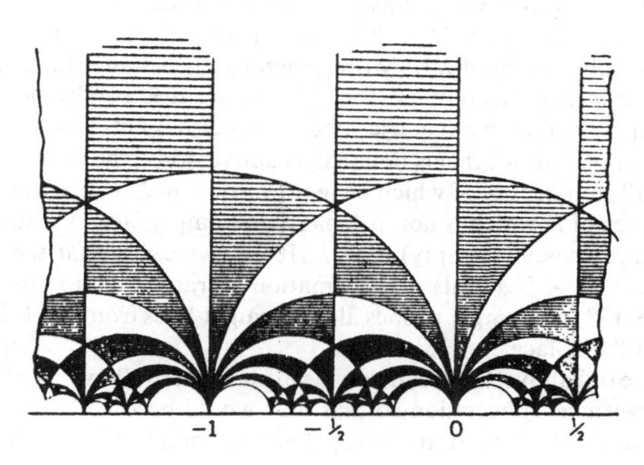

Figure 9.2c

(Figures 9.2a and 9.2c are reproduced from F. Klein and R. Fricke, Vorlesungen über die Theorie der elliptischen Modulfunctionen, Band I, B. G. Teubner, Leipzig, 1890, by permission of the publisher.)

the simplicial topology. It can be shown, however, that the simplicial topology agrees with the usual topology on the open subspace X of X^*.]

Now let T be the simplicial complement of ∂X^* in the barycentric subdivision K of X^*, i.e., T is the largest subcomplex of K disjoint from ∂X^*. (Explicitly, T consists of all simplices of K none of whose vertices are in ∂X^*.) The subcomplex T is shown in Fig. 9.2b. See also Fig. 9.2c, where the entire barycentric subdivision K is shown in the upper half plane model for X. The reader is invited to trace out the subcomplex T in this picture. [Note: The vertices of the large triangle in Fig. 9.2a correspond to the points -1, 0, ∞ in Fig. 9.2c; the barycenter of this triangle corresponds to the point $(-1 + i\sqrt{3})/2$ in Fig. 9.2c.]

I claim that this simplicial complement T is, in a canonical way, a deformation retract of the geometric complement $X = X^* - \partial X^*$. Indeed, one deforms X to T by pushing away from ∂X^* along straight lines (in the hyperbolic or simplicial sense). More precisely, any $x \in X$ lies in a closed 2-simplex σ of K with one vertex v in ∂X^* and the opposite face τ in T. Since $x \neq v$, there is a well-defined ray from v to x, and the deformation moves x along this ray away from v until it hits τ.

Note that T and the deformation are described purely in terms of the simplicial structure on X^*, so they are compatible with the action of $SL_2(\mathbb{Z})$ and its subgroup Γ. It follows that T/Γ is a deformation retract of X/Γ, and it has dimension $1 = 2(2 - 1)/2$, as required.

Before proceeding to the general case, we make some remarks about the construction above. Let K be a simplicial complex. A subcomplex L is said to be *full* if every simplex of K having all of its vertices in L is itself in L. For example, the subcomplex ∂X^* above is not full in X^*, but it is full in the barycentric subdivision of X^*. (More generally, if K is an arbitrary complex and L an arbitrary subcomplex, then the barycentric subdivision of L is a full subcomplex of the barycentric subdivision of K.) Let T be the simplicial complement of L in K; this is defined, as above, to be the subcomplex consisting of all simplices of K which have no vertices in L. If L is full in K, then every simplex of K which is not in L has a non-empty face in T such that the opposite face (possibly empty) is in L. It follows easily that the geometric complement $K - L$ admits a deformation retraction onto the simplicial complement T; one simply pushes along straight lines from the L-face of any simplex to the T-face.

What we did above, then, can now be explained as follows: We first passed to the barycentric subdivision so that ∂X^* would become full, and we then had a deformation retraction of the geometric complement of ∂X^* onto the simplicial complement.

The generalization to $SL_n(\mathbb{Z})$ for arbitrary n is based on a theory due to Voronoi [1907]. Voronoi constructs an enlargement X^* of X, obtained by adjoining certain positive *semi*-definite quadratic forms, namely, those whose nullspace admits a basis consisting of vectors in $\mathbb{Q}^n$. He gives X^* an explicit decomposition into convex cells which are permuted by the action of $SL_n(\mathbb{Z})$. The subspace $\partial X^* = X^* - X$ is a subcomplex. The cells of X are not necessarily simplices, but X^* admits a barycentric subdivision K which is simplicial and which inherits a simplicial action of $SL_n(\mathbb{Z})$. It follows, as above, that $X = X^* - \partial X^*$ admits a deformation retraction (compatible with the $SL_n(\mathbb{Z})$-action) onto the simplicial complement T of ∂X^* in K. One sees by looking at Voronoi's construction that ∂X^* contains the entire

$(n-2)$-skeleton of X^*, and it follows easily that T has codimension $\geq n-1$. Thus

$$\dim T/\Gamma = \dim T \leq \frac{n(n+1)}{2} - 1 - (n-1) = \frac{n(n-1)}{2},$$

as required. Finally, Voronoi proves in addition that X^* has only finitely many cells mod the $SL_n(\mathbb{Z})$-action, so T/Γ is in fact a finite complex and Γ is of type FL.

Remark. In case $n = 2$, the subcomplex T is a tree, being a 1-dimensional deformation retract of the contractible space X. This tree was first introduced by Serre [1977a], I.4.2, who used it to give an easy proof of the classical theorem expressing $SL_2(\mathbb{Z})$ as an amalgamation $\mathbb{Z}_4 *_{\mathbb{Z}_2} \mathbb{Z}_6$. The existence of an analogue of T for $n > 2$ (i.e., a Γ-invariant, contractible subcomplex of X of dimension $n(n-1)/2$ with T/Γ compact) was proved by Soulé [1978] for $n = 3$ and by Ash [1977] in general. Ash, in fact, proved an analogous result for a more general class of groups, using a generalization of Voronoi's theory. The cohomological dimension of such groups had previously been computed by Borel and Serre [1974], using different methods which we will describe in Example 5 below.

The groups in Examples 2 and 3 are examples of "arithmetic" groups. Before looking at general arithmetic groups, let's see what can be said about arbitrary discrete subgroups of Lie groups.

4. Let G be a Lie group with only finitely many connected components, let K be a maximal compact subgroup, and let $d = \dim G - \dim K$. If Γ is any torsion-free discrete subgroup of G, I claim that

$$\operatorname{cd} \Gamma \leq d,$$

with equality if and only if Γ is *co-compact* in G, i.e., if and only if G/Γ is compact. Moreover, in the co-compact case Γ is of type FL and one has $H^i(\Gamma, \mathbb{Z}\Gamma) = 0$ for $i \neq d$ and $H^d(\Gamma, \mathbb{Z}\Gamma) = \mathbb{Z}$ (possibly with non-trivial Γ-action). To prove these assertions we need only recall (cf. §II.4, Example 8) that $\Gamma \backslash G/K$ is a $K(\Gamma, 1)$-manifold (without boundary) of dimension d; moreover, $\Gamma \backslash G/K$ is compact if and only if Γ is co-compact. The assertions now follow at once from 8.1 and 8.2.

5. We now specialize Example 4 to the "arithmetic" case. We will confine ourselves to a brief outline of this theory, referring to the survey paper of Serre [1979] for more details and further references. Let G be a linear algebraic group defined over $\mathbb{Q}$. In concrete terms, this simply means that G is a subgroup of some general linear group GL_n, defined by polynomial equations (with rational coefficients) in the n^2 matrix entries. For example, G could be the group SL_n (defined by the single equation $\det(a_{ij}) = 1$) or the

strict upper triangular group (defined by the equations $a_{ii} = 1$, $a_{ij} = 0$ for $i > j$). The real matrices satisfying the given polynomial equations then form a Lie group $G(\mathbb{R})$ (e.g., $SL_n(\mathbb{R})$), which is known to have only finitely many components; and the integral matrices satisfying the equations form a discrete subgroup $G(\mathbb{Z}) \subseteq G(\mathbb{R})$ (e.g., $SL_n(\mathbb{Z}) \subset SL_n(\mathbb{R})$). The group $G(\mathbb{Z})$ is said to be an *arithmetic group*. More generally, if $\Gamma \subseteq G(\mathbb{Q})$ is a subgroup *commensurable* with $G(\mathbb{Z})$, i.e., such that $\Gamma \cap G(\mathbb{Z})$ is of finite index in both Γ and $G(\mathbb{Z})$, then Γ is said to be an *arithmetic subgroup* of $G(\mathbb{Q})$. In particular, any subgroup of $G(\mathbb{Z})$ of finite index is arithmetic.

Suppose now that Γ is a *torsion-free* arithmetic group. As in Example 4, we have a contractible manifold $X = G(\mathbb{R})/K$ on which Γ acts freely, so that X/Γ is a $K(\Gamma, 1)$-manifold. It turns out, however, that Γ is usually *not* co-compact in $G(\mathbb{R})$, and hence that X/Γ is non-compact. In fact, there is a "$\mathbb{Q}$-rank" $l \geq 0$ attached to the algebraic group G, and Γ is co-compact in $G(\mathbb{R})$ if and only if $l = 0$.

[Here is the precise definition of l, for the benefit of the reader familiar with algebraic groups: l is the rank of a maximal $\mathbb{Q}$-split torus in G/RG, where RG is the radical of G. If G is the strict upper triangular group, for example, then $RG = G$, so $l = 0$ and Γ is co-compact, as we saw in Example 2. If $G = SL_n$ ($n \geq 2$), on the other hand, then $RG = \{1\}$ and a maximal split torus in $G/RG = G$ is given by the diagonal matrices; thus $l = n - 1 > 0$ and Γ is not co-compact, as we saw in Example 3.]

In spite of this failure of Γ to be co-compact, however, Borel and Serre [1974] show that Γ is of type FL and compute $H^*(\Gamma, \mathbb{Z}\Gamma)$. Their method is to replace X by a contractible manifold $\bar{X}$ *with boundary* (with X as its interior if G is semi-simple) on which $G(\mathbb{Q})$ operates. They show that the action of Γ is free and proper and that the quotient $\bar{X}/\Gamma$ is compact. Thus $\bar{X}/\Gamma$ is a compact $K(\Gamma, 1)$-manifold, so that Γ is of type FL by 8.1b. Moreover, the construction of $\bar{X}$ is explicit enough that Borel and Serre are able to identify the homotopy type of the boundary: namely, $\partial\bar{X}$ has the homotopy type of a countable bouquet of $(l - 1)$-spheres, where l is the $\mathbb{Q}$-rank mentioned above. [If $l = 0$, this simply means that $\partial\bar{X} = \varnothing$.] Letting $d = \dim \bar{X}$, we conclude from 8.2 that $H^i(\Gamma, \mathbb{Z}\Gamma) = 0$ for $i \neq d - l$, and that $H^{d-l}(\Gamma, \mathbb{Z}\Gamma)$ is a free abelian group of countable rank if $l > 0$ and of rank 1 if $l = 0$. In particular,

$$\operatorname{cd} \Gamma = d - l.$$

If $G = SL_n$, for example, then $d = n(n + 1)/2 - 1$ and $l = n - 1$, so $\operatorname{cd} \Gamma = d - l = n(n - 1)/2$, as we saw in Example 3.

6. Let S be a finite set of prime numbers and let $\mathbb{Z}_S \subset \mathbb{Q}$ be the localization of $\mathbb{Z}$ obtained by inverting the elements of S. If G is as in example 5, then a subgroup of $G(\mathbb{Q})$ is called *S-arithmetic* if it is commensurable with $G(\mathbb{Z}_S)$. Borel and Serre [1976] have shown that the results of example 5 extend to torsion-free S-arithmetic groups, provided G is reductive.

7. Serre [1971] noted the following remarkable consequence of example 6: If Γ is any finitely generated torsion-free subgroup of $GL_n(\mathbb{Q})$, then cd $\Gamma < \infty$. [Sketch of proof: Note that $\Gamma \subseteq GL_n(\mathbb{Z}_S)$ for some S, and apply example 6 to the S-arithmetic group $GL_n(\mathbb{Z}_S)$ to conclude that any torsion-free subgroup of the latter has finite cohomological dimension.]

8. The result of example 7 is no longer true if $\mathbb{Q}$ is replaced by a field K of transcendence degree ≥ 1 over $\mathbb{Q}$. For one can find finitely generated torsion-free subgroups of $GL_n(K)$ which contain unipotent subgroups that are free abelian of infinite rank. [A subgroup U of $GL_n(K)$ is *unipotent* if every element of U has all n of its eigenvalues equal to 1. According to a famous theorem of Kolchin, this holds iff U is conjugate to a subgroup of the strict upper triangular group.] But Alperin and Shalen [1982] have shown that this is essentially the only type of counter-example. More precisely, they show that if Γ is a finitely generated torsion-free subgroup of $GL_n(K)$, where K is an arbitrary field of characteristic 0, then cd $\Gamma < \infty$ iff there is a finite upper bound on the Hirsch ranks of the finitely generated unipotent subgroups of Γ.

Remark. The reader may have noticed that all of the examples of groups of type *FL* which we gave in this section have the following remarkable property: There is an integer n (necessarily equal to cd Γ) such that $H^i(\Gamma, \mathbb{Z}\Gamma) = 0$ for $i \neq n$ and $H^n(\Gamma, \mathbb{Z}\Gamma)$ is free abelian. The significance of this, as we will see in the next section, is that it implies that these groups satisfy a *duality* relation, somewhat analogous to Poincaré duality for manifolds.

EXERCISE

Let G and Γ be as in Example 2. Let $C = \{(a_{ij}) \in G: |a_{ij}| \leq 1$ for all $i, j\}$. Show that $G = \Gamma \cdot C$ and deduce that Γ is co-compact in G. [Hint: Consider the elementary matrices in Γ, and interpret left multiplication by such a matrix as an elementary row operation.]

10 Duality Groups

For any group Γ of type *FP*, we have seen (6.9) that there is an isomorphism

$$H^n(\Gamma, M) \approx H_0(\Gamma, D \otimes M)$$

for any Γ-module M, where $n = \text{cd}\,\Gamma$, D is the right Γ-module $H^n(\Gamma, \mathbb{Z}\Gamma)$, and $D \otimes M = D \otimes_{\mathbb{Z}} M$ with the diagonal Γ-action. In this section we will study those groups Γ for which the isomorphism above extends to isomorphisms

$$H^i(\Gamma, M) \approx H_{n-i}(\Gamma, D \otimes M)$$

for all i. These groups were first considered by Bieri and Eckmann [1973], who proved the following characterization of them:

(10.1) Theorem. *The following conditions are equivalent for a group Γ of type FP:*

(i) *There exist an integer n and a Γ-module D such that*

$$H^i(\Gamma, M) \approx H_{n-i}(\Gamma, D \otimes M)$$

for all Γ-modules M and all integers i.

(ii) *There is an integer n such that $H^i(\Gamma, \mathbb{Z}\Gamma \otimes A) = 0$ for all $i \neq n$ and all abelian groups A.*

(iii) *There is an integer n such that $H^i(\Gamma, \mathbb{Z}\Gamma) = 0$ for all $i \neq n$ and $H^n(\Gamma, \mathbb{Z}\Gamma)$ is torsion-free (as an abelian group).*

(iv) *There are natural isomorphisms*

$$H^i(\Gamma, -) \approx H_{n-i}(\Gamma, D \otimes -),$$

where $n = \operatorname{cd} \Gamma$ and $D = H^n(\Gamma, \mathbb{Z}\Gamma)$, which are compatible with the connecting homomorphisms in the long exact homology and cohomology sequences associated to a short exact sequence of modules.

PROOF. (i) $\Rightarrow$ (ii): Apply (i) with M an induced module $\mathbb{Z}\Gamma \otimes A$. It is easy to see that $D \otimes M$ is also induced (cf. III.5.7). Since induced modules are H_*-acyclic, it follows from (i) that $H^i(\Gamma, M) \approx H_{n-i}(\Gamma, D \otimes M) = 0$ for $i \neq n$.

(ii) $\Rightarrow$ (iii): Applying (ii) with $A = \mathbb{Z}$, we find $H^i(\Gamma, \mathbb{Z}\Gamma) = 0$ for $i \neq n$. Now apply (ii) with $A = \mathbb{Z}_k$ $(k > 0)$, and use the cohomology exact sequence associated to the short exact sequence

$$0 \to \mathbb{Z}\Gamma \xrightarrow{k} \mathbb{Z}\Gamma \to \mathbb{Z}\Gamma \otimes \mathbb{Z}_k \to 0.$$

This yields

$$0 = H^{n-1}(\Gamma, \mathbb{Z}\Gamma \otimes \mathbb{Z}_k) \to H^n(\Gamma, \mathbb{Z}\Gamma) \xrightarrow{k} H^n(\Gamma, \mathbb{Z}\Gamma),$$

showing that $H^n(\Gamma, \mathbb{Z}\Gamma)$ has no k-torsion.

(iii) $\Rightarrow$ (iv): Note first that the integer n in (iii) is necessarily equal to $\operatorname{cd} \Gamma$ by 6.7. We now give three different proofs that (iii) $\Rightarrow$ (iv).

First proof: We will use the general nonsense of §III.7, which the reader may want to review before proceeding further. Consider the cohomological functor $H^*(\Gamma, -)$. Since $\operatorname{cd} \Gamma = n < \infty$, we can re-index and regard $H^*(\Gamma, -)$ as a *homological* functor; indeed, an exact sequence $0 \to M' \to M \to M'' \to 0$ of Γ-modules yields an exact sequence

$$\cdots \to H^{n-1}(\Gamma, M'') \to H^n(\Gamma, M') \to H^n(\Gamma, M) \to H^n(\Gamma, M'') \to 0,$$

so we obtain a homological functor T by setting $T_i = H^{n-i}(\Gamma, -)$. The hypothesis (iii) implies (via 5.2) that $H^i(\Gamma, F) = 0$ for all $i \neq n$ and all free

$\mathbb{Z}\Gamma$-modules F. Consequently, T_i is effaceable for $i > 0$. Now consider the functor $H_*(\Gamma, D \otimes -)$, where $D = H^n(\Gamma, \mathbb{Z}\Gamma)$. This is also a homological functor. For if $0 \to M' \to M \to M'' \to 0$ is exact, then so is $0 \to D \otimes M' \to D \otimes M \to D \otimes M'' \to 0$ (since D is torsion-free by hypothesis), hence we obtain a long exact homology sequence. Moreover, $H_i(\Gamma, D \otimes -)$ is effaceable for $i > 0$ because $D \otimes M$ is induced if M is induced. Since $T_0 \approx H_0(\Gamma, D \otimes -)$ by 6.9, Theorem III.7.3 yields an isomorphism $T \approx H_*(\Gamma, D \otimes -)$ of homological functors, whence (iv).

Second proof: This is similar to the first proof, except that we will use cap products instead of general nonsense to get the map $T \to H_*(\Gamma, D \otimes -)$. By exercise 5d of §6, there is a "fundamental class" $z \in H_n(\Gamma, D)$ such that cap product with z gives an isomorphism $H^n(\Gamma, -) \xrightarrow{\approx} H_0(\Gamma, D \otimes -)$. Since the cap product map

$$\cap z \colon T_i = H^{n-i}(\Gamma, -) \to H_i(\Gamma, D \otimes -)$$

is compatible (up to sign) with connecting homomorphisms in long exact sequences, it follows by dimension-shifting that this map is an isomorphism for all i.

Third proof: Let $P = (P_i)_{0 \le i \le n}$ be a finite projective resolution of $\mathbb{Z}$ over $\mathbb{Z}\Gamma$ of length n, and consider the dual complex $\bar{P} = \mathscr{H}om_\Gamma(P, \mathbb{Z}\Gamma)$. Since $H^i(\Gamma, \mathbb{Z}\Gamma) = 0$ for $i \neq n$, $\bar{P}$ provides a projective resolution of $D = H^n(\Gamma, \mathbb{Z}\Gamma)$:

$$\cdots \to \bar{P}^{n-1} \to \bar{P}^n \to D \to 0;$$

more precisely, the n-th suspension $\Sigma^n \bar{P}$ is a projective resolution of D. In view of the duality isomorphism

$$\mathscr{H}om_\Gamma(P, M) \approx \bar{P} \otimes_\Gamma M$$

of I.8.3, it follows that

$$(*) \qquad H^i(\Gamma, M) \approx H_{-i}(\bar{P} \otimes_\Gamma M) = H_{n-i}(\Sigma^n \bar{P} \otimes_\Gamma M) = \operatorname{Tor}^\Gamma_{n-i}(D, M)$$

for any Γ-module M. Since D is torsion free, we can now apply III.2.2 to obtain

$$(**) \qquad \operatorname{Tor}^\Gamma_{n-i}(D, M) \approx H_{n-i}(\Gamma, D \otimes M).$$

The reader can easily verify that $(*)$ and $(**)$ are natural and compatible with connecting homomorphisms, whence (iv).

(iv) $\Rightarrow$ (i): Trivial. □

If Γ satisfies the conditions of the theorem then Γ is said to be a *duality group* and the Γ-module $D = H^n(\Gamma, \mathbb{Z}\Gamma)$ is called the *dualizing module* of Γ. If, in addition, D is infinite cyclic (as an abelian group), then Γ is said to be a *Poincaré duality group*, and in this case Γ is said to be *orientable* if Γ acts trivially on D and *non-orientable* otherwise. Note that if Γ is an orientable Poincaré duality group then (iv) takes the familiar form

$$H^i(\Gamma, M) \approx H_{n-i}(\Gamma, M),$$

as in Poincaré duality for closed, orientable manifolds.

Remarks

1. It is sometimes useful to note that if Γ is a duality group, then there are duality isomorphisms $H^i(\Gamma, M) \xrightarrow{\approx} H_{n-i}(\Gamma, D \otimes M)$ given by cap product with a fixed element $z \in H_n(\Gamma, D)$. This was made explicit in the second proof above, and it also follows from the third proof, exactly as in the proof of VI.7.2.

2. Bieri and Eckmann originally defined a duality group to be a group Γ, *not necessarily of type FP*, such that there are cap product isomorphisms

$$\cap z\colon H^i(\Gamma, M) \xrightarrow{\approx} H_{n-i}(\Gamma, D \otimes M)$$

for all i and M, where n is a fixed integer, D a fixed Γ-module, and z a fixed element of $H_n(\Gamma, D)$. It is clear from the previous remark that a duality group in our sense is also one in theirs. Conversely, if Γ is a duality group in their sense, then naturality of the cap product implies that there are *natural* isomorphisms

$$H^i(\Gamma, -) \approx H_{n-i}(\Gamma, D \otimes -).$$

Since $D \otimes -$ and $H_*(\Gamma, -)$ commute with direct limits, it follows that $H^*(\Gamma, -)$ commutes with direct limits. It also follows that $\operatorname{cd} \Gamma \leq n < \infty$. Thus Γ is of type FP by 4.8, and hence Γ is a duality group in our sense. It is also worth noting that the integer n which occurs in the Bieri–Eckmann definition is necessarily equal to $\operatorname{cd} \Gamma$ (this is clear from the proof of 10.1) and that the module D is necessarily isomorphic to $H^n(\Gamma, \mathbb{Z}\Gamma)$. Indeed, the given duality isomorphism with $i = n$ and $M = \mathbb{Z}\Gamma$ gives

$$H^n(\Gamma, \mathbb{Z}\Gamma) \approx H_0(\Gamma, D \otimes \mathbb{Z}\Gamma) \approx D \otimes_{\mathbb{Z}\Gamma} \mathbb{Z}\Gamma \approx D,$$

and this is compatible with the Γ-action by naturality of the duality isomorphism.

EXAMPLES

1. If Γ is a group such that there exists a closed $K(\Gamma, 1)$-manifold Y, then, as one would expect, Γ is a Poincaré duality group and is orientable if and only if Y is orientable. This follows immediately from 8.1b, 8.2, and the criterion 10.1(iii) for duality. [Alternatively, use Poincaré duality in Y, with local coefficients, to verify directly that Γ satisfies the Bieri–Eckmann definition of duality group which we quoted above, with D equal to the orientation module Ω associated to Y.] For example, the free abelian group $\mathbb{Z}^n$ is an orientable Poincaré duality group. [Take Y to be the n-dimensional torus.] More generally, any finitely generated, torsion-free, nilpotent group

is an orientable Poincaré duality group by example 2 of §9. [Alternatively, one can give a purely algebraic proof, cf. Bieri [1976], Theorem 9.10.] Still more generally, if G is an arbitrary Lie group with only finitely many connected components and Γ is a discrete, co-compact, torsion-free subgroup, then Γ is a Poincaré duality group and is orientable if G is connected (cf. example 4 of §9).

Remark. It is not known whether every Poincaré duality group admits a closed $K(\Gamma, 1)$-manifold. This is even unknown in the 2-dimensional case. But Eckmann and Müller [1980] have shown that every 2-dimensional Poincaré duality group Γ which has the homology of a surface other than S^2 or P^2 is in fact isomorphic to the fundamental group of that surface. It follows that if there is a 2-dimensional counter-example (i.e., a 2-dimensional Poincaré duality group Γ for which there is no closed $K(\Gamma, 1)$-manifold, then there is even a counter-example Γ such that $H_*(\Gamma) \approx H_*(S^2)$. Such a Γ would be very interesting indeed. In particular, the method of J. Cohen [1972] shows that Γ would not be of type FL.

2. If Γ is a free group on k generators, $1 \leq k < \infty$, then Γ is a 1-dimensional duality group. Indeed, we know that $\operatorname{cd} \Gamma = 1$ and that Γ is of type FL, so we need only check that $H^0(\Gamma, \mathbb{Z}\Gamma \otimes A) = 0$ for any A; but this is easily seen to hold for any infinite group Γ, cf. §III.5, exercise 4a. If $k = 1$, then Γ is an orientable Poincaré duality group by example 1 (or by directly computing $H^*(\Gamma, \mathbb{Z}\Gamma)$ from the resolution I.4.5). If $k > 1$, however, then Γ is not a Poincaré duality group; for we have

$$\mathbb{Z}^k \approx H^1(\Gamma, \mathbb{Z}) \approx H_0(\Gamma, D \otimes \mathbb{Z}) = H_0(\Gamma, D),$$

so D cannot be infinite cyclic. (In fact, it is not hard to show that D is a free abelian group of countable rank, cf. exercise 2 below.)

3. The results stated in example 1 of §9 show that knot groups are 2-dimensional duality groups (but not Poincaré duality groups).

4. The results of Borel and Serre quoted in example 5 of §9 show that torsion-free arithmetic groups are duality groups (but not Poincaré duality groups except in the rank 0 case.)

5. If Γ is a finitely presented group of cohomological dimension 2 which cannot be decomposed as a free product, then Γ is a duality group. Indeed, Γ is of type FP by the comments following 6.1, so it suffices to show that $H^i(\Gamma, \mathbb{Z}\Gamma \otimes A) = 0$ for $i < 2$ and all A. This is clear for $i = 0$. For $i = 1$, it is not hard to see (cf. Swan [1969], §3) that $H^1(\Gamma, \mathbb{Z}\Gamma)$ is a free abelian group of rank $e - 1$, where e is the number of "ends" of Γ, and that $H^1(\Gamma, \mathbb{Z}\Gamma \otimes A) \approx H^1(\Gamma, \mathbb{Z}\Gamma) \otimes A$. But $e = 1$ (and hence $H^1(\Gamma, \mathbb{Z}\Gamma \otimes A) = 0$) under our hypotheses, because of the following theorem of Stallings [1968]: A finitely generated, torsion-free group with more than one end is either infinite cyclic or a free product. (See also Scott–Wall [1979] for the theory of ends and a proof of Stallings's theorem.)

6. Further examples of duality groups can be constructed from known examples by forming extensions and amalgamations. The precise results are somewhat complicated, and we refer to Bieri [1976] for details. [Warning: These group theoretic constructions do *not* always preserve the property of being a duality group. For example, a free product of duality groups is not a duality group unless the groups are free, cf. exercise 3 below.]

Remark. The examples above suggest that if Γ is a duality group then the dualizing module D is either infinite cyclic (in which case Γ is a Poincaré duality group) or free abelian of infinite rank. It is not known whether this is true, the problem being the freeness of D. It is known, however, as a consequence of results of Farrell [1975], that if D is not infinite cyclic then D has infinite rank (i.e., $\dim_{\mathbb{Q}}(D \otimes \mathbb{Q}) = \infty$). A proof can be found in Bieri [1976], §9.5, along with other interesting properties of D.

We close this section by proving one useful property of duality groups:

(10.2) Proposition. *Let Γ be a torsion-free group and Γ' a subgroup of finite index. Then Γ is a duality group if and only if Γ' is a duality group. Moreover, if Γ and Γ' are duality groups then they have the same dualizing module.* (*More precisely, if D is the dualizing module for Γ, then the dualizing module for Γ' is isomorphic to the module* $\mathrm{Res}^{\Gamma}_{\Gamma'} D$ *obtained by restricting the action of Γ to Γ'.*)

PROOF. By 6.6, Γ is of type FP if and only if Γ' is of type FP. Since $H^*(\Gamma, \mathbb{Z}\Gamma) \approx H^*(\Gamma', \mathbb{Z}\Gamma')$ by Shapiro's lemma, the first part of the proposition follows from the criterion 10.1(iii) for duality. To prove the second part, we must check that the isomorphism $H^n(\Gamma, \mathbb{Z}\Gamma) \approx H^n(\Gamma', \mathbb{Z}\Gamma')$ is an isomorphism of right Γ'-modules. By straightforward definition-checking (cf. §III.8, exercise 2), one sees that this isomorphism is the composite

$$H^n(\Gamma, \mathbb{Z}\Gamma) \xrightarrow{\text{res}} H^n(\Gamma', \mathbb{Z}\Gamma) \to H^n(\Gamma', \mathbb{Z}\Gamma'),$$

where the second map is induced by the coefficient homomorphism

$$\mathbb{Z}\Gamma \approx \mathbb{Z}\Gamma \otimes_{\mathbb{Z}\Gamma'} \mathbb{Z}\Gamma' \underset{(\text{III}.5.9)}{\xrightarrow{\approx}} \mathrm{Hom}_{\Gamma'}(\mathbb{Z}\Gamma, \mathbb{Z}\Gamma') \xrightarrow{\text{eval at } 1} \mathbb{Z}\Gamma'.$$

One checks that this homomorphism is given explicitly by

$$\gamma \mapsto \begin{cases} \gamma & \gamma \in \Gamma' \\ 0 & \text{otherwise,} \end{cases}$$

and this is clearly compatible with the right Γ'-action. Since res: $H^n(\Gamma, \mathbb{Z}\Gamma) \to H^n(\Gamma', \mathbb{Z}\Gamma)$ is obviously compatible with the right Γ'-action (in fact, with the right Γ-action), the proof is complete. □

EXERCISES

1. If Γ is an n-dimensional duality group with dualizing module D, prove that D is a $\mathbb{Z}\Gamma$-module of type FP (i.e., it admits a finite projective resolution) and of projective dimension n. [Hint: Look at the third proof of (iii) ⇒ (iv) in 10.1.]

2. Let Γ be a free group of finite rank > 1. Prove that the dualizing module $H^1(\Gamma, \mathbb{Z}\Gamma)$ is a free abelian group of infinite rank. [Hint: Use exercise 5 of §5 and argue by induction on the rank of Γ. Alternatively, use the theory of ends.]

3. Show that a duality group of cohomological dimension > 1 cannot be decomposed as a free product. [Hint: Exercise 5 of §5.]

4. If Γ is an n-dimensional duality group whose dualizing module D is $\mathbb{Z}$-free, prove that

$$H_i(\Gamma, M) \approx H^{n-i}(\Gamma, \text{Hom}(D, M))$$

for any Γ-module M. [Hint: The starting point is exercise 5b of §6. Now imitate any of the three proofs that (iii) $\Rightarrow$ (iv) in 10.1. If you imitate the first or second proof, you will need to use exercise 4b of §5.]

5. Show that if we are willing to use chain complex coefficients (cf. §VII.5) then *every* group of type *FP* behaves like a duality group. Namely, there is a "dualizing complex" $\mathscr{D}$ such that

$$H^i(\Gamma, \mathscr{C}) \approx H_{n-i}(\Gamma, \mathscr{C} \otimes \mathscr{D})$$

and

$$H_i(\Gamma, \mathscr{C}) \approx H^{n-i}(\Gamma, \mathscr{H}om(\mathscr{D}, \mathscr{C}))$$

for every complex $\mathscr{C}$ of Γ-modules. [Hint: Let $\mathscr{D}$ be the dual (suitably re-indexed) of a finite projective resolution of $\mathbb{Z}$ over $\mathbb{Z}\Gamma$. Look at the definitions of $H^i(\Gamma, \mathscr{C})$ and $H_i(\Gamma, \mathscr{C})$, and apply exercise 1 of §VI.6.]

*6. (a) Prove the following result of Strebel [1977]: If Γ is a Poincaré duality group and $\Gamma' \subset \Gamma$ is a subgroup of infinite index, then cd $\Gamma' <$ cd Γ. [Hint: Use exercise 4 of §6.] This result is motivated by the special case (which may in fact be the general case) given in the exercise of §8.

(b) Give an example which shows that the result of (a) does not hold for arbitrary duality groups.

11 Virtual Notions

Up to now this chapter has been concerned primarily with torsion-free groups. Indeed, we observed in §2 that groups with torsion necessarily have cd $= \infty$, and hence they certainly cannot satisfy the *FL* or *FP* conditions or be duality groups. On the other hand, we have seen examples of groups with torsion (such as $SL_n(\mathbb{Z})$) which have torsion-free subgroups of finite index that satisfy these finiteness conditions. The purpose of this section is to introduce a convenient language for describing this situation.

In general we will say that a group *virtually* has a given property if some subgroup of finite index has that property. For instance, Γ is *virtually torsion-free* if Γ has a torsion-free subgroup of finite index. In this case it follows from

Serre's theorem (3.1) that all such subgroups have the same cohomological dimension (which may be finite or infinite); for if Γ' and Γ'' are two torsion-free subgroups of finite index, then $\Gamma' \cap \Gamma''$ has finite index in both Γ' and Γ'', so Serre's theorem gives $\operatorname{cd} \Gamma' = \operatorname{cd}(\Gamma' \cap \Gamma'') = \operatorname{cd} \Gamma''$. The common cohomological dimension of the torsion-free subgroups of finite index is called the *virtual cohomological dimension* of Γ and is denoted $\operatorname{vcd} \Gamma$.

Similarly, Γ is said to be of *type VFP* (resp. *VFL*) if some subgroup of finite index is of type *FP* (resp. *FL*). If Γ is of type *VFP* then it follows from 6.6 that *every* torsion-free subgroup of finite index is of type *FP*. The corresponding statement for groups of type *VFL* is not known, so one is led to introduce the following apparently stronger condition: Γ is said to be of *type WFL* if Γ is virtually torsion-free and *every* torsion-free subgroup of finite index is of type *FL*.

To summarize, we have introduced four "virtual" finiteness conditions in this section, which are related as follows:

$$WFL \Rightarrow VFL \Rightarrow VFP \Rightarrow (\operatorname{vcd} < \infty).$$

It is not known whether the first two implications are reversible. Note also that $VFP \Rightarrow FP_\infty$ by 5.1.

We saw in §7 that $\operatorname{cd} \Gamma < \infty$ if and only if there exists a finite dimensional, contractible, free Γ-*CW*-complex. To state the "virtual" analogue of this, we introduce the following weakening of the notion of "free action": We say that a Γ-complex is *proper* if the isotropy group Γ_σ is finite for every cell σ. (This is consistent with the notion of "proper action" introduced in §II.4, Example 6, but we will not need to know this.) Note that a proper Γ-complex is Γ'-free for every torsion-free subgroup $\Gamma' \subseteq \Gamma$. Conversely, if X is a Γ-complex which is Γ'-free for some subgroup Γ' of finite index (i.e., if X is *virtually free*), then X is proper. Thus, if Γ is virtually torsion-free, a Γ-complex is proper if and only if it is virtually free.

We can now state the topological interpretation of finite virtual cohomological dimension:

(11.1) Theorem. *Let Γ be a virtually torsion-free group. Then* $\operatorname{vcd} \Gamma < \infty$ *if and only if there exists a finite dimensional, contractible, proper Γ-complex X. If there exists such an X which has only finitely many cells* mod Γ, *then Γ is of type WFL.*

PROOF. The second assertion and the "if" part of the first assertion are obvious from the remarks above. The "only if" part follows from the proof of Serre's theorem (3.1). Indeed, the entire proof of that theorem, except for the last sentence, is valid for any group Γ with $\operatorname{vcd} \Gamma < \infty$, and it yields the desired X. □

There remain several open questions in connection with this theorem. It is not known, for example, whether X can be taken to have dimension equal to vcd Γ. Clearly we have vcd $\Gamma \leq \dim X$ for any X as in 11.1; but the proof above *always* yields an X with $\dim X >$ vcd Γ, except in the trivial case where Γ is finite and X is a point. It is also not known whether the converse of the second assertion of 11.1 is true. In particular, the proof above will always yield an X with infinitely many cells mod Γ, unless Γ is finite. Thus we are very far from understanding the relation between our algebraically defined finiteness conditions and the analogous topologically defined conditions.

For future reference, we record now some facts which can be deduced from the proof of 11.1:

(11.2) Addendum. *If* vcd $\Gamma < \infty$ *then the complex* X *in* 11.1 *can be taken to be simplicial (with simplicial* Γ*-action) and to have the following property: For every finite subgroup* $H \subset \Gamma$*, the fixed-point set* X^H *is non-empty and contractible.*

PROOF. The assertion about the fixed-point sets follows easily from the proof of 11.1, cf. §3, exercise. The fact that X can be taken to be simplicial requires some preliminaries concerning ordered simplicial complexes.

An *ordered simplicial complex* is a simplicial complex together with a partial ordering on its vertices, such that the vertices of any simplex are linearly ordered: $v_0 < v_1 < \cdots < v_n$. The canonical example of an ordered simplicial complex is the barycentric subdivision K' of an arbitrary simplicial complex K. To see that K' is ordered, recall that the vertices of K' are the barycenters of the simplices of K, hence there is an ordering on the vertices of K' corresponding to the obvious ordering (by the face relation) on the simplices of K; moreover, the simplices of K' are precisely the finite sets of barycenters which are linearly ordered (cf. Spanier [1966], 3.3, or Schubert [1968], III.2.6), so K' is indeed an ordered simplicial complex.

Next we recall the definition of the *simplicial product* $K \times L$ of two ordered simplicial complexes K and L. The set of vertices vert$(K \times L)$ is defined by vert$(K \times L) =$ vert$(K) \times$ vert(L), with the product ordering: $(v, w) \leq (v', w') \Leftrightarrow v \leq v'$ and $w \leq w'$. A simplex of $K \times L$ is defined to be a linearly ordered set of vertices $(v_0, w_0) < \cdots < (v_n, w_n)$ such that $\{v_0, \ldots, v_n\}$ is a simplex of K (possibly of dimension $< n$) and $\{w_0, \ldots, w_n\}$ is a simplex of L. It is a classical fact (cf. Eilenberg–Steenrod [1952], II.8.9, or Milnor [1957b]) that this abstract product complex $K \times L$ provides a triangulation of the geometric product of K and L, i.e., that $|K \times L| \approx |K| \times |L|$, where $|\ \ |$ denotes geometric realization. [Here, as in the proof of 3.1, some care needs to be taken in topologizing $|K| \times |L|$.]

Look now at the proof of 3.1. The $K(\Gamma', 1)$-complex Y' can be taken to be simplicial by 7.3. Its universal cover X' is then simplicial (with simplicial Γ'-action), cf. Spanier [1966], Theorem 3.8.3, or Schubert [1968], §III.6.9. Passing to the barycentric subdivision if necessary, we can assume that X' is an *ordered* simplicial complex and that the Γ'-action preserves the ordering. The co-induction construction given in the proof of 3.1 can therefore be done in the category of ordered simplicial complexes using the product described above, and the resulting X will then be simplicial. □

EXAMPLES

1. vcd $\Gamma = 0$ if and only if Γ is finite.

2. If $\Gamma = \Gamma_1 *_A \Gamma_2$ where Γ_1 and Γ_2 are finite, then Γ is of type WFL and vcd $\Gamma \leq 1$. More generally, the same results are true whenever Γ is the fundamental group of a finite graph of finite groups, cf. §VII.9. This is an easy consequence of Serre's theory of groups acting on trees, which provides a contractible 1-complex on which Γ acts properly and with finite quotient. For details see Serre [1977a], II.2.6, Prop. 11. [Note: If we drop the requirement that the graph of finite groups be finite and require instead that there be a bound on the orders of the finite groups, then it is still true that vcd $\Gamma \leq 1$, but Γ will not necessarily be of type WFL.] Conversely, if Γ is a group such that vcd $\Gamma \leq 1$, then the Stallings–Swan theorem (§2, Example 2) shows that Γ has a free subgroup of finite index, and it is then known that Γ is the fundamental group of a graph of finite groups. This is due to Karrass–Pietrowski–Solitar [1973] in case Γ is finitely generated (in which case the graph can be taken to be finite) and to D. E. Cohen [1972] and Scott [1974] in the general case. Proofs can also be found in Dunwoody [1979] and Scott–Wall [1979].

3. If Γ is a finitely generated one-relator group, then Γ is of type WFL and vcd $\Gamma \leq 2$. Indeed, Γ is virtually torsion-free by Fischer–Karrass–Solitar [1972]. And Lyndon [1950] established an exact sequence

$$0 \to \mathbb{Z}[\Gamma/C] \to F \to \mathbb{Z}\Gamma \to \mathbb{Z} \to 0,$$

where F is a free $\mathbb{Z}\Gamma$-module of finite rank and C is a finite cyclic subgroup of Γ. Our assertions follow at once, since $\mathbb{Z}[\Gamma/C]$ is clearly a free $\mathbb{Z}\Gamma'$-module of finite rank for every torsion-free subgroup $\Gamma' \subseteq \Gamma$ of finite index. [Note that we have in this case a contractible, proper, 2-dimensional Γ-complex with finite quotient. This follows from the topological version of Lyndon's theorem, which we stated in Exercise 2c of §II.5.]

4. Let G be a Lie group and let X be the associated homogeneous space G/K, where K is a maximal compact subgroup. Let Γ be a discrete subgroup of G, and assume that Γ is virtually torsion-free. [This is automatic if, for instance, Γ is a subgroup of some $GL_n(\mathbb{Z})$, cf. §II.4, Exercise 3.] Then it follows from Example 4 of §9 that

$$\text{vcd } \Gamma \leq \dim X,$$

with equality if and only if Γ is co-compact in G. In the latter case, Γ is of type WFL. More interestingly, Γ is of type WFL in the arithmetic case discussed in

Example 5 of §9, even though Γ is usually not co-compact, and the Borel–Serre results which we stated there give a precise formula for vcd Γ. For instance

$$\operatorname{vcd} SL_n(\mathbb{Z}) = \frac{n(n-1)}{2}.$$

Finally, a group Γ is said to be a *virtual duality group* if some subgroup of finite index is a duality group. The arithmetic groups provide a large and interesting class of virtual duality groups.

(11.3) Proposition. *A group Γ is a virtual duality group if and only if the following two conditions are satisfied:*

(a) *Γ is of type VFP.*
(b) *There is an integer n such that $H^i(\Gamma, \mathbb{Z}\Gamma) = 0$ for $i \neq n$ and $H^n(\Gamma, \mathbb{Z}\Gamma)$ is torsion-free.*

In this case every torsion-free subgroup of finite index is a duality group with dualizing module $H^n(\Gamma, \mathbb{Z}\Gamma)$.

This is an easy consequence of Shapiro's lemma, as in the proof of 10.2.

EXERCISES

1. Suppose Γ is virtually torsion-free and let Γ' be an arbitrary subgroup. Show that vcd $\Gamma' \leq$ vcd Γ, with equality if $(\Gamma : \Gamma') < \infty$.

*2. Study the behavior of the virtual finiteness properties with respect to extensions, amalgamations, etc. [Hint: Choose a torsion-free subgroup of finite index and apply 2.4b, Exercise 4 of §2, and Exercise 8 of §6.] *Warning*: These group-theoretic constructions do *not* preserve the property of being virtually torsion-free, cf. Schneebeli [1978]. So the results you state will have to include a hypothesis that the group under consideration is virtually torsion-free.

CHAPTER IX

Euler Characteristics

1 Ranks of Projective Modules: Introduction

Let Γ be a finite group and P a finitely generated projective $\mathbb{Z}\Gamma$-module. There are several ways that one might try to associate a "rank" to P. One candidate for this, which we will denote by $\varepsilon(P)$ or $\varepsilon_\Gamma(P)$, is defined via extension of scalars with respect to the augmentation map $\mathbb{Z}\Gamma \to \mathbb{Z}$. Namely, we consider $P_\Gamma = \mathbb{Z} \otimes_{\mathbb{Z}\Gamma} P$, which is a finitely generated projective $\mathbb{Z}$-module; since projective $\mathbb{Z}$-modules are free, we can set

(1.1) $$\varepsilon(P) = \mathrm{rk}_{\mathbb{Z}}(P_\Gamma),$$

the right hand side being the rank in the naive sense (i.e., the cardinality of a basis). Note that 1.1 makes sense even if Γ is infinite.

Our second definition of a "rank" of P, which we will denote $\rho(P)$ or $\rho_\Gamma(P)$, is based on the fact that P itself is finitely generated and projective as a $\mathbb{Z}$-module; hence we can set

(1.2) $$\rho(P) = \mathrm{rk}_{\mathbb{Z}}(P)/|\Gamma|.$$

Here, of course, we do need Γ to be finite.

Both ε and ρ give the "right" answer n if P is the free module $\mathbb{Z}\Gamma^n$. Other than that, however, it is not obvious that they have anything to do with one another. In fact, it is not even obvious that the rational number $\rho(P)$ is an integer. But it turns out that $\varepsilon = \rho$ for all finite groups Γ and all finitely generated projectives P. This fact, which is due to Swan, plays a crucial role in our theory of Euler characteristics. We will therefore devote the first few sections of the chapter to a proof of this equality.

Following Bass [1976, 1979], our proof will go as follows: In §2 we will introduce a third notion of rank, due to Hattori and Stallings and defined for finitely generated projectives over an arbitrary ring. In §3 we will use standard commutative algebra to give a concrete interpretation of the Hattori–Stallings rank in case the ring is commutative. This interpretation, together with formal properties of the Hattori–Stallings rank over group rings, will lead in §4 to a proof that $\varepsilon = \rho$.

Exercise

The purpose of this exercise is to show that some restriction on a ring A is needed if there is to be a "reasonable" $\mathbb{Z}$-valued rank for finitely generated projective A-modules. Suppose there is a $\mathbb{Z}$-valued function r on finitely generated projective A-modules, satisfying:

(i) $r(P \oplus Q) = r(P) + r(Q)$.
(ii) $r(P) > 0$ if $P \neq 0$.
(iii) $r(A) = 1$.

Prove then that A is *indecomposable*, i.e., that A cannot be decomposed as the direct sum of two non-zero left ideals. [Hint: A decomposition $A = I \oplus J$ yields an equation $1 = r(I) + r(J)$ in $\mathbb{Z}$.]

2 The Hattori–Stallings Rank

If F is a finitely generated free $\mathbb{Z}$-module, then the rank of F is equal to the trace of the identity endomorphism of F; indeed, this is just a restatement of the obvious fact that the trace of the $n \times n$ identity matrix is n. With this as motivation, Hattori [1965] and Stallings [1965b] defined the "rank" of an arbitrary finitely generated projective module over an arbitrary ring to be the trace of the identity map. In order to make sense out of this, we need to begin by developing a theory of traces for endomorphisms of projectives which are not necessarily free over rings which are not necessarily commutative.

Let A be an arbitrary ring, F a finitely generated free A-module, and $\alpha: F \to F$ an endomorphism. One would like to define $\operatorname{tr}(\alpha)$ as the sum of the diagonal entries of the matrix of α relative to a basis for F. Unfortunately, this sum is not in general independent of the choice of basis for F, unless A is commutative. (One can already see this for 1×1 matrices.)

There is, however, a well-defined trace in the quotient $T(A) = A/[A, A]$, where $[A, A]$ is the additive subgroup of A generated by all commutators $ab - ba$, $a, b \in A$. [Warning: $[A, A]$ is not an ideal, in general, so $T(A)$ is simply an abelian group, not a ring.] Namely, we define $\operatorname{tr}(\alpha) \in T(A)$ for α an

$n \times n$ matrix to be $\sum_{i=1}^{n} \bar{\alpha}_{ii}$, where $\bar{\alpha}_{ii}$ is the image of α_{ii} in $T(A)$. It is trivial to verify that

$$\text{(2.1)} \qquad \operatorname{tr}(\alpha\beta) = \operatorname{tr}(\beta\alpha)$$

whenever α is an $m \times n$ matrix and β is an $n \times m$ matrix (so that $\alpha\beta$ and $\beta\alpha$ are both square); and from this it follows that $\operatorname{tr}(\beta\alpha\beta^{-1}) = \operatorname{tr}(\beta^{-1}\beta\alpha) = \operatorname{tr}(\alpha)$ for any invertible β, so the trace of an endomorphism is indeed a well-defined element of $T(A)$, independent of the choice of basis.

We would like now to extend the trace to endomorphisms of finitely generated *projective* modules. I claim that there is a unique way to do this so that 2.1 continues to hold for maps $P \underset{\beta}{\overset{\alpha}{\leftrightarrows}} P'$. For let $\alpha: P \to P$ be an endomorphism of a finitely generated projective, and express P as a direct summand of a finitely generated free module F via maps $P \underset{\iota}{\overset{\pi}{\leftrightarrows}} F$ with $\pi\iota = \mathrm{id}_P$. Then α gives rise to an endomorphism $\iota\alpha\pi$ of F. [In concrete terms, this is simply α on the summand P and 0 on the complementary summand $\ker \iota\pi$.] If we want 2.1 to hold, then we are forced to define

$$\text{(2.2)} \qquad \operatorname{tr}(\alpha) = \operatorname{tr}(\iota\alpha\pi);$$

for 2.1 would imply $\operatorname{tr}(\iota\alpha\pi) = \operatorname{tr}(\alpha\pi\iota) = \operatorname{tr}(\alpha \circ \mathrm{id})$.

It remains to show that the right side of 2.2 is independent of the choice of $P \leftrightarrows F$ and that 2.1 continues to hold. Suppose we have maps

$$F \underset{\pi}{\overset{\iota}{\leftrightarrows}} P \underset{\beta}{\overset{\alpha}{\leftrightarrows}} P' \underset{\iota'}{\overset{\pi'}{\leftrightarrows}} F',$$

where F and F' are free, $\pi\iota = \mathrm{id}_P$, and $\pi'\iota' = \mathrm{id}_{P'}$. Then we can apply 2.1 to matrices representing the composites

$$F \underset{\iota'\beta\pi}{\overset{\iota\alpha\pi'}{\leftrightarrows}} F'$$

to conclude that $\operatorname{tr}(\iota\alpha\pi'\iota'\beta\pi) = \operatorname{tr}(\iota'\beta\pi\iota\alpha\pi')$. Since $\pi'\iota'$ and $\pi\iota$ are identity maps, this yields

$$\operatorname{tr}(\iota\alpha\beta\pi) = \operatorname{tr}(\iota'\beta\alpha\pi').$$

Taking $P' = P$ and $\beta = \mathrm{id}$, we conclude that the right side of 2.2 is indeed independent of the choice of $P \leftrightarrows F$. And taking P' and β arbitrary, we conclude that 2.1 holds for arbitrary $P \underset{\beta}{\overset{\alpha}{\leftrightarrows}} P'$. This proves the claim. The resulting trace $\operatorname{tr}(\alpha)$ will sometimes be denoted $\operatorname{tr}_A(\alpha)$.

Next we discuss the behavior of traces with respect to *extension and restriction of scalars.*

(2.3) Proposition. *Let $\varphi: A \to B$ be a ring homomorphism and let $T(\varphi): T(A) \to T(B)$ be the induced map. If $\alpha: P \to P$ is an endomorphism of a finitely generated projective A-module, then $B \otimes_A \alpha: B \otimes_A P \to B \otimes_A P$ is an endomorphism of a finitely generated projective B-module, and we have*

$$\operatorname{tr}_B(B \otimes_A \alpha) = T(\varphi)(\operatorname{tr}_A(\alpha)).$$

PROOF. It suffices to consider the case where P is free. Then $B \otimes_A P$ is also free, and a matrix representing $B \otimes_A \alpha$ is obtained by applying φ to the entries of a matrix representing α. □

(2.4) Proposition. *Let $\varphi: A \to B$ be a ring homomorphism such that B is finitely generated and projective as a left A-module. Then there is a map*

$$\mathrm{tr}_{B/A}: T(B) \to T(A)$$

with the following property: Let $\alpha: P \to P$ be an endomorphism of a finitely generated projective B-module; then P is also finitely generated and projective as an A-module, and

$$\mathrm{tr}_A(\alpha) = \mathrm{tr}_{B/A}(\mathrm{tr}_B(\alpha)).$$

The proof will use:

(2.5) Lemma. *Let A be an arbitrary ring and let P be a finitely generated projective A-module which is a direct sum $P_1 \oplus \cdots \oplus P_n$. Let $\alpha: P \to P$ be an endomorphism with components $\alpha_{ij}: P_i \to P_j$. Then $\mathrm{tr}(\alpha) = \sum_1^n \mathrm{tr}(\alpha_{ii})$.*

PROOF. This reduces easily to the case where each P_i is free, in which case the result follows from an examination of the matrix of α. □

PROOF OF 2.4. For any $b \in B$, the right multiplication map $\mu_b: B \to B$ is an endomorphism of the left A-module B, so it has a trace $\mathrm{tr}_A(\mu_b) \in T(A)$. Consider the additive map $B \to T(A)$ given by $b \mapsto \mathrm{tr}_A(\mu_b)$. Since $\mu_{b_1 b_2} = \mu_{b_2}\mu_{b_1}$, this map annihilates commutators $b_1 b_2 - b_2 b_1$. There is therefore an induced map $T(B) \to T(A)$, which is the desired $\mathrm{tr}_{B/A}$. To prove the formula $\mathrm{tr}_A(\alpha) = \mathrm{tr}_{B/A}(\mathrm{tr}_B(\alpha))$, note first that it holds if $P = B$ (in which case α is necessarily of the form μ_b). In view of Lemma 2.5, it continues to hold if P is a direct sum of copies of B, i.e., if P is a free B-module. In the general case, choose $P \overset{\pi}{\underset{\iota}{\leftrightarrows}} F$ with $\pi\iota = \mathrm{id}_P$, where F is free over B. Then $\mathrm{tr}(\iota\alpha\pi) = \mathrm{tr}(\alpha\pi\iota) = \mathrm{tr}(\alpha)$ over both A and B, so the formula for $\alpha: P \to P$ follows from the formula for $\iota\alpha\pi: F \to F$. □

We can now define the *Hattori–Stallings rank* of P, denoted $R(P)$ or $R_A(P)$, by

$$R(P) = \mathrm{tr}(\mathrm{id}_P).$$

Note that $R(P)$ is not a number, but rather an element of the somewhat mysterious abelian group $T(A)$. In the next two sections, however, we will give concrete interpretations of $R(P)$ for some interesting classes of rings. Before proceeding to this, the reader is strongly urged to look at the exercises below; the results of exercises 5, 6, and 7 will be needed in §4.

EXERCISES

1. Let P be a finitely generated projective (left) A-module and let P^* be its dual. Then P^* is a right A-module and we have $P^* \otimes_A P \approx \mathrm{Hom}_A(P, P)$ by I.8.3b. Prove that there is a commutative diagram

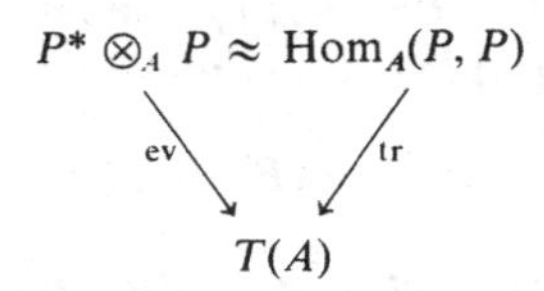

where ev, the "evaluation map," is given by $\mathrm{ev}(u \otimes x) = \overline{u(x)}$. In some treatments of the theory of traces, this diagram is used to define tr, cf. Bass [1976, 1979]. [Hint: Consider $P \overset{x}{\underset{u}{\leftrightarrows}} A$.]

2. (a) Let C be a finite projective chain complex over A and let $\tau: C \to C$ be a chain map with components $\tau_i: C_i \to C_i$. Define the "Lefschetz number" $L(\tau) \in T(A)$ by $L(\tau) = \sum (-1)^i \mathrm{tr}(\tau_i)$. Prove that $L(\tau_0) = L(\tau_1)$ if $\tau_0 \simeq \tau_1$. [Hint: Write down the definition of homotopy and use 2.1.]

   (b) Deduce that one can define a trace map $\mathrm{tr}: \mathrm{Hom}_A(M, M) \to T(A)$ whenever M is an A-module of type FP by setting $\mathrm{tr}(\alpha) = L(\tau)$, where τ is any lifting of α to a finite projective resolution of M.

3. Suppose A is an algebra over a commutative ring k (e.g., A is always an algebra over its center). Show that $\mathrm{Hom}_A(P, P)$ and $T(A)$ are k-modules and that $\mathrm{tr}: \mathrm{Hom}_A(P, P) \to T(A)$ is k-linear. More generally, show that the same is true if P is replaced by a module M of type FP as in exercise 2b.

4. Show by checking definitions that $R(P)$ is equal to the trace of an idempotent matrix defining P. More precisely, let F be a finitely generated free module and $e: F \to F$ an (idempotent) projection operator of F onto a direct summand isomorphic to P; then $R(P) = \mathrm{tr}(e)$.

5. Let Γ be a group and $\varphi: \mathbb{Z}\Gamma \to \mathbb{Z}$ the augmentation map. Show that the numerical "rank" $\varepsilon_\Gamma(\ )$ defined in §1 can be obtained from the Hattori–Stallings rank $R_{\mathbb{Z}\Gamma}(\ )$ by the formula
$$\varepsilon_\Gamma(P) = T(\varphi)(R_{\mathbb{Z}\Gamma}(P)).$$
[Apply 2.3 with $\alpha = \mathrm{id}_P$ and note that $T(\varphi)$ takes values in $\mathbb{Z}$ because $T(\mathbb{Z}) = \mathbb{Z}$.]

6. Let Γ be a finite group.

   (a) Look at the definition of $\mathrm{tr}_{\mathbb{Z}\Gamma/\mathbb{Z}}: T(\mathbb{Z}\Gamma) \to T(\mathbb{Z}) = \mathbb{Z}$ in the proof of 2.4 and deduce that $\mathrm{tr}_{\mathbb{Z}\Gamma/\mathbb{Z}}(\bar{1}) = |\Gamma|$ and that $\mathrm{tr}_{\mathbb{Z}\Gamma/\mathbb{Z}}(\bar{\gamma}) = 0$ for $1 \neq \gamma \in \Gamma$. Here $\bar{\gamma}$ is the image of γ in $T(\mathbb{Z}\Gamma)$. Consequently, there is a well-defined homomorphism $\tau: T(\mathbb{Z}\Gamma) \to \mathbb{Z}$ such that $\tau(\bar{1}) = 1$ and $\tau(\bar{\gamma}) = 0$ for $1 \neq \gamma \in \Gamma$, and one has $\mathrm{tr}_{\mathbb{Z}\Gamma/\mathbb{Z}} = |\Gamma| \cdot \tau$.
   (b) Deduce from 2.4 applied to the inclusion $\mathbb{Z} \hookrightarrow \mathbb{Z}\Gamma$ (with $\alpha = \mathrm{id}_P$) that $\mathrm{rk}_{\mathbb{Z}}(P) = |\Gamma| \cdot \tau(R_{\mathbb{Z}\Gamma}(P))$ for any finitely generated projective $\mathbb{Z}\Gamma$-module P. Hence the "rank" ρ defined in §1 is given by
$$\rho_\Gamma(P) = \tau(R_{\mathbb{Z}\Gamma}(P)).$$

7. Exercise 6b shows that $\rho_\Gamma(P) \in \mathbb{Z}$. On the other hand, it is obvious from the definition of ρ that $\rho_\Gamma(P) > 0$ if $P \neq 0$. Show, therefore, that $\mathbb{Z}\Gamma$ is indecomposable when Γ is finite. [Use the exercise of §1.]

8. Show that the definition of τ in exercise 6a makes sense even if Γ is infinite, i.e., there is a well-defined homomorphism $\tau\colon T(\mathbb{Z}\Gamma) \to \mathbb{Z}$ such that $\tau(\bar{1}) = 1$ and $\tau(\bar{\gamma}) = 0$ for $1 \neq \gamma \in \Gamma$. We can therefore define a $\mathbb{Z}$-valued "rank" ρ over $\mathbb{Z}\Gamma$ by $\rho(P) = \tau(R(P))$. In view of exercise 6b, this agrees with the ρ of §1 if Γ is finite. Moreover, it turns out for arbitrary Γ that $\rho(P) > 0$ if $P \neq 0$. [This non-trivial result is due to Kaplansky; proofs can be found in Montgomery [1969] and in Passman [1977], Theorem 1.8 and Exercise 9 of Chapter 1.] Consequently, you can show as in exercise 7 that $\mathbb{Z}\Gamma$ is indecomposable for *any* group Γ.

3 Ranks over Commutative Rings

If A is an integral domain with field of fractions K, then there is an easy way to define a $\mathbb{Z}$-valued rank for finitely generated projective A-modules. Namely, we set

$$\operatorname{rk}_A(P) = \dim_K(K \otimes_A P).$$

Moreover, the Hattori–Stallings rank $R_A(P)$ is simply given by

$$R_A(P) = \operatorname{rk}_A(P) \cdot 1, \tag{3.1}$$

where $1 \in A = T(A)$ is the identity element. [To prove this, we may regard both sides of the equation as elements of K. The left hand side is then equal to $R_K(K \otimes_A P)$ by the compatibility of traces with extension of scalars (2.3), and this in turn is equal to $\dim_K(K \otimes_A P) \cdot 1 = \operatorname{rk}_A(P) \cdot 1$ since $K \otimes_A P$ is a free K-module.]

Our goal in this section is to prove a generalization of 3.1 to indecomposable commutative rings which are not necessarily domains. [This restriction to indecomposable rings should not surprise the reader in view of the exercise of §1. There are, however, interesting results along these lines for decomposable rings; see the exercises below.]

Since A is no longer assumed to be a domain, we do not have a field of fractions to work with. We do, however, have a localization $A_\mathfrak{p}$ for every prime ideal $\mathfrak{p}$, and we can use this instead. [See Atiyah–Macdonald [1969], Chapter 3, for the definitions and elementary facts concerning localization.] The ring $A_\mathfrak{p}$ is a *local ring* (it has a unique maximal ideal), and we will prove:

(3.2) Proposition. *Any finitely generated projective over a local ring is free.*

We can therefore define the *rank of P at $\mathfrak{p}$* to be the rank in the naive sense of the free $A_\mathfrak{p}$-module $P_\mathfrak{p} = A_\mathfrak{p} \otimes_A P$. Let Spec A be the set of prime ideals of A. Then what we have, so far, is a *function* $\mathscr{R}k_A(P)$: Spec $A \to \mathbb{Z}$ whose value at $\mathfrak{p}$ is the rank of P at $\mathfrak{p}$.

Spec A has a well-known topology (the "Zariski topology"), whose definition we will recall below, and we will prove:

(3.3) Proposition. *The function* $\mathscr{R}k_A(P)$ *is locally constant.*

Finally, we will establish the (not surprising) geometric interpretation of indecomposability:

(3.4) Proposition. *If A is indecomposable then* Spec A *is connected.*

The function $\mathscr{R}k_A(P)$ is therefore constant if A is indecomposable, and we set $\mathrm{rk}_A(P)$ equal to its constant value.

Accepting all this for the moment, we can prove the desired generalization of 3.1:

(3.5) Theorem. *Let A be a commutative indecomposable ring, let P be a finitely generated projective A-module, and let* $\mathrm{rk}_A(P)$ *be the rank of P at any prime ideal* $\mathfrak{p}$. *Then the Hattori–Stallings rank $R_A(P)$ is given by* $R_A(P) = \mathrm{rk}_A(P) \cdot 1$.

Proof. The compatibility of traces with extension of scalars implies, as in the proof of 3.1, that $R_A(P)$ and $\mathrm{rk}_A(P) \cdot 1$ have the same image in $A_\mathfrak{p}$ for every $\mathfrak{p}$; hence they are equal. □

It remains to prove 3.2, 3.3, and 3.4. For 3.2 we will need "Nakayama's lemma":

(3.6) Lemma. *Let A be a local ring with maximal ideal $\mathfrak{m}$ and residue field $k = A/\mathfrak{m}$, and let M be a finitely generated A-module. If $k \otimes_A M = 0$ then $M = 0$.*

(Note the similarity between this and VI.8.4. Both results are special cases of a general "Nakayama's lemma.")

Proof. Let $m_1, \ldots, m_r$ be a minimal set of generators of M, and suppose $r \geq 1$. Since $k \otimes_A M = M/\mathfrak{m}M$, our hypothesis says that $M = \mathfrak{m}M$. In particular, we can write $m_1 = \sum_1^r a_i m_i$ with $a_i \in \mathfrak{m}$, so $(1 - a_1)m_1 = \sum_2^r a_i m_i$. But $1 - a_1$ is invertible since it is not in the unique maximal ideal of A; thus M is generated by $m_2, \ldots, m_r$. This contradicts the minimality of $m_1, \ldots, m_r$, so r must be 0. □

Proof of 3.2. This is similar to the proof of VI.8.5. Let A be local with residue field k, let P be a finitely generated projective A-module, and let

$$r = \dim_k(k \otimes_A P).$$

Then we can find a map $f: A^r \to P$ such that

$$k \otimes f: k^r = k \otimes A^r \to k \otimes P$$

is an isomorphism. Then $k \otimes \operatorname{coker} f = \operatorname{coker}(k \otimes f) = 0$, so $\operatorname{coker} f = 0$ by 3.6. We now have a short exact sequence

$$0 \to \ker f \to A^r \xrightarrow{f} P \to 0,$$

which splits because P is projective. The sequence therefore remains exact after being tensored with k, so $k \otimes \ker f = 0$. Now $\ker f$ is finitely generated, being a direct summand of A^r, so we can apply 3.6 again to conclude that $\ker f = 0$. Thus f is an isomorphism. □

The next step is to topologize Spec A. For any subset $S \subseteq A$, we define a subset $V(S) \subseteq \operatorname{Spec} A$ by

$$V(S) = \{\mathfrak{p} : \mathfrak{p} \supseteq S\}.$$

(3.7) Lemma.

(i) *If I is the ideal generated by S, then $V(S) = V(I)$.*
(ii) $V(S) = \operatorname{Spec} A$ *if and only if every element of S is nilpotent.*
(iii) *If I is an ideal, then $V(I) = \varnothing$ if and only if $I = A$.*
(iv) *If I and J are ideals, then $V(IJ) = V(I) \cup V(J)$.*
(v) *If $\{I_\alpha\}$ is a family of ideals and I is their sum, then $V(I) = \bigcap_\alpha V(I_\alpha)$.*

Proof. The proofs are routine verifications which are left for the reader, except for the "only if" part of (ii), which is proved as follows: Suppose $a \in S$ and a is not nilpotent. Then the localization $A[a^{-1}]$ is not the zero ring and hence contains a maximal ideal $\mathfrak{m}$. The contraction of $\mathfrak{m}$ to A is then a prime ideal not containing a, so $V(S) \neq \operatorname{Spec} A$. □

It follows from (ii)–(v) that we can topologize Spec A by declaring that the closed sets are the sets of the form $V(I)$, where I is an ideal of A. In view of (i), every $V(S)$ is then closed.

Proof of 3.3. Suppose $P_\mathfrak{p}$ is free of rank r over $A_\mathfrak{p}$. We must show that $P_\mathfrak{q}$ is free of rank r over $A_\mathfrak{q}$ for all $\mathfrak{q}$ in some neighborhood of $\mathfrak{p}$. It is easy to see that we can find r elements of P whose images in $P_\mathfrak{p}$ form a basis for $P_\mathfrak{p}$. In other words, we can find a map $f: F \to P$, where F is a free A-module of rank r, such that $f_\mathfrak{p}: F_\mathfrak{p} \to P_\mathfrak{p}$ is an isomorphism. Therefore $(\operatorname{coker} f)_\mathfrak{p} = 0$, so every element of $\operatorname{coker} f$ is annihilated by an element of $A - \mathfrak{p}$. But $\operatorname{coker} f$ is finitely generated, so there is a single element $s \in A - \mathfrak{p}$ such that $s \cdot \operatorname{coker} f = 0$. Thus if we localize A by inverting s, we obtain a surjection $f[s^{-1}]: F[s^{-1}] \to P[s^{-1}]$ which becomes an isomorphism when localized at $\mathfrak{p}$. (This makes sense since $A_\mathfrak{p}$ can be viewed as a localization of $A[s^{-1}]$.)

$P[s^{-1}]$ being projective over $A[s^{-1}]$, this surjection splits, and $\ker f[s^{-1}]$ is therefore finitely generated. Since it becomes zero when localized at $\mathfrak{p}$, it is annihilated by some element of $A[s^{-1}] - \mathfrak{p}[s^{-1}]$ and hence by some $t \in A - \mathfrak{p}$. Letting $u = st$, we conclude that $F[u^{-1}] \xrightarrow{\approx} P[u^{-1}]$. But then $F_\mathfrak{q} \xrightarrow{\approx} P_\mathfrak{q}$ for all $\mathfrak{q}$ such that $u \notin \mathfrak{q}$, since $A_\mathfrak{q}$ is a localization of $A[u^{-1}]$ for such $\mathfrak{q}$. Thus $P_\mathfrak{q}$ is free of rank r for all $\mathfrak{q}$ in the neighborhood Spec $A - V(u)$ of $\mathfrak{p}$. □

Proof of 3.4. Note first that a ring A is decomposable if and only if it contains an idempotent element $e \neq 0, 1$; for direct sum decompositions of the A-module A correspond to 1×1 idempotent matrices, which are the same as idempotent elements. What we must prove, then, is that if Spec A is disconnected then A contains an idempotent $e \neq 0, 1$.

Suppose Spec A is disconnected, say Spec $A = V(I) \coprod V(J)$ where I and J are proper ideals. Since an element e is idempotent if and only if $e(1 - e) = 0$, what we are looking for is a decomposition of 1 as $e + f$, where $ef = 0$ and $e, f \neq 1$. Now we know (cf. 3.7(v)) that $V(I + J) = V(I) \cap V(J) = \varnothing$, so $I + J = A$ by 3.7(iii). Thus $e_1 + f_1 = 1$ for some $e_1 \in I, f_1 \in J$. We also know (3.7(iv)) that $V(IJ) = V(I) \cup V(J) =$ Spec A, so every element of IJ is nilpotent by 3.7(ii). In particular, $e_1 f_1$ is nilpotent. I claim now that we can get our desired decomposition $e + f = 1$ by raising both sides of the equation $e_1 + f_1 = 1$ to a high enough power. Indeed, for any integer $n > 0$, we can group the terms in the binomial expansion of $(e_1 + f_1)^{2n-1}$ to obtain $1 = (e_1 + f_1)^{2n-1} = e + f$, where $e \in Ae_1^n \subseteq I$ and $f \in Af_1^n \subseteq J$ (hence $e, f \neq 1$). Since $e_1 f_1$ is nilpotent, we will then have $ef = 0$ for large enough n. □

Exercises

*1. Generalize 3.4 as follows: For any commutative ring A, the idempotents of A are in 1–1 correspondence with the subsets of Spec A which are both open and closed.

*2. Generalize 3.5 as follows: Let A be an arbitrary commutative ring and let P be a finitely generated projective A-module. For any integer $i \geq 0$ let $U_i \subseteq$ Spec A be $\mathscr{R}\!k_A(P)^{-1}(i)$. Let $e_i \in A$ be the idempotent corresponding to U_i (cf. exercise 1). Then $e_i = 0$ for all but finitely many i and $R_A(P) = \sum_{i \geq 0} ie_i$.

*3. In view of exercise 4 of §2, the result of exercise 2 can be stated in the following concrete terms: Let E be an idempotent matrix over a commutative ring A; then there are idempotent elements $e_i \in A$ such that $\mathrm{tr}(E) = \sum_{i \geq 0} ie_i$. It is natural to ask whether the e_i can be computed explicitly in terms of the entries of the matrix E. The purpose of this exercise is to give an affirmative answer, due to Goldman [1961] (see also Almkvist [1973]), which gives a formula for e_i in terms of the coefficients of the characteristic polynomial of E.

(a) Let E be an idempotent $n \times n$ matrix and let $\varphi(t) = \det(1 + tE) = 1 + \mathrm{tr}(E)t + \cdots + \det(E)t^n$. Let P be the corresponding projective module $\mathrm{Im}(E)$, and let $(e_i)_{i \geq 0}$ be as in exercise 2. Show that $\varphi(t - 1) = \sum_{i \geq 0} e_i t^i$. [Hint: At a prime ideal where P has rank i, both sides of this equation are equal to t^i.]

(b) Writing $\varphi(t) = \sum_i a_i t^i$, deduce that

$$e_i = \sum_j (-1)^j \binom{i+j}{i} a_{i+j}.$$

If $n = 2$, for example, then $e_2 = \det(E)$, $e_1 = \operatorname{tr}(E) - 2\det(E)$, and $e_0 = 1 - \operatorname{tr}(E) + \det(E)$.

4 Ranks over Group Rings; Swan's Theorem

For simplicity we will confine ourselves to *integral* group rings $\mathbb{Z}\Gamma$, although it will be obvious that some of what we do in this section extends to more general group rings $k\Gamma$.

In order to understand the Hattori–Stallings rank $R(P)$ (also denoted $R_\Gamma(P)$) over $\mathbb{Z}\Gamma$, we must first understand the group $T(\mathbb{Z}\Gamma)$. This group is the quotient of the additive group of $\mathbb{Z}\Gamma$ by the subgroup generated by the commutators $[\gamma, \gamma'] = \gamma\gamma' - \gamma'\gamma$ $(\gamma, \gamma' \in \Gamma)$. Now the commutators of this form are precisely the elements $\gamma\gamma_1\gamma^{-1} - \gamma_1$ $(\gamma, \gamma_1 \in \Gamma)$; for we have $[\gamma, \gamma'] = \gamma(\gamma'\gamma)\gamma^{-1} - (\gamma'\gamma)$ and $\gamma\gamma_1\gamma^{-1} - \gamma_1 = [\gamma, \gamma_1\gamma^{-1}]$. It follows easily that $T(\mathbb{Z}\Gamma)$ can be identified with the free abelian group on the conjugacy classes of elements of Γ. Thus any element of $T(\mathbb{Z}\Gamma)$ has the form

$$\sum_{\gamma \in \mathscr{C}} t(\gamma) \cdot [\gamma],$$

where $\mathscr{C}$ is a set of representatives for the conjugacy classes in Γ, $[\gamma]$ denotes the conjugacy class of γ, and $t\colon \Gamma \to \mathbb{Z}$ is a function which is constant on each conjugacy class and is zero for almost all conjugacy classes. In concrete terms, then, the trace of a matrix over $\mathbb{Z}\Gamma$ is obtained by summing the diagonal elements and grouping together the terms corresponding to conjugate group elements.

Now let P be a finitely generated projective $\mathbb{Z}\Gamma$-module. Its Hattori–Stallings rank $R_\Gamma(P) \in T(\mathbb{Z}\Gamma)$ has an expansion

$$R_\Gamma(P) = \sum_{\gamma \in \mathscr{C}} R_\Gamma(P)(\gamma) \cdot [\gamma].$$

Thus $R_\Gamma(P)$ can be viewed as a family of integers $R_\Gamma(P)(\gamma)$ $(\gamma \in \mathscr{C})$ associated to P. From this point of view, the restriction formula 2.4 takes the following form:

(4.1) Proposition. *Let Γ be a group and Γ' a subgroup of finite index. For any $\gamma \in \Gamma'$ there is an integer $n(\gamma) > 0$ such that for every finitely generated projective $\mathbb{Z}\Gamma$-module P,*

$$R_{\Gamma'}(P)(\gamma) = n(\gamma) \cdot R_\Gamma(P)(\gamma).$$

Moreover, $n(1) = (\Gamma : \Gamma')$.

(See exercise 2 below for a precise formula for $n(\gamma)$.)

PROOF. We need to compute $\mathrm{tr} = \mathrm{tr}_{\mathbb{Z}\Gamma/\mathbb{Z}\Gamma'}: T(\mathbb{Z}\Gamma) \to T(\mathbb{Z}\Gamma')$. Fix $\gamma \in \Gamma$, and recall that $\mathrm{tr}([\gamma])$ is the trace over $\mathbb{Z}\Gamma'$ of the right multiplication map $\mu_\gamma: \mathbb{Z}\Gamma \to \mathbb{Z}\Gamma$ (cf. proof of 2.4). Let S be a set of coset representatives for $\Gamma' \backslash \Gamma$. Then S is a basis for $\mathbb{Z}\Gamma$ as a left $\mathbb{Z}\Gamma'$-module, and μ_γ can be described as follows in terms of this basis: Given $s \in S$, we have $\mu_\gamma(s) = s\gamma = \gamma' t$, where $t \in S$ is the representative of the coset $\Gamma' s\gamma$ and $\gamma' = s\gamma t^{-1} \in \Gamma'$. Thus we get a non-zero contribution to the trace of μ_γ if and only if $t = s$, i.e., if and only if $s\gamma s^{-1} \in \Gamma'$, and this contribution is then $\gamma' = s\gamma s^{-1}$. We therefore have

$$\mathrm{tr}([\gamma]) = \sum_{s \in S} [s\gamma s^{-1}]_{\Gamma'},$$

where

$$[s\gamma s^{-1}]_{\Gamma'} = \begin{cases} \text{the } \Gamma'\text{-conjugacy class of } s\gamma s^{-1} \text{ if } s\gamma s^{-1} \in \Gamma' \\ 0 \quad \text{otherwise.} \end{cases}$$

For any $\gamma \in \Gamma'$, let $n(\gamma)$ be the number of cosets $\Gamma' s$ such that $[s\gamma s^{-1}]_{\Gamma'} = [\gamma]_{\Gamma'}$. Note that $n(\gamma) \geq 1$ and that $n(1) = (\Gamma : \Gamma')$. Let $\mathscr{C}$ (resp. $\mathscr{C}'$) be a set of representatives for the Γ-conjugacy classes (resp. Γ'-conjugacy classes). Then the formula of the previous paragraph yields the following formula for $\mathrm{tr}: T(\mathbb{Z}\Gamma) \to T(\mathbb{Z}\Gamma')$:

$$\mathrm{tr}\left(\sum_{\gamma \in \mathscr{C}} r(\gamma)[\gamma] \right) = \sum_{\gamma \in \mathscr{C}'} n(\gamma) r(\gamma) [\gamma]_{\Gamma'}.$$

The proposition is now immediate from 2.4. □

For example, if we take Γ finite and $\Gamma' = \{1\}$, 4.1 simply says $\mathrm{rk}_{\mathbb{Z}}(P) = |\Gamma| \cdot R_\Gamma(P)(1)$; thus the "rank" $\rho_\Gamma(P)$ introduced in §1 satisfies

$$\rho_\Gamma(P) = R_\Gamma(P)(1). \tag{4.2}$$

[Note: We have just rederived the result of exercise 6 of §2.] On the other hand, our other "rank" $\varepsilon_\Gamma(P)$ can also be expressed in terms of $R_\Gamma(P)$. Indeed, we have already done this in exercise 5 of §2, and the result of that exercise, when translated into the notation of the present section, is

$$\varepsilon_\Gamma(P) = \sum_{\gamma \in \mathscr{C}} R_\Gamma(P)(\gamma). \tag{4.3}$$

We can now prove our main result, which can be viewed as saying that the three "ranks" ε, ρ, and R that we have been talking about for finite Γ are all essentially the same:

(4.4) Theorem (Swan [1960b]). *If Γ is finite and P is a finitely generated projective $\mathbb{Z}\Gamma$-module, then $R_\Gamma(P)(\gamma) = 0$ for $\gamma \neq 1$. Thus there is an integer r such that $R_\Gamma(P) = r \cdot [1]$, and one has $\varepsilon_\Gamma(P) = \rho_\Gamma(P) = r$.*

PROOF. The second assertion follows from the first in view of 4.2 and 4.3. To prove the first assertion, we may replace Γ by the cyclic subgroup generated by γ; for the restriction formula 4.1 shows that this does not change the question as to whether $R_\Gamma(P)(\gamma) = 0$. In particular, we may assume that $\mathbb{Z}\Gamma$ is commutative. Also, as we pointed out in exercise 7 of §2, it follows from formal properties of $\rho_\Gamma(\ )$ that $\mathbb{Z}\Gamma$ is indecomposable. Theorem 3.5 is therefore applicable and shows that $R_\Gamma(P)$ is an integral multiple of $[1]$, as required. □

For emphasis, we go back to the definitions of ρ and ε and state explicitly what it means for them to be equal:

(4.5) Corollary. *Let Γ be a finite group and P a finitely generated projective $\mathbb{Z}\Gamma$-module. Then*

$$\mathrm{rk}_{\mathbb{Z}}(P) = |\Gamma| \cdot \mathrm{rk}_{\mathbb{Z}}(P_\Gamma).$$

Remark. Using some elementary representation theory (cf. exercise 3 below), one can restate Swan's theorem as follows: If Γ is finite and P is a finitely generated projective $\mathbb{Z}\Gamma$-module, then $\mathbb{Q} \otimes_{\mathbb{Z}} P$ is a free $\mathbb{Q}\Gamma$-module. It is in this form that Swan stated and proved the theorem. The formulation and proof in terms of Hattori–Stallings ranks are due to Bass [1976, 1979]. Bass went on to conjecture that 4.4 remains valid for arbitrary groups Γ. [To make sense out of the second assertion of 4.4 for arbitrary Γ, one should take 4.2 as a *definition* when Γ is infinite, as we essentially did in exercise 8 of §2.] He proved the conjecture for a large class of *torsion-free* groups Γ. For infinite groups with torsion, however, the conjecture is still completely open.

EXERCISES

1. Where did the proof of 4.4 use the assumption that Γ was finite?

2. Show that the integer $n(\gamma)$ in 4.1 is equal to $(C_\Gamma(\gamma): C_{\Gamma'}(\gamma))$, where $C_\Gamma(\gamma)$ (resp. $C_{\Gamma'}(\gamma)$) is the centralizer of γ in Γ (resp. Γ').

3. Let Γ be finite, let k be a field of characteristic zero, and let V be a $k\Gamma$-module which is finite dimensional over k. Then V is finitely generated and projective (cf. §I.8, exercise 5), hence it has a Hattori–Stallings rank $\sum R(\gamma) \cdot [\gamma]$, where $R: \Gamma \to k$ is a central function, i.e., a function which is constant on conjugacy classes. On the other hand, there is a classical way to use traces to associate a central function $\chi: \Gamma \to k$ to V, called the *character* of V. Namely, $\chi(\gamma)$ is the trace over k of the action of γ on V. The purpose of this exercise is to relate χ and R and to draw some consequences from this.

   (a) Show that the Hattori–Stallings rank of V is equal to

$$\frac{1}{|\Gamma|} \sum_{\gamma \in \Gamma} \chi(\gamma^{-1}) \cdot [\gamma] = \sum_{\gamma \in \mathscr{C}} \frac{\chi(\gamma^{-1})}{|C_\Gamma(\gamma)|} \cdot [\gamma],$$

i.e., that $R(\gamma) = \chi(\gamma^{-1})/|C_\Gamma(\gamma)|$. [Method 1: There is a canonical way of writing V as a direct summand of an induced module; namely, we have $k\Gamma \otimes_k V \underset{\pi}{\overset{\iota}{\leftrightarrows}} V$, where $\pi(\gamma \otimes v) = \gamma v$ and

$$\iota(v) = \frac{1}{|\Gamma|} \sum_{\gamma \in \Gamma} \gamma \otimes \gamma^{-1} v.$$

Choose a basis (e_i) for V over k, so that $(1 \otimes e_i)$ is a basis for $k\Gamma \otimes V$ over $k\Gamma$. You can now write down the matrix of $\iota\pi$ in terms of the matrices $a_{ij}(\gamma)$ of the elements of Γ acting on V. Method 2: It's enough to prove this when k is algebraically closed and V is irreducible. In this case, the element

$$e = \frac{n}{|\Gamma|} \sum_{\gamma \in \Gamma} \chi(\gamma^{-1})\gamma,$$

where $n = \dim_k V$, is the idempotent of $k\Gamma$ which projects $k\Gamma$ onto its summand of type V (cf. Serre [1977b], 2.6). Since V occurs n times in $k\Gamma$, the image of e in $T(k\Gamma)$ is the Hattori–Stallings rank of the sum of n copies of V. Method 3: By the restriction formula (in the precise form provided by exercise 2), we may assume Γ is cyclic, so that the map $\mu_\gamma\colon V \to V$ given by the action of γ is $k\Gamma$-linear for each $\gamma \in \Gamma$. Then $\mathrm{tr}_{k\Gamma}(\mu_\gamma) = \gamma \cdot \mathrm{tr}_{k\Gamma}(\mathrm{id}) = \gamma \cdot R_{k\Gamma}(V)$ by exercise 3 of §2. In particular, the coefficient $\mathrm{tr}_{k\Gamma}(\mu_\gamma)(1)$ of $[1]$ is equal to $R_{k\Gamma}(V)(\gamma^{-1})$. On the other hand, the analogue of 4.1 for traces gives $\chi(\gamma) = \mathrm{tr}_k(\mu_\gamma) = |\Gamma| \cdot \mathrm{tr}_{k\Gamma}(\mu_\gamma)(1)$.]

(b) Using the fact (known from representation theory) that the module V is determined up to isomorphism by its character, show that two finitely generated $k\Gamma$-modules are isomorphic if and only if they have the same Hattori–Stallings rank.

(c) Deduce the reformulation of 4.4 given in the remark above.

4. Take 4.2 as a definition, and prove that $\rho_{\Gamma'}(P) = (\Gamma : \Gamma')\rho_\Gamma(P)$ for any group Γ, subgroup Γ' of finite index, and finitely generated projective $\mathbb{Z}\Gamma$-module P.

5. With Γ, Γ', and P as in exercise 4, prove that $\varepsilon_{\Gamma'}(P) = (\Gamma : \Gamma')\varepsilon_\Gamma(P)$. [Hint: Suppose first that Γ' is normal in Γ. Let $\bar{\Gamma} = \Gamma/\Gamma'$ and let $\bar{P} = P_{\Gamma'} = \mathbb{Z}\bar{\Gamma} \otimes_{\mathbb{Z}\Gamma} P$ (cf. §II.2, exercise 3). $\bar{P}$ is a finitely generated projective $\mathbb{Z}\bar{\Gamma}$-module, and the desired result is obtained by applying 4.5 to it. In the general case, choose a subgroup $\Gamma'' \subseteq \Gamma'$ of finite index such that Γ'' is normal in Γ; this is possible because Γ' has only finitely many conjugates in Γ, so their intersection is still of finite index and is normal. Now compare $\varepsilon_{\Gamma''}$ to both $\varepsilon_{\Gamma'}$ and ε_Γ.]

5 Consequences of Swan's Theorem

In this section, finally, the reader will begin to see some Euler characteristics. We will then be ready in §6 to discuss Euler characteristics of *groups*.

For any finitely generated abelian group A, we define the *rank* of A (sometimes called the *torsion-free rank* of A) by

$$\mathrm{rk}_{\mathbb{Z}}(A) = \dim_{\mathbb{Q}}(\mathbb{Q} \otimes_{\mathbb{Z}} A).$$

If A is free this obviously agrees with the naive rank [cardinality of a basis] which we have been using throughout this chapter for free modules; in general, $\mathrm{rk}_{\mathbb{Z}}(A)$ as defined above is equal to the rank in this naive sense of the "free part" of A, i.e., of the quotient of A by its torsion subgroup. In particular, $\mathrm{rk}_{\mathbb{Z}}(A) = 0$ if and only if A is finite.

Let C be a non-negative chain complex of abelian groups. We say that C is *finite dimensional* if $C_i = 0$ for sufficiently large i. If, in addition, each C_i is finitely generated, then C is called *finite*. Suppose that C is finite dimensional and that $H_* C$ is finitely generated; then we define the *Euler characteristic* $\chi(C)$ by

$$\chi(C) = \sum_{i \geq 0} (-1)^i \, \mathrm{rk}_{\mathbb{Z}}(H_i C).$$

In case C is the cellular chain complex $C(X)$ of a finite dimensional CW-complex X, we write $\chi(X)$ instead of $\chi(C(X))$. Thus

$$\chi(X) = \sum_{i \geq 0} (-1)^i \, \mathrm{rk}_{\mathbb{Z}}(H_i X).$$

If X is actually finite, then $\chi(X)$ is equal to the classical Euler characteristic $\sum_i (-1)^i n_i$, where n_i is the number of i-cells of X. This follows from:

(5.1) Proposition. *If C is a finite chain complex, then*

$$\chi(C) = \sum_i (-1)^i \, \mathrm{rk}_{\mathbb{Z}}(C_i).$$

See, for instance, Spanier [1966], 4.3.14, or Dold [1972], V.5.2, for the (easy) proof.

It is also convenient sometimes to compute Euler characteristics using "mod p" homology. This is justified by:

(5.2) Proposition. *Let C be a finite dimensional free chain complex over $\mathbb{Z}$ with $H_* C$ finitely generated. Let p be a prime number. Then*

$$\chi(C) = \sum_i (-1)^i \dim_{\mathbb{Z}_p}(H_i(C \otimes \mathbb{Z}_p)).$$

PROOF. Let $r_i = \dim_{\mathbb{Z}_p}((H_i C)_p)$ and $s_i = \dim_{\mathbb{Z}_p}({}_p(H_i C))$. [Recall that $A_p = A \otimes \mathbb{Z}_p = A/pA$ and that ${}_pA = \{a \in A : pa = 0\} = \mathrm{Tor}(A, \mathbb{Z}_p)$ for any abelian group A.] Using the universal coefficient theorem (or the long exact homology sequence associated to the short exact sequence $0 \to C \xrightarrow{p} C \to C \otimes \mathbb{Z}_p \to 0$), one finds that $\dim_{\mathbb{Z}_p}(H_i(C \otimes \mathbb{Z}_p)) = r_i + s_{i-1}$. On the other hand, since $H_i C$ is a direct sum of cyclic groups, one sees easily that $r_i = \mathrm{rk}_{\mathbb{Z}}(H_i C) + s_i$. Thus $\dim_{\mathbb{Z}_p}(H_i(C \otimes \mathbb{Z}_p)) = \mathrm{rk}_{\mathbb{Z}}(H_i C) + s_i + s_{i-1}$, and the proposition follows at once. □

We now prove the main result of this section:

(5.3) Theorem. *Let G be a finite group and let C be a finite dimensional chain complex of projective $\mathbb{Z}G$-modules. If $H_*(C)$ is finitely generated then so is $H_*(C_G)$, and*

$$\chi(C) = |G| \cdot \chi(C_G).$$

(In other words, C behaves like a finite free complex as far as Euler characteristics are concerned.)

PROOF. In case C is a *finite* projective complex, this is immediate from 4.5, since χ can then be computed on the chain level (5.1). The general case is reduced to the finite case by means of Lemma 5.4 below. □

Recall that a ring R is *noetherian* if every submodule of a finitely generated module is finitely generated. For example, $\mathbb{Z}G$ is noetherian if G is finite, since a finitely generated $\mathbb{Z}G$-module is finitely generated as a $\mathbb{Z}$-module.

(5.4) Lemma. *Let R be a noetherian ring and C a finite dimensional chain complex of projective R-modules. If $H_* C$ is finitely generated over R, then C is homotopy equivalent to a finite projective complex.*

PROOF. First we construct, step-by-step, a free complex F of finite type which admits a weak equivalence $\tau: F \to C$. The inductive step is carried out as follows: Suppose we have constructed the n-skeleton of F together with a chain map

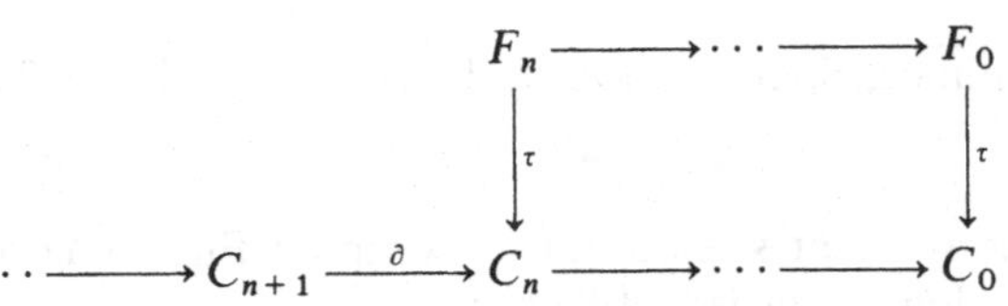

such that $H_i\tau$ is an isomorphism for $i < n$ and an epimorphism for $i = n$. Then $\ker(H_n\tau)$ is finitely generated since F_n is finitely generated and R is noetherian. Let $x_1, \ldots, x_r \in F_n$ be cycles representing generators of $\ker(H_n\tau)$, let $y_1, \ldots, y_r \in C_{n+1}$ be chains such that $\partial y_i = \tau x_i$, and let $z_1, \ldots, z_s \in C_{n+1}$ be cycles representing generators of the finitely generated module $H_{n+1}C$. Then we can take F_{n+1} to be free with basis $e_1, \ldots, e_r, f_1, \ldots, f_s$ and set $\partial e_i = x_i$, $\partial f_i = 0$, $\tau e_i = y_i$, and $\tau f_i = z_i$. The resulting chain map

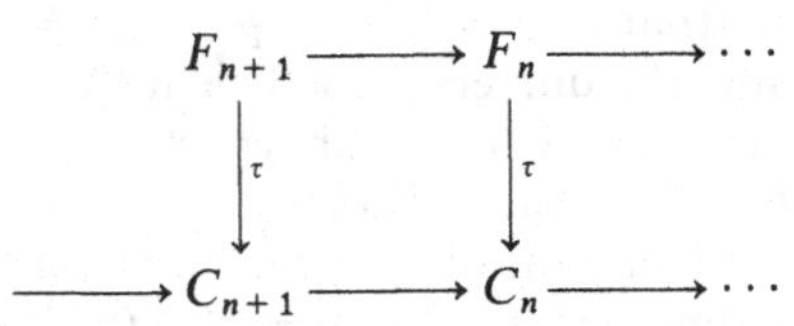

induces an H_i-isomorphism for $i < n + 1$ and an H_{n+1}-epimorphism. This completes the inductive construction of F and τ. Note that the weak equivalence τ is actually a homotopy equivalence by I.8.4.

Now let n be an integer such that $C_i = 0$ for $i > n$, and consider the quotient $\bar{F}$ of F defined by

$$\bar{F}_i = \begin{cases} F_i & i < n \\ F_n/B_n & i = n \\ 0 & i > n, \end{cases}$$

B_n being the module of n-boundaries. Then we still have a weak equivalence $\bar{F} \to C$, and the lemma will clearly follow if we can show that F_n/B_n is projective. For any R-module M, we have

$$H^{n+1}(\mathcal{H}om_R(F, M)) \approx H^{n+1}(\mathcal{H}om_R(C, M)) = 0$$

since $F \simeq C$ and $C_{n+1} = 0$. Examining cocycles and coboundaries in $\mathcal{H}om_R(F, M)$ as in the proof of VIII.2.1, we conclude easily that B_n is a direct summand of F_n, and hence that F_n/B_n is projective. □

To illustrate the significance of Theorem 5.3, we give two special cases.

(5.5) Corollary. *Let G be a finite group and let X be a finite dimensional free G-CW-complex with $H_* X$ finitely generated. Then $H_*(X/G)$ is finitely generated and $\chi(X) = |G| \cdot \chi(X/G)$.*

Proof. Apply 5.3 to the cellular chain complex of X. □

This result is obvious, of course, if X is finite, since we can then compute Euler characteristics by counting cells.

The second special case involves group cohomology. If Γ is a group such that $H_i \Gamma$ is finitely generated for all i and finite for sufficiently large i, then we set

$$\tilde{\chi}(\Gamma) = \sum_i (-1)^i \operatorname{rk}_{\mathbb{Z}}(H_i \Gamma).$$

[The notation $\tilde{\chi}$ here is intended to suggest that this is not always the "right" Euler characteristic. We will explain this in §7.]

(5.6) Corollary. *Let Γ be a group with $\operatorname{cd} \Gamma < \infty$ and let Γ' be a normal subgroup of finite index. If $H_* \Gamma'$ is finitely generated then so is $H_* \Gamma$, and $\tilde{\chi}(\Gamma') = (\Gamma : \Gamma') \cdot \tilde{\chi}(\Gamma)$.*

Proof. Let P be a projective resolution of finite length of $\mathbb{Z}$ over $\mathbb{Z}\Gamma$, let $G = \Gamma/\Gamma'$, and let $C = P_{\Gamma'} = \mathbb{Z}G \otimes_{\mathbb{Z}\Gamma} P$ (cf. §II.2, exercise 3). Then C is a finite dimensional complex of projective $\mathbb{Z}G$-modules with $H_* C = H_* \Gamma'$ and $H_*(C_G) = H_* \Gamma$; now apply 5.3. □

Finally, we will need a mod p analogue of 5.3:

(5.7) Theorem. *Let G be a finite p-group for some prime p, let k be a field of characteristic p, and let C be a finite dimensional complex of projective kG-modules. If $H_* C$ is of finite rank over k, then so is $H_*(C_G)$, and $\chi(C) = |G| \cdot \chi(C_G)$, where $\chi(\ \ )$ now denotes $\sum (-1)^i \dim_k(H_i(\ \ ))$.*

PROOF. This is identical to that of 5.3, with one major simplification; namely, the mod p analogue of 4.5 is now true for trivial reasons, since all projective kG-modules are free (VI.8.5). □

One also has, obviously, mod p analogues of the corollaries 5.5 and 5.6 of 5.3.

EXERCISES

1. Let C and G be as in 5.3. For each $g \in G$, let $L(g)$ be the Lefschetz number of g acting on C, i.e.,
$$L(g) = \sum_i (-1)^i \operatorname{tr}_{\mathbb{Q}} (g \text{ acting on } H_i C \otimes \mathbb{Q}).$$
Prove that $L(g) = 0$ for $g \neq 1$, and state a topological corollary analogous to 5.5. [Hint: We may assume C is finite, so that $L(g)$ can be computed on the chain level. Exercise 3 of §4 now describes $L(g)$ in terms of Hattori–Stallings ranks, and the latter are computed by Swan's theorem (4.4).]

2. Let Γ and Γ' be as in 5.6. Let ψ_i be the character associated to the representation of the finite group Γ/Γ' acting on $H_i(\Gamma', \mathbb{Q})$. Show that $\sum_i (-1)^i \psi_i$ is an integral multiple of the character of the regular representation of Γ/Γ'. [The regular representation, by definition, is the free $\mathbb{Q}[\Gamma/\Gamma']$-module of rank 1; its character takes the value 0 on all non-identity elements of Γ/Γ'.]

6 Euler Characteristics of Groups: The Torsion-Free Case

Euler characteristics of groups have been defined by a number of people under a number of different finiteness conditions. For our purposes the following condition is the most convenient one to impose: A group Γ is said to be of *finite homological type* if (i) $\operatorname{vcd} \Gamma < \infty$ and (ii) for every Γ-module M which is finitely generated as an abelian group, $H_i(\Gamma, M)$ is finitely generated for all i. [Warning: This definition differs slightly from that given in Brown [1974].] Note that if Γ is torsion-free, then (i) implies that $\operatorname{cd} \Gamma < \infty$ (cf. §VIII.3).

Remark. The main examples of groups of finite homological type are the groups of type VFP, and the reader may prefer to replace "finite homological type" by "type VFP" in what follows; this results in occasional simplifications of the proofs. On the other hand, we will see that one cannot restrict attention to groups of type VFL, even though all known examples of groups of type VFP are in fact of type VFL.

(6.1) Lemma. *If Γ is a group and Γ' is a subgroup of finite index, then Γ is of finite homological type if and only if Γ' is of finite homological type.*

PROOF. It is obvious that $\operatorname{vcd} \Gamma < \infty$ if and only if $\operatorname{vcd} \Gamma' < \infty$, so we need only check that (ii) holds for Γ if and only if it holds for Γ'. The "only if" part follows from Shapiro's lemma, since $\operatorname{coind}_{\Gamma'}^{\Gamma} M$ is finitely generated over $\mathbb{Z}$ if M is. Conversely, suppose Γ' satisfies (ii). We know that Γ' has a subgroup Γ'' of finite index which is normal in Γ (cf. §4, hint to exercise 5), and Γ'' still satisfies (ii) by what we have just proved. Replacing Γ' by Γ'', then, we may assume Γ' is normal in Γ. For any Γ-module M which is finitely generated over $\mathbb{Z}$, the Hochschild–Serre spectral sequence

$$E_{pq}^2 = H_p(\Gamma/\Gamma', H_q(\Gamma', M)) \Rightarrow H_{p+q}(\Gamma, M)$$

now shows that $H_i(\Gamma, M)$ is finitely generated for all i, as required. □

Suppose now that Γ is of finite homological type and *torsion-free*. We then define the *Euler characteristic* $\chi(\Gamma)$ by

$$\chi(\Gamma) = \sum_i (-1)^i \operatorname{rk}_{\mathbb{Z}}(H_i \Gamma).$$

[Thus $\chi(\Gamma) = \tilde{\chi}(\Gamma)$ in this case, where $\tilde{\chi}$ is as in §5.] In case Γ is of type FP, one has an analogue of the combinatorial formula for the Euler characteristic of a finite CW-complex. Namely, if P is a finite projective resolution of $\mathbb{Z}$ over $\mathbb{Z}$, then

$$\chi(\Gamma) = \sum_i (-1)^i \varepsilon(P_i), \tag{6.2}$$

where ε is the "rank" defined in §1. [To prove this, we need only note that $H_*\Gamma$ can be computed from the finite chain complex P_Γ; the assertion now follows from 5.1.] We also have, trivially, the topological interpretation $\chi(\Gamma) = \chi(K(\Gamma, 1))$.

Here are a few examples where $\chi(\Gamma)$ can easily be computed.

EXAMPLES

1. If Γ is a free group of rank $n \geq 0$, then there is a $K(\Gamma, 1)$ with one vertex and n 1-cells, hence $\chi(\Gamma) = 1 - n$.

2. If Γ is a free abelian group of rank $n \geq 1$, then the n-torus $S^1 \times \cdots \times S^1$ is a $K(\Gamma, 1)$ with Euler characteristic $\chi(S^1) \cdots \chi(S^1) = 0$, hence $\chi(\Gamma) = 0$.

3. Let Γ be the commutator subgroup of $SL_2(\mathbb{Z})$. The latter is well-known to be an amalgamation $\mathbb{Z}_4 *_{\mathbb{Z}_2} \mathbb{Z}_6$. (See Serre [1977a], I.4.2, for an indication of an easy proof of this; see also example 3 of §VIII.9 above.) In particular, it follows that $SL_2(\mathbb{Z})_{ab}$ is of order $4 \cdot 6/2 = 12$, so that Γ is of index 12 in $SL_2(\mathbb{Z})$. One can also check, using the normal form for words in an amalgamated free product, that Γ is free of rank 2 (cf. Serre [1977a], I.1.3, proof of Prop. 4). Hence $\chi(\Gamma) = 1 - 2 = -1$ by example 1.

We will see more examples later.

The following property of $\chi(\quad)$ is fundamental:

(6.3) Theorem. *If Γ is torsion-free and of finite homological type and Γ' is a subgroup of finite index, then*

$$\chi(\Gamma') = (\Gamma : \Gamma') \cdot \chi(\Gamma).$$

PROOF. It suffices to prove this when Γ' is normal in Γ, in which case it follows from Corollary 5.6. □

The proof of 6.3 simplifies slightly if Γ is of type FP; in this case we can use 6.2 and exercise 5 of §4 instead of 5.6. And the proof simplifies drastically if Γ is of type FL, since exercise 5 of §4 is trivial for free modules.

Theorem 6.3 has an interesting application to the theory of group extensions:

(6.4) Corollary. *Let $1 \to \Gamma \to E \to G \to 1$ be a group extension such that Γ is torsion-free and of finite homological type and G is of prime order p. If $p \nmid \chi(\Gamma)$, then the extension splits.*

(If Γ is free on n generators, for instance, then such an extension must split whenever $p \nmid (n - 1)$.)

PROOF. The group E necessarily has torsion; for if it were torsion-free, then 6.3 would be applicable and would yield $p \mid \chi(\Gamma)$, contrary to our hypothesis. Let $\tilde{G}$ be a non-trivial finite subgroup of E. Since Γ is torsion-free, $\tilde{G} \cap \Gamma = \{1\}$; thus $\tilde{G}$ maps injectively to G. But $|G|$ is prime, so $\tilde{G}$ maps isomorphically and provides a splitting. □

Remarks

1. Even if we are interested in 6.4 only for groups Γ of type FL, the proof still requires a theory of Euler characteristics for groups of type FP. For we need to apply 6.3 to E in the first part of the proof (assuming E is torsion-free); and all we know about E, given that Γ is of type FL, is that it is of type FP if it is torsion-free.

2. We will give a substantial improvement of 6.4 later (cf. 9.4).

7 Extension to Groups with Torsion

From the topological point of view, it may seem very reasonable to restrict attention to torsion-free groups. Indeed, topologists usually consider Euler characteristics only for finite dimensional complexes, and we know that Γ must be torsion-free if there is a finite dimensional $K(\Gamma, 1)$. From the algebraic point of view, however, this seems less reasonable; for it forces us to exclude certain groups Γ (like $SL_2(\mathbb{Z})$) just because they have torsion, even though Γ may have subgroups of finite index for which χ is defined.

There is a simple solution to this problem, based on an idea due to Wall [1961]. Namely, if Γ is an arbitrary group of finite homological type, then we choose a torsion-free subgroup Γ' of finite index and set

$$\chi(\Gamma) = \frac{\chi(\Gamma')}{(\Gamma:\Gamma')} \in \mathbb{Q}.$$

This is motivated by 6.3, and, in fact, 6.3 is precisely what is needed to justify this definition, i.e., to show that the right-hand side is independent of the choice of Γ'. For suppose Γ'' is another such subgroup and let $\Gamma_0 = \Gamma' \cap \Gamma''$; then

$$\frac{\chi(\Gamma')}{(\Gamma:\Gamma')} \overset{(6.3)}{=} \frac{\chi(\Gamma_0)/(\Gamma':\Gamma_0)}{(\Gamma:\Gamma')} = \frac{\chi(\Gamma_0)}{(\Gamma:\Gamma_0)}$$

and similarly $\chi(\Gamma'')/(\Gamma:\Gamma'') = \chi(\Gamma_0)/(\Gamma:\Gamma_0)$, which proves our assertion.

If Γ is finite, for instance, then we can take $\Gamma' = \{1\}$ and we find

(7.1) $$\chi(\Gamma) = \frac{1}{|\Gamma|}.$$

Or if $\Gamma = SL_2(\mathbb{Z})$, then we can take Γ' to be the commutator subgroup (cf. example 3 of §6), and we find

(7.2) $$\chi(SL_2(\mathbb{Z})) = -\frac{1}{12}.$$

These examples show that the rational number $\chi(\Gamma)$ is not generally an integer. In particular, $\chi(\Gamma)$ need not be equal to the naive Euler characteristic $\tilde{\chi}(\Gamma)$ defined in §5, which is always an integer. [Note that $\tilde{\chi}(\Gamma)$ is indeed defined if Γ is of finite homological type. For we know $H_i\Gamma$ is finitely generated for all i, and a transfer argument (III.10.1) shows that $\mathrm{rk}_{\mathbb{Z}}(H_i\Gamma) = 0$ for $i > \mathrm{vcd}\,\Gamma$.] This failure of $\chi(\Gamma)$ to be integral raises a number of interesting questions, which we will take up in §9.

In our study of Euler characteristics, we will attempt to get information about $\chi(\Gamma)$ by studying actions of Γ on CW-complexes. For this purpose, the *equivariant Euler characteristic* $\chi_\Gamma(X)$ is a fundamental tool. It is defined as follows: Suppose X is a Γ-complex such that (i) every isotropy group Γ_σ

is of finite homological type and (ii) X has only finitely many cells mod Γ; then we set

$$\chi_\Gamma(X) = \sum_{\sigma \in \mathscr{E}} (-1)^{\dim \sigma} \chi(\Gamma_\sigma),$$

where $\mathscr{E}$ is a set of representatives for the cells of X mod Γ. Note that $\chi(\Gamma) = \chi_\Gamma(\text{pt.})$, so the equivariant Euler characteristic can be viewed as a generalization of the ordinary Euler characteristic.

We now close this section by listing some useful properties of $\chi(\ \ )$ and $\chi_\Gamma(\ \ )$.

(7.3) Proposition. (a) *If Γ is torsion-free and of finite homological type then*

$$\chi(\Gamma) = \tilde{\chi}(\Gamma) = \sum_i (-1)^i \dim_{\mathbb{Z}_p}(H_i(\Gamma, \mathbb{Z}_p))$$

for any prime p.

(b) *If Γ is of finite homological type and Γ' is a subgroup of finite index, then*

$$\chi(\Gamma') = (\Gamma : \Gamma') \cdot \chi(\Gamma).$$

(b′) *More generally, if $\chi_\Gamma(X)$ is defined and $\Gamma' \subseteq \Gamma$ is a subgroup of finite index, then $\chi_{\Gamma'}(X)$ is defined and*

$$\chi_{\Gamma'}(X) = (\Gamma : \Gamma') \cdot \chi_\Gamma(X).$$

(c) *Suppose that $\chi_\Gamma(X)$ is defined and that each Γ_σ is torsion-free. Then the equivariant homology $H_*^\Gamma(X)$ is finitely generated and*

$$\chi_\Gamma(X) = \tilde{\chi}_\Gamma(X),$$

where $\tilde{\chi}_\Gamma(X) = \sum_i (-1)^i \operatorname{rk}_{\mathbb{Z}}(H_i^\Gamma(X))$.

(d) *Let $1 \to \Gamma' \to \Gamma \to \Gamma'' \to 1$ be a short exact sequence of groups with Γ' and Γ'' of finite homological type. If Γ is virtually torsion-free, then Γ is of finite homological type and*

$$\chi(\Gamma) = \chi(\Gamma') \cdot \chi(\Gamma'').$$

(e) *Let $\Gamma = \Gamma_1 *_A \Gamma_2$, where Γ_1, Γ_2, and A are of finite homological type. If Γ is virtually torsion-free then Γ is of finite homological type and*

$$\chi(\Gamma) = \chi(\Gamma_1) + \chi(\Gamma_2) - \chi(A).$$

(e′) *More generally, suppose X is a contractible Γ-complex such that $\chi_\Gamma(X)$ is defined. If Γ is virtually torsion-free, then Γ is of finite homological type and*

$$\chi(\Gamma) = \chi_\Gamma(X).$$

PROOF. The first equality of (a) is true by definition, and the second follows from 5.2. (b) is immediate from the definition of χ. To prove (b′), fix a cell σ of X and consider the set $\Gamma\sigma$ of cells which are equivalent to σ mod Γ. Since $\Gamma\sigma$ is in 1-1 correspondence with Γ/Γ_σ, it decomposes into finitely many

Γ'-orbits, represented by the cells $\gamma\sigma$ where γ ranges over a set S of representatives for $\Gamma'\backslash\Gamma/\Gamma_\sigma$. Note that $\Gamma'_{\gamma\sigma} = \Gamma' \cap \Gamma_{\gamma\sigma} = \Gamma' \cap \gamma\Gamma_\sigma\gamma^{-1}$, which is conjugate in Γ to $\gamma^{-1}\Gamma'\gamma \cap \Gamma_\sigma$. The contribution of $\Gamma\sigma$ to the sum defining $\chi_{\Gamma'}(X)$ is therefore

$$\begin{aligned}\sum_{\gamma\in S}(-1)^{\dim \gamma\sigma}\chi(\Gamma' \cap \gamma\Gamma_\sigma\gamma^{-1}) &= (-1)^{\dim\sigma}\sum_{\gamma\in S}\chi(\gamma^{-1}\Gamma'\gamma \cap \Gamma_\sigma)\\ &= (-1)^{\dim\sigma}\sum_{\gamma\in S}(\Gamma_\sigma : \gamma^{-1}\Gamma'\gamma \cap \Gamma_\sigma)\cdot\chi(\Gamma_\sigma)\\ &= (-1)^{\dim\sigma}(\Gamma : \Gamma')\cdot\chi(\Gamma_\sigma).\end{aligned}$$

(The last equality here is obtained by decomposing $\Gamma'\backslash\Gamma$ into orbits under the right action of Γ_σ.) (b′) now follows at once.

To prove (c), consider the equivariant homology spectral sequence (VII.7.7)

$$E^1_{pq} = \bigoplus_{\sigma\in\mathscr{E}_p} H_q(\Gamma_\sigma, \mathbb{Z}_\sigma) \Rightarrow H^\Gamma_{p+q}(X),$$

where $\mathscr{E}_p$ is a set of representatives for the p-cells of X mod Γ. Since Γ_σ is torsion-free and of finite homological type, we know that $H_*(\Gamma_\sigma)$ is finitely generated. Hence $H^\Gamma_*(X)$ is finitely generated, and we can compute $\tilde{\chi}_\Gamma(X)$ from the E^1-term of the spectral sequence (cf. VII.2.7). Assuming for the moment that Γ_σ acts trivially on $\mathbb{Z}_\sigma$ for all σ, we find

$$\begin{aligned}\tilde{\chi}_\Gamma(X) &= \sum_{p,q}\sum_{\sigma\in\mathscr{E}_p}(-1)^{p+q}\operatorname{rk}_{\mathbb{Z}}(H_q(\Gamma_\sigma))\\ &= \sum_p\sum_{\sigma\in\mathscr{E}_p}(-1)^p\chi(\Gamma_\sigma)\\ &= \chi_\Gamma(X).\end{aligned}$$

In the general case, we need only note that all Euler characteristics can be computed from homology with $\mathbb{Z}_2$-coefficients (cf. (a) above). Since Γ_σ acts trivially on $(\mathbb{Z}_2)_\sigma$, the result follows as above from the equivariant homology spectral sequence with $\mathbb{Z}_2$-coefficients.

(d) Let $\Gamma_0 \subseteq \Gamma$ be a torsion-free subgroup of finite index whose image Γ''_0 in Γ'' is torsion-free, and let $\Gamma'_0 = \Gamma_0 \cap \Gamma'$. I claim that $(\Gamma:\Gamma_0) = (\Gamma':\Gamma'_0)\cdot(\Gamma'':\Gamma''_0)$; for $(\Gamma:\Gamma_0) = (\Gamma:\Gamma'\Gamma_0)\cdot(\Gamma'\Gamma_0:\Gamma_0)$, and the well-known isomorphism laws of group theory imply that $(\Gamma:\Gamma'\Gamma_0) = (\Gamma'':\Gamma''_0)$ and $(\Gamma'\Gamma_0:\Gamma_0) = (\Gamma':\Gamma'_0)$. We may therefore replace the given exact sequence by $1 \to \Gamma'_0 \to \Gamma_0 \to \Gamma''_0 \to 1$, i.e., we may assume that Γ', Γ, and Γ'' are torsion-free. Then $\operatorname{cd}\Gamma < \infty$ by VIII.2.4b, and the Hochschild–Serre spectral sequence

$$E^2_{pq} = H_p(\Gamma'', H_q(\Gamma', M)) \Rightarrow H_{p+q}(\Gamma, M)$$

shows that Γ is of finite homological type. Now take $M = \mathbb{Z}_2$. Since $H_*(\Gamma', \mathbb{Z}_2)$ is finite, there is a subgroup $\Gamma''_0 \subseteq \Gamma''$ of finite index which acts trivially on it. Replacing Γ'' by Γ''_0 and Γ by the inverse image of Γ''_0, we may assume that Γ''

acts trivially on $H_*(\Gamma', \mathbb{Z}_2)$. Then $E^2_{pq} = H_p(\Gamma'', \mathbb{Z}_2) \otimes_{\mathbb{Z}_2} H_q(\Gamma', \mathbb{Z}_2)$. Computing Euler characteristics from this spectral sequence, we conclude that $\chi(\Gamma) = \chi(\Gamma') \cdot \chi(\Gamma'')$.

Finally, (e) follows from (e′) applied to the tree associated to $\Gamma_1 *_A \Gamma_2$ (cf. appendix to Chapter II), so it remains only to prove (e′). In view of (b′), it suffices to prove (e′) when Γ is torsion-free. Since X is contractible, we have $H_*(\Gamma, M) \approx H^\Gamma_*(X, M)$ for any Γ-module M (VII.7.3), and similarly for cohomology. A spectral sequence argument (cf. proof of (c)) then shows that $H^i(\Gamma, M) = 0$ for $i > \dim X + \max\{\operatorname{cd} \Gamma_\sigma\}$ and that $H_*(\Gamma, M)$ is finitely generated if M is. Hence F is of finite homological type. Moreover,

$$\begin{aligned} \chi(\Gamma) &= \tilde{\chi}(\Gamma) && \text{because } \Gamma \text{ is torsion-free} \\ &= \tilde{\chi}_\Gamma(X) && \text{because } H_*\Gamma \approx H^\Gamma_*(X) \\ &= \chi_\Gamma(X) && \text{by (c).} \end{aligned}$$

□

As an example of (e), consider $SL_2(\mathbb{Z}) \approx \mathbb{Z}_4 *_{\mathbb{Z}_2} \mathbb{Z}_6$. Then (e) gives

$$\chi(SL_2(\mathbb{Z})) = \tfrac{1}{4} + \tfrac{1}{6} - \tfrac{1}{2} = -\tfrac{1}{12},$$

in agreement with 7.2.

Exercises

1. The purpose of this exercise is to outline an alternative method of defining a "rational Euler characteristic" under suitable finiteness hypotheses. This is due to Bass [1976], Chiswell [1976b] and Stallings [unpublished]. For any group Γ and any finitely generated projective $\mathbb{Q}\Gamma$-module P, we can define a "rational rank" $\rho(P)$ [or $\rho_\Gamma(P)$] by $\rho(P) = R_{\mathbb{Q}\Gamma}(P)(1) \in \mathbb{Q}$. [This is motivated by 4.2.] Note that $\rho_{\Gamma'}(P) = (\Gamma : \Gamma')\rho_\Gamma(P)$ if $\Gamma' \subseteq \Gamma$ is a subgroup of finite index (cf. 4.1).

   (a) Γ is said to be of type *FP* over $\mathbb{Q}$ if $\mathbb{Q}$ admits a finite projective resolution over $\mathbb{Q}\Gamma$. In this case we choose such a resolution P and set $e(\Gamma) = \sum(-1)^i \rho(P_i)$. Show that $e(\Gamma)$ is well-defined and that $e(\Gamma') = (\Gamma : \Gamma')e(\Gamma)$ if $\Gamma' \subseteq \Gamma$ is a subgroup of finite index.

   (b) Show that any group of type *VFP* is of type *FP* over $\mathbb{Q}$. [Hint: If P is a $\mathbb{Q}\Gamma$-module which is projective over $\mathbb{Q}\Gamma'$, where $(\Gamma : \Gamma') < \infty$, then P is projective. This can be proved by an averaging argument.]

   (c) If Γ is of type *VFL*, show that $e(\Gamma) = \chi(\Gamma)$.

   [**Remark.** It is not known whether $e(\Gamma) = \chi(\Gamma)$ for arbitrary groups Γ of type *VFP*.]

2. (a) If $e(\Gamma) \neq 0$, show that $C(\Gamma)$, the center of Γ, is finite. [Consider the "complete Euler characteristic" $E(\Gamma) \in T(\mathbb{Q}\Gamma)$, defined by $E(\Gamma) = \sum(-1)^i R_{\mathbb{Q}\Gamma}(P_i)$, where (P_i) is a finite projective resolution of $\mathbb{Q}$ over $\mathbb{Q}\Gamma$. Note that $E(\Gamma)$ is simply $\operatorname{tr}_{\mathbb{Q}\Gamma}(\operatorname{id}_{\mathbb{Q}})$, as defined in exercise 2 of §2, and that $E(\Gamma)(1) = e(\Gamma)$. If $\gamma \in C(\Gamma)$, then we find (using exercise 3 of §2) that $E(\Gamma) = \operatorname{tr}(\operatorname{id}_{\mathbb{Q}}) = \operatorname{tr}(\gamma \cdot \operatorname{id}_{\mathbb{Q}}) = \gamma \cdot \operatorname{tr}(\operatorname{id}_{\mathbb{Q}}) = \gamma \cdot E(\Gamma)$. Hence $e(\Gamma) = E(\Gamma)(\gamma^{-1})$.]

   (b) Deduce the following theorem of Gottlieb [1965]: If Y is a finite $K(\Gamma, 1)$-complex with $\chi(Y) \neq 0$, then $C(\Gamma) = \{1\}$.

Remark. This proof of Gottlieb's theorem is due to Stallings [1965b], as reformulated by Bass [1976].

*3. Suppose Γ is torsion-free and of finite homological type and let M be a Γ-module which is finitely generated as an abelian group. Show that

$$\sum_i (-1)^i \operatorname{rk}_{\mathbb{Z}}(H_i(\Gamma, M)) = \chi(\Gamma) \cdot \operatorname{rk}_{\mathbb{Z}}(M).$$

*4. If Γ is of finite homological type and M is a Γ-module which is finitely generated as an abelian group, show that $H^i(\Gamma, M)$ is finitely generated for all i. [Hint: Reduce to the case where M is $\mathbb{Z}$-free. In this case let $M^* = \operatorname{Hom}(M, \mathbb{Z})$ and note that $\mathcal{H}om_\Gamma(P, M) \approx \mathcal{H}om(P \otimes_\Gamma M^*, \mathbb{Z})$.]

*5. Let $\Gamma \subset SL_2(\mathbb{Z})$ be a torsion-free subgroup of finite index k. Prove that Γ is a free group on $1 + k/12$ generators. Example: The principal congruence subgroup of level 3 has index $24 = |SL_2(\mathbb{Z}_3)|$, hence it is free on 3 generators.

8 Euler Characteristics and Number Theory

The study of groups of integral matrices is intimately related to algebraic number theory. For example, if one tries to classify elements of order p in $GL_n(\mathbb{Z})$ up to conjugacy, one quickly finds that ideal classes of the p-th cyclotomic field come into play. Recall now that many integral matrix groups are of type VFL (cf. §VIII.11). We therefore have rational numbers $\chi(\Gamma)$ associated to such groups, and it is natural to ask whether these numbers have any number-theoretic significance. The remarkable answer is that $\chi(\Gamma)$, for many arithmetic groups Γ, can be expressed in terms of values of zeta-functions. In this section we will give a brief survey of results of this type. See Serre [1971, 1979] for more details and a guide to the literature.

The starting point for the computation of $\chi(\Gamma)$ is the Gauss-Bonnet theorem of differential geometry, which says the following: Suppose Y is a closed Riemannian manifold. Then there is a measure μ on Y, constructed canonically from the curvature associated to the Riemannian metric, such that

$$\chi(Y) = \mu(Y).$$

[*Note*: μ is not in general a *positive* measure. If Y is a surface, for example, then $\mu = K \cdot dA/2\pi$, where K is the Gaussian curvature function and dA is the canonical "area" measure on Y.] We call μ the *Gauss–Bonnet measure* on Y.

If Y is a $K(\Gamma, 1)$ and X is its universal cover, then we can rewrite this formula as

$$\chi(\Gamma) = \mu(X/\Gamma). \tag{8.1}$$

Moreover, we can interpret the right-hand side as the measure of a fundamental domain for the action of Γ on X, relative to the Gauss–Bonnet measure on X.

We now specialize to the arithmetic case. Suppose G is a semi-simple algebraic group defined over $\mathbb{Q}$ and Γ is a *torsion-free* arithmetic subgroup of $G(\mathbb{Q}) \subseteq G(\mathbb{R})$. Let K be a maximal compact subgroup of $G(\mathbb{R})$ and let $X = G(\mathbb{R})/K$. Since K is compact, it is easy to see that X admits a Riemannian metric invariant under the action of $G(\mathbb{R})$; hence there is a $G(\mathbb{R})$-invariant Gauss-Bonnet measure μ on X, and 8.1 holds if Γ is co-compact in $G(\mathbb{R})$. Using the compactness of K again, we can lift μ to a left-invariant measure (still called μ) on $G(\mathbb{R})$; 8.1 then takes the form

$$\chi(\Gamma) = \mu(\Gamma \backslash G(\mathbb{R})). \tag{8.2}$$

The advantage of this formulation is that it remains valid even if Γ has torsion. For both sides of 8.2 get multiplied by $(\Gamma : \Gamma')$ if we replace Γ by a torsion-free subgroup Γ' of finite index.

We derived 8.2 under the assumption that Γ was co-compact in $G(\mathbb{R})$, and, as we saw in §VIII.9, this situation is far from typical. A remarkable theorem of Harder [1971], however, says that 8.2 remains valid even if Γ is not co-compact. Harder went on to explicitly calculate the measure μ for an interesting family of groups G, the so-called "Chevalley groups." [There is one such group for every simple Lie algebra over $\mathbb{C}$. Thus we have the special linear groups SL_n, the symplectic groups Sp_{2n}, the special orthogonal groups SO_n, and five exceptional groups G_2, F_4, E_6, E_7, and E_8.]

For such G there is a canonical Haar measure (i.e., positive invariant measure) μ_a on $G(\mathbb{R})$, called the *arithmetic measure*, with the property that $\mu_a(G(\mathbb{R})/G(\mathbb{Z}))$ is expressible in terms of values of the zeta-function. On the other hand the Gauss–Bonnet measure μ, being invariant, must have the form $\mu = c\mu_a$ for some $c \in \mathbb{R}$. What Harder did was to calculate the scalar c, so that 8.2 would yield an explicit formula for $\chi(G(\mathbb{Z}))$ in terms of values of the zeta-function. We will give some examples of this formula below, following a brief review of the Riemann zeta-function.

Recall that $\zeta(s)$ is defined for $\mathrm{Re}(s) > 1$ by

$$\zeta(s) = \sum_{n=1}^{\infty} \frac{1}{n^s} = \prod_{p \text{ prime}} \frac{1}{1 - p^{-s}}.$$

Thus $\zeta(s)$ is an analytic function in the half-plane $\mathrm{Re}(s) > 1$, and one extends it to a meromorphic function in the entire complex plane. Moreover, the function $\xi(s) = \pi^{-s/2}\Gamma(s/2)\zeta(s)$ satisfies the functional equation $\xi(s) = \xi(1 - s)$. We will be interested in the values of $\zeta(s)$ at negative integers. At *even* negative integers, this is trivial to compute:

$$\zeta(-2k) = 0 \qquad \text{if} \quad k \geq 1.$$

For $\xi(s)$ is holomorphic at $s = 2k + 1$, hence also at $s = 1 - (2k + 1) = -2k$. But $\Gamma(s/2)$ has a pole at $s = -2k$, so $\zeta(s)$ must have a zero there. At *odd* negative integers, on the other hand, $\zeta(s)$ is non-zero and can be described in terms of the Bernoulli numbers, whose definition we now recall.

The function $z/(e^z - 1)$ has a power series expansion of the form

$$\frac{z}{e^z - 1} = 1 - \frac{z}{2} + \sum_{k=1}^{\infty} \frac{B_{2k}}{(2k)!} z^{2k},$$

where $B_{2k} \in \mathbb{Q}$. The numbers B_{2k} which arise in this way are called the *Bernoulli numbers*. Here are the first few of them, with their numerators and denominators factored into primes:

$$B_2 = \frac{1}{2 \cdot 3} \qquad B_4 = -\frac{1}{2 \cdot 3 \cdot 5} \qquad B_6 = \frac{1}{2 \cdot 3 \cdot 7}$$

$$B_8 = -\frac{1}{2 \cdot 3 \cdot 5} \qquad B_{10} = \frac{5}{2 \cdot 3 \cdot 11} \qquad B_{12} = -\frac{691}{2 \cdot 3 \cdot 5 \cdot 7 \cdot 13}$$

$$B_{14} = \frac{7}{2 \cdot 3} \qquad B_{16} = -\frac{3617}{2 \cdot 3 \cdot 5 \cdot 17} \qquad B_{18} = \frac{43867}{2 \cdot 3 \cdot 7 \cdot 19}$$

$$B_{20} = -\frac{283 \cdot 617}{2 \cdot 3 \cdot 5 \cdot 11} \qquad B_{22} = \frac{11 \cdot 131 \cdot 593}{2 \cdot 3 \cdot 23} \qquad B_{24} = -\frac{103 \cdot 2294797}{2 \cdot 3 \cdot 5 \cdot 7 \cdot 13}$$

It is well-known that the Bernoulli numbers arise when one computes $\zeta(2k)$ for an integer $k \geq 1$:

$$\zeta(2k) = -\frac{(-1)^k 2^{2k-1} B_{2k}}{(2k)!} \pi^{2k}.$$

Using the functional equation to rewrite this as a formula for $\zeta(1 - 2k)$, we find that the factorial and the powers of -1, 2, and π disappear; the result is

$$\zeta(1 - 2k) = -\frac{B_{2k}}{2k}. \tag{8.3}$$

We can now state a few examples of Harder's formula for $\chi(G(\mathbb{Z}))$. First, taking $G = SL_2$, we have

$$\chi(SL_2(\mathbb{Z})) = \zeta(-1). \tag{8.4}$$

Since $\zeta(-1) = -B_2/2 = -\frac{1}{12}$, this agrees with 7.2. The group SL_2 is part of two infinite families of groups, namely, the special linear groups SL_n and the symplectic groups Sp_{2n}. [One has $Sp_2 = SL_2$.] The generalizations of 8.4 to these families are

$$\chi(SL_n(\mathbb{Z})) = \prod_{k=2}^{n} \zeta(1 - k) \tag{8.5}$$

and

$$\chi(Sp_{2n}(\mathbb{Z})) = \prod_{k=1}^{n} \zeta(1-2k). \tag{8.6}$$

Of course 8.5 involves the factor $\zeta(-2)$ as soon as $n \geq 3$, so it simply says $\chi(SL_n(\mathbb{Z})) = 0$ for $n \geq 3$. But 8.6 involves only non-zero values of $\zeta(s)$ and hence is more interesting.

For our last example we take G to be the exceptional Chevalley group E_7. Harder's formula in this case turns out to involve ζ at $-1, -5, -7, -9, -11, -13$, and -17, hence it involves B_k for $k = 2, 6, 8, 10, 12, 14$, and 18. The precise formula turns out to be

$$\chi(E_7(\mathbb{Z})) = -\frac{691 \cdot 43867}{2^{21} \cdot 3^9 \cdot 5^2 \cdot 7^3 \cdot 11 \cdot 13 \cdot 19}. \tag{8.7}$$

Finally, we mention that Harder's formulas (such as 8.5 and 8.6) are valid with $\mathbb{Z}$ replaced by the ring of integers in a number field F and ζ replaced by the Dedekind ζ-function ζ_F associated to F. And Serre has shown that one can generalize even further, namely, to rings $\mathcal{O}_S$ of "S-integers". Here S is a finite set of primes of F and $\mathcal{O}_S \subset F$ is the subring consisting of the elements of F which are integral at all primes not in S. One then uses the ζ-function $\zeta_{F,S}$ obtained from ζ_F by omitting from its infinite product expansion the factors corresponding to the primes in S. If $F = \mathbb{Q}$ and S consists of a single prime p, for example, then $\zeta_{F,S}(s) = \zeta(s)(1 - p^{-s})$ and the generalization of 8.4 gives

$$\chi(SL_2(\mathbb{Z}[1/p])) = \zeta_{F,S}(-1) = \frac{p-1}{12}.$$

9 Integrality Properties of $\chi(\Gamma)$

Let Γ be a group of finite homological type (e.g., a group of type VFP). We know that $\chi(\Gamma) = \tilde{\chi}(\Gamma) \in \mathbb{Z}$ if Γ is torsion-free, so the non-integrality of $\chi(\Gamma)$ in general is somehow due to the presence of torsion in Γ. The rest of this chapter will be devoted to proving various precise versions of this vague statement.

The most obvious integrality statement that one can make in general is that $(\Gamma : \Gamma') \cdot \chi(\Gamma) \in \mathbb{Z}$ if Γ' is a torsion-free subgroup of finite index; for $(\Gamma : \Gamma') \cdot \chi(\Gamma) = \chi(\Gamma') = \tilde{\chi}(\Gamma') \in \mathbb{Z}$. Consequently,

$$d \cdot \chi(\Gamma) \in \mathbb{Z}, \tag{9.1}$$

where $d = \gcd\{(\Gamma : \Gamma') : \Gamma'$ is torsion-free and of finite index$\}$. This simple observation, in conjunction with Harder's deep results (§8), already has non-trivial applications. For example, Serre [1971] used it to prove integrality results about values of the zeta-function of a number field. Another application (also due to Serre) concerns the torsion in the exceptional Chevalley groups. It is based on:

(9.2) Lemma. *The prime divisors of d are precisely the primes p such that Γ has p-torsion.*

PROOF. Let H be a finite subgroup of Γ. If $\Gamma' \subseteq \Gamma$ is a torsion-free subgroup, then H acts freely on Γ/Γ', so $(\Gamma : \Gamma')$, if finite, is divisible by $|H|$. Hence $|H|$ divides d. In particular, it follows that d is divisible by any prime p such that Γ has p-torsion. Conversely, suppose p is a prime such that Γ has no p-torsion. Let $\Gamma_0 \subseteq \Gamma$ be a torsion-free normal subgroup of finite index and let Γ' be an intermediate subgroup ($\Gamma_0 \subseteq \Gamma' \subseteq \Gamma$) such that Γ'/Γ_0 is a p-Sylow subgroup of Γ/Γ_0. Then $(\Gamma' : \Gamma_0)$ is a power of p and $(\Gamma : \Gamma')$ is relatively prime to p. By the first part of the proof, any finite subgroup of Γ' must be a p-group, so Γ' is torsion-free. But then $d \mid (\Gamma : \Gamma')$, so $p \nmid d$. □

Looking at formula 8.7 for $\chi(E_7(\mathbb{Z}))$, for example, we conclude from 9.1 and 9.2 that $E_7(\mathbb{Z})$ must have p-torsion for $p = 2, 3, 5, 7, 11, 13, 19$.

The proof of 9.2 shows that d is divisible by the order of every finite subgroup of Γ, hence d is divisible by the least common multiple m of the orders of the finite subgroups. Moreover, it is clear from 9.2 that d and m have the same prime divisors. But in general $d \neq m$. If $\Gamma = SL_3(\mathbb{Z})$, for instance, it can be shown that $d = 2^4 \cdot 3$ whereas $m = 2^3 \cdot 3$.

It is reasonable to ask, then, whether 9.1 can be improved to the statement $m \cdot \chi(\Gamma) \in \mathbb{Z}$. This question was raised by Serre, and it turns out to have an affirmative answer:

(9.3) Theorem (Brown [1974]). *Let Γ be a group of finite homological type and let m be the least common multiple of the orders of the finite subgroups of Γ. Then $m \cdot \chi(\Gamma) \in \mathbb{Z}$. Consequently, if a prime power p^a divides the denominator of $\chi(\Gamma)$, then Γ has a subgroup of order p^a.*

The second assertion follows easily from the first and the Sylow theorems. The proof of the first assertion requires some topological ideas involving finite group actions; it will be given in the next section.

Using 9.3 one can improve Serre's integrality results on the values of the zeta function (cf. Brown [1974], §9). One can also obtain good estimates of the amount of torsion in some of the exceptional Chevalley groups. For instance, 9.3 and 8.7 show that $E_7(\mathbb{Z})$ has subgroups of order 2^{21}, 3^9, etc. See Serre [1979] for more information on this. A third application is the following improvement of 6.4:

(9.4) Corollary. *Let $1 \to \Gamma \to E \to G \to 1$ be a group extension such that Γ is torsion-free and of finite homological type and G is a p-group for some prime p. If $p \nmid \chi(\Gamma)$, then the extension splits.*

Proof. We have $\chi(E) = \chi(\Gamma)/|G|$, and this fraction is in lowest terms since $p \nmid \chi(\Gamma)$. Theorem 9.3 therefore implies that E has a finite subgroup $\tilde{G}$ with $|\tilde{G}| = |G|$. Since Γ is torsion-free, $\tilde{G}$ maps isomorphically to G and provides a splitting. □

10 Proof of Theorem 9.3; Finite Group Actions

We begin by interpreting Theorem 9.3 topologically. Since vcd $\Gamma < \infty$, we know from VIII.11.1 that there is a finite dimensional contractible Γ-complex X with finite isotropy groups. Moreover, I claim that X can be taken to be an *admissible* Γ-complex, in the sense that the following condition holds: For each cell σ of X, the isotropy group Γ_σ fixes σ pointwise. To see this, we need only recall that X can be taken to be simplicial (VIII.11.2); passing to the barycentric subdivision if necessary, we can assume that X is an *ordered* simplicial complex and that Γ is order-preserving. Then clearly no element of Γ can permute the vertices of a simplex non-trivially, so the admissibility condition holds.

Let $\Gamma' \subseteq \Gamma$ be a torsion-free normal subgroup of finite index. Then Γ' acts freely on X and the quotient $Y = X/\Gamma'$ is a $K(\Gamma', 1)$. Thus

$$\chi(\Gamma) = \frac{\tilde{\chi}(\Gamma')}{(\Gamma : \Gamma')} = \frac{\chi(Y)}{(\Gamma : \Gamma')},$$

where $\chi(Y) = \sum (-1)^i \operatorname{rk}_{\mathbb{Z}}(H_i Y)$. What we are trying to prove, then, is that

$$\frac{m}{(\Gamma : \Gamma')} \cdot \chi(Y) \in \mathbb{Z},$$

where $m = \operatorname{lcm}\{|H| : H \text{ is a finite subgroup of } \Gamma\}$. In other words, we want to prove that

$$k \mid \chi(Y),$$

where k is the integer $(\Gamma : \Gamma')/m$.

Let G be the finite quotient Γ/Γ'. Note that the action of Γ on X induces an action of G on Y. Moreover, I claim that the isotropy group G_τ of any cell τ of Y is simply $\pi(\Gamma_\sigma)$, where σ is any cell of X lying over τ and $\pi : \Gamma \to G$ is the quotient map. For suppose $g\tau = \tau$ and choose $\gamma \in \Gamma$ such that $\pi(\gamma) = g$. Then $\gamma\sigma$ and σ both lie over τ, so $\gamma'\gamma\sigma = \sigma$ for some $\gamma' \in \Gamma'$. Since $\pi(\gamma'\gamma) = g$, this shows that $G_\tau \subseteq \pi(\Gamma_\sigma)$. The opposite inclusion is trivially true, whence the claim. (Incidentally, π actually maps Γ_σ isomorphically onto G_τ; for $\Gamma_\sigma \cap \Gamma'$ is finite and torsion-free, hence trivial.)

Two facts follow from this. First, the action of G on Y is admissible. Second, every isotropy group G_τ has order dividing m, hence the cardinality $|G|/|G_\tau|$ of the orbit of τ is divisible by $k = |G|/m$.

Theorem 9.3 now follows from:

(10.1) Theorem. *Let G be a finite group and let Y be a finite dimensional admissible G-complex such that $H_* Y$ is finitely generated. If k is an integer which divides the cardinality of every G-orbit, then $k \mid \chi(Y)$.*

Note that it does not matter whether we interpret G-orbit to mean G-orbit of cells or G-orbit of points; for the admissibility condition implies that $G_\tau = G_y$ for any interior point y of τ.

Note also that the theorem is trivially true if Y is finite, since $\chi(Y)$ can then be computed by counting cells. The point of 10.1 is that it is true under fairly mild *homological* finiteness hypotheses on Y. In fact, one can even prove a version of 10.1 for spaces Y which are not CW-complexes (cf. Brown [1979], Theorem 7.4), but the proof then requires more machinery than is needed for 10.1.

PROOF OF 10.1. For any subgroup $H \subseteq G$, let $Y_H = \{y \in Y: G_y = H\}$. The admissibility condition implies that Y_H is a union of (open) cells, namely, those cells τ such that $G_\tau = H$. Note that $g \cdot Y_H = Y_{gHg^{-1}}$ for any $g \in G$. In particular, Y_H is stable under the action of the normalizer $N(H)$ of H in G. Moreover, $N(H)/H$ acts freely on Y_H. We can now easily explain the idea of the proof. Y is the disjoint union (set-theoretically) of the subspaces Y_H, so one might expect

$$(*) \qquad \chi(Y) = \sum_{H \subseteq G} \chi(Y_H).$$

Now $Y_{gHg^{-1}} = g \cdot Y_H \approx Y_H$, so we can group together the terms in $(*)$ corresponding to conjugate subgroups H, and we obtain

$$(**) \qquad \chi(Y) = \sum_{H \in \mathscr{C}} (G: N(H))\chi(Y_H),$$

where $\mathscr{C}$ is a set of representatives for the conjugacy classes of subgroups of G which occur as isotropy groups in Y. [We have used the elementary fact that $(G: N(H))$ is the number of conjugates of H in G.] Since $N(H)/H$ acts freely on Y_H, 5.5 should now imply that $\chi(Y_H)$ is divisible by $(N(H): H)$, so that $(G: N(H))\chi(Y_H)$ is divisible by $(G: H)$. But $(G: H)$ is the cardinality of an orbit, so every term on the right-hand side of $(**)$ is divisible by k; hence $k \mid \chi(Y)$.

We now must clarify and justify the steps in this "proof." In the first place, Y_H is not in general a subcomplex of Y, since it need not be closed. (A boundary point of a cell can have a bigger isotropy group than the interior points.) But each fixed-point set Y^H is a subcomplex, and hence Y_H is a difference of subcomplexes:

$$Y_H = Y^H - Y^{>H},$$

where $Y^{>H} = \bigcup_{H' \supsetneq H} Y^{H'}$. Let's *assume* temporarily that each Y^H has finitely generated homology. Then the same is true of $Y^{>H}$ by a Mayer-Vietoris argument, since the family $\{Y^{H'}: H' \supsetneq H\}$ is closed under intersection. Hence $H_*(Y^H, Y^{>H})$ is finitely generated, and I claim that (*) becomes true if we replace $\chi(Y_H)$ by $\chi(Y^H, Y^{>H})$.

To prove this, let $H_1, \ldots, H_n$ be the subgroups of G, ordered so that $|H_i| \geq |H_{i+1}|$. We filter Y by subcomplexes

$$\varnothing = Y_0 \subseteq Y_1 \subseteq \cdots \subseteq Y_n = Y,$$

defined inductively by

$$Y_i = Y_{i-1} \cup Y^{H_i}.$$

Note that $Y_{i-1} \cap Y^{H_i} = Y^{>H_i}$. We therefore have an excision isomorphism

$$H_*(Y_i, Y_{i-1}) \approx H_*(Y^{H_i}, Y^{>H_i}).$$

[Note that, because we are dealing with subcomplexes, there is no problem justifying the excision. Indeed, we already have an excision isomorphism on the level of cellular chain complexes.] Hence

$$\begin{aligned} \chi(Y) &= \sum_{i=1}^{n} \chi(Y_i, Y_{i-1}) \\ &= \sum_{i=1}^{n} \chi(Y^{H_i}, Y^{>H_i}), \end{aligned}$$

as claimed.

The remainder of our initial heuristic proof is now valid, with Y_H replaced everywhere by $(Y^H, Y^{>H})$. One needs, of course, to apply a relative version of 5.5 to $N(H)/H$ acting on $(Y^H, Y^{>H})$, but this presents no problem—simply apply 5.3 to the cellular chain complex $C(Y^H, Y^{>H})$, which is a finite dimensional complex of free $\mathbb{Z}[N(H)/H]$-modules.

If we now drop the assumption that $H_*(Y^H)$ is finitely generated for each H, then we can argue as follows. Suppose first that G is a p-group for some prime p. Since $H_* Y$ is finitely generated, we can compute $\chi(Y)$ from $H_*(Y; \mathbb{Z}_p)$. We now repeat the argument above using $H_*(\ \ ; \mathbb{Z}_p)$ and using 5.7 instead of 5.3. This time, however, we have a *theorem* (VII.10.5 and exercise 2 of §VII.10) which says that $H_*(Y^H; \mathbb{Z}_p)$ is finitely generated, so we do not need to assume this.

Finally, we can easily reduce the general case to the p-group case: To prove 10.1, it suffices to show that if p^a is a prime power which divides k then $p^a | \chi(Y)$. Let $G(p)$ be a p-Sylow subgroup of G. By hypothesis $p^a | (G: G_\tau) = |G|/|G_\tau|$ for any cell τ of Y. Looking at the p-part of $|G|$ and $|G_\tau|$, we conclude that p^a divides $|G(p)|/|P|$, where P is any p-subgroup of G_τ. In particular, p^a divides $|G(p)|/|G(p)_\tau|$, which is the cardinality of a typical $G(p)$-orbit. It now follows from the p-group case that $p^a | \chi(Y)$. □

EXERCISE

Use the method of proof of 10.1 to prove the following "Lefschetz fixed-point theorem": Let G be a finite group and Y an admissible finite dimensional G-complex. Assume that each fixed-point set Y^g ($g \in G$) has finitely generated homology. Prove that the Lefschetz number $L(g)$ is given by

$$L(g) = \chi(Y^g).$$

[Hint: You may assume G is cyclic, generated by g. Then g is non-trivial in $N(H)/H = G/H$ for every $H \subsetneqq G$, so $L(g \mid Y_H) = 0$ by exercise 1 of §5. Now use the Lefschetz number analogue of (*) and note that every term is zero except the one corresponding to $H = G$.]

Remark. One can use this theorem to prove the following theorem about Euler characteristics of groups (cf. Brown [1982]): Let Γ be a group such that the centralizer $C(\gamma)$ is of finite homological type for every $\gamma \in \Gamma$ of finite order (including $\gamma = 1$). Let $\mathscr{C}$ be a set of representatives for the conjugacy classes of elements of finite order. Then $\mathscr{C}$ is finite, and

$$\tilde{\chi}(\Gamma) = \sum_{\gamma \in \mathscr{C}} \chi(C(\gamma)).$$

Consequently, we have the following formula, which expresses the deviation between $\chi(\Gamma)$ and $\tilde{\chi}(\Gamma)$ in terms of the torsion in Γ:

$$\tilde{\chi}(\Gamma) = \chi(\Gamma) + \sum_{\gamma \in \mathscr{C}'} \chi(C(\gamma)),$$

where $\mathscr{C}' = \mathscr{C} - \{1\}$.

11 The Fractional Part of $\chi(\Gamma)$

Under suitable hypotheses on Γ it is possible to deduce from the proof of 9.3 precise formulas expressing $\chi(\Gamma) - \tilde{\chi}(\Gamma)$ in terms of the torsion in Γ. Since $\tilde{\chi}(\Gamma) \in \mathbb{Z}$, such a formula can be regarded as a formula for the "fractional part" of the rational number $\chi(\Gamma)$. One formula of this type was just written down, following the exercise of §10. Another formula, which can be found in Brown [1974], has the form

$$\chi(\Gamma) = \tilde{\chi}(\Gamma) + \sum_{H} \frac{c(H)}{|H|},$$

where H ranges over the non-trivial finite subgroups of Γ (up to conjugacy) and $c(H)$ is an integer whose definition is too complicated to repeat here.

Rather than state and prove this formula, we will use similar methods to prove some formulas which are less precise but easier to prove and apply.

If S is a partially ordered set, then there is an obvious way to construct from S an ordered simplicial complex $|S|$. Namely, the vertices of $|S|$ are the elements of S, and the simplices of $|S|$ are the linearly ordered finite subsets $s_0 < \cdots < s_n$ of S. [We have already seen one example of this construction: If K is any simplicial complex and S is the set of simplices of K, ordered by the face relation, then $|S|$ is simply the barycentric subdivision of K, cf. §VIII.11.] If Γ acts on S (and preserves the ordering), then there is an induced action of Γ on $|S|$, and we set

$$\chi_\Gamma(S) = \chi_\Gamma(|S|)$$

if the equivariant Euler characteristic on the right-hand side is defined. More generally, we will use the functor $S \mapsto |S|$ to assign arbitrary topological concepts to partially ordered sets. For instance we will write $C_*(S)$ and $H_*(S)$ instead of $C_*(|S|)$ and $H_*(|S|)$, and we will say that S is contractible or acyclic if $|S|$ is.

(11.1) Theorem. *Let Γ be a group of finite homological type. Assume that Γ has only finitely many finite subgroups (up to conjugacy) and that the normalizer $N(H)$ of each finite subgroup H has finite homological type. Let S be the set of non-trivial finite subgroups of Γ, regarded as a partially ordered set under inclusion, with Γ-action by conjugation. Then $\chi_\Gamma(S)$ is defined and*

$$\chi(\Gamma) \equiv \chi_\Gamma(S) \bmod \mathbb{Z}.$$

(In concrete terms, this gives the fractional part of $\chi(\Gamma)$ in terms of the Euler characteristics of subgroups of Γ of the form $N(H_0) \cap \cdots \cap N(H_n)$, where $H_0 \subset \cdots \subset H_n$ is a chain of non-trivial finite subgroups.)

PROOF. To show $\chi_\Gamma(S)$ is defined, we must show that $|S|$ has only finitely many simplices σ mod Γ and that each Γ_σ has finite homological type. Let $\mathscr{F}$ be a (finite) set of representatives for the conjugacy classes of finite subgroups of Γ. Then any simplex $H_0 \subset \cdots \subset H_n$ of $|S|$ is equivalent mod Γ to one such that $H_n \in \mathscr{F}$. But there are clearly only finitely many such simplices, since H_n is finite. To prove the assertion about the isotropy groups, note that $N(H)$ for any $H \in S$ permutes the simplices $H_0 \subset \cdots \subset H_n$ with $H_n = H$. Since there are only finitely many such simplices, it follows that each such simplex σ has $(N(H): N(H)_\sigma) < \infty$, hence $N(H)_\sigma$ is of finite homological type. But $N(H)_\sigma = \Gamma_\sigma$, so the first part of the theorem is proved.

Now let X, Γ', G, and Y be as in the proof of Theorem 9.3. By VIII.11.2 we may assume that X^H is contractible for each $H \in S$. Let X_0 be the part of X where the action of Γ is not free, i.e.

$$X_0 = \bigcup_{H \in S} X^H,$$

and let $Y_0 = X_0/\Gamma' \subseteq Y$. The analysis of the isotropy groups G_τ that we gave in the proof of 9.3 shows that Y_0 is the part of Y where the G-action is not free. [Y_0 is in fact equal to the subcomplex called Y_{n-1} in the proof of 10.1.]

I claim now that $H_*(Y_0)$ is finitely generated, and hence so is $H_*(Y, Y_0)$. Accepting this for the moment, we can apply 5.3 to $C(Y, Y_0)$ since G acts freely in $Y - Y_0$. It follows that $\chi(Y, Y_0)$ is divisible by $|G|$. Thus

$$\begin{aligned}\chi(\Gamma) &= \frac{\chi(Y)}{|G|}\\ &= \frac{\chi(Y_0)}{|G|} + \frac{\chi(Y, Y_0)}{|G|}\\ &\equiv \frac{\chi(Y_0)}{|G|} \bmod \mathbb{Z}.\end{aligned}$$

To complete the proof, then, we will prove that $H_*(Y_0)$ is finitely generated and that $\chi(Y_0)/|G| = \chi_\Gamma(S)$. Note first that $H_*(Y_0) \approx H_*^{\Gamma'}(X_0)$ since Γ' acts freely on X_0 (cf. VII.7.8). Next, we apply the following observation to X_0:

(11.2) Lemma. *Let Z be an admissible proper Γ-complex such that each Γ_σ is non-trivial and each Z^H ($H \in S$) is acyclic, where S is the set of non-trivial finite subgroups of Γ. Then Z is homologically equivalent to S, in the following sense: There is a chain complex C of Γ-modules which admits Γ-maps $C(Z) \leftarrow C \rightarrow C(S)$ inducing homology isomorphisms.*

(*Note*: By "acyclic" we mean "having the homology of a point." In particular, acyclic $\Rightarrow$ non-empty.)

The proof is a straightforward analogue of that of VII.4.4. Details will be given in the next section. [The intuitive idea is as follows: We have $Z = \bigcup_{H \in S} Z^H$, and the order relation on the index set S completely determines the intersection pattern of the acyclic complexes Z^H. Namely, $Z^{H_1} \cap Z^{H_2} \neq \varnothing$ if and only if H_1 and H_2 have a least upper bound H in S, in which case $Z^{H_1} \cap Z^{H_2} = Z^H$. It is not surprising, then that we can use $|S|$ instead of the nerve of the covering in VII.4.4.]

Returning to the proof of 11.1, we conclude (cf. VII.5.2) that $H_*(Y_0) \approx H_*^{\Gamma'}(X_0) \approx H_*^{\Gamma'}(S)$. In view of 7.3c and b′, it follows that $H_*(Y_0)$ is finitely generated and that

$$\frac{\chi(Y_0)}{|G|} = \frac{\tilde{\chi}_{\Gamma'}(S)}{|G|} = \chi_\Gamma(S). \qquad \square$$

With very little effort, we can generalize 11.1 to the following result:

(11.3) Proposition. *Let Γ and S be as in* 11.1. *Let Z be an admissible Γ-complex such that $\chi_\Gamma(Z)$ is defined. If Z^H is acyclic for each $H \in S$, then*

$$\chi_\Gamma(Z) \equiv \chi_\Gamma(S) \bmod \mathbb{Z}.$$

(This reduces to 11.1 if Z is a point.)

Proof. Let X and Γ' be as above, and apply the proof of 11.1 with X replaced by $X \times Z$. Note that the latter is finite dimensional, admissible, and proper, and that $(X \times Z)^H = X^H \times Z^H$ is acyclic for each $H \in S$. Moreover,

$$H_*((X \times Z)/\Gamma') \approx H_*^{\Gamma'}(X \times Z) \approx H_*^{\Gamma'}(Z)$$

since X is contractible. We therefore obtain, as in the proof of 11.1,

$$\begin{aligned}\chi_\Gamma(Z) &= \frac{\tilde{\chi}_{\Gamma'}(Z)}{(\Gamma : \Gamma')} \\ &= \frac{\chi((X \times Z)/\Gamma')}{(\Gamma : \Gamma')} \\ &\equiv \frac{\chi((X \times Z)_0/\Gamma')}{(\Gamma : \Gamma')} \mod \mathbb{Z} \\ &= \chi_\Gamma(S).\end{aligned}$$

□

Combining 11.1 and 11.3, we see that

$$\chi(\Gamma) \equiv \chi_\Gamma(Z) \mod \mathbb{Z} \tag{11.4}$$

under the hypotheses of 11.3.

Finally, we will need the following "local" version of 11.4, for which the hypotheses on Γ can be weakened.

For any prime p, let $\mathbb{Z}_{(p)} = \{a/b : a, b \in \mathbb{Z}, p \nmid b\}$.

(11.5) Proposition. *Let Γ be a group of finite homological type and let Z be an admissible Γ-complex such that $\chi_\Gamma(Z)$ is defined. If p is a prime such that Z^H is acyclic for every non-trivial finite p-subgroup of Γ, then*

$$\chi(\Gamma) \equiv \chi_\Gamma(Z) \mod \mathbb{Z}_{(p)}.$$

Proof. We continue to use the notation of the proof of 11.1. Assume first that Γ/Γ' is a p-group. Then every finite subgroup of Γ is a p-group, so our hypothesis says that Z^H is acyclic for every $H \in S$. We now repeat the proofs of 11.1 and 11.3, but taking all homology groups to have $\mathbb{Z}_p$-coefficients. We no longer have hypotheses to guarantee that $\chi_\Gamma(S)$ is defined, but instead we can use VII.10.5 as in the proof of 10.1 to conclude that $H_*(Y_0; \mathbb{Z}_p)$ is finitely generated. Arguing as in the proof of 11.1 (and using 5.7 instead of 5.3), we find that $H_*^{\Gamma'}(S; \mathbb{Z}_p)$ is finitely generated and that

$$\chi(\Gamma) \equiv \frac{\tilde{\chi}_{\Gamma'}(S)}{(\Gamma : \Gamma')} \mod \mathbb{Z},$$

where the equivariant Euler characteristic on the right is now understood to be defined in terms of $H_*^{\Gamma'}(S; \mathbb{Z}_p)$. Similarly,

$$\chi_\Gamma(Z) \equiv \frac{\tilde{\chi}_{\Gamma'}(S)}{(\Gamma : \Gamma')} \bmod \mathbb{Z},$$

so

$$\chi(\Gamma) \equiv \chi_\Gamma(Z) \bmod \mathbb{Z}.$$

In the general case, where Γ/Γ' is not necessarily a p-group choose a subgroup $\Gamma_p \supseteq \Gamma'$ such that Γ_p/Γ' is a p-Sylow subgroup of Γ/Γ'. By what we have just proved,

$$\chi(\Gamma_p) \equiv \chi_{\Gamma_p}(Z) \bmod \mathbb{Z},$$

so that

$$(\Gamma : \Gamma_p) \cdot \chi(\Gamma) \equiv (\Gamma : \Gamma_p) \cdot \chi_\Gamma(Z) \bmod \mathbb{Z}.$$

Since $(\Gamma : \Gamma_p)$ is relatively prime to p, this proves the proposition. □

Knowing $\chi(\Gamma) \bmod \mathbb{Z}_{(p)}$ is equivalent to knowing the "p-fractional part" of $\chi(\Gamma)$, i.e., that part of the partial fractions decomposition of $\chi(\Gamma)$ with denominator a power of p. Thus 11.5 says that $\chi(\Gamma)$ and $\chi_\Gamma(Z)$ have the same p-fractional part. We will give some substance to this result in §13 by giving an interesting example of a complex Z satisfying the hypotheses of 11.5.

12 Acyclic Covers; Proof of Lemma 11.2

The proof of Lemma 11.2 will be based on the following variant of VII.4.4:

(12.1) Proposition. *Let X be a CW-complex and $(X_s)_{s\in S}$ a family of subcomplexes indexed by a partially ordered set S. Assume that $X_s \supseteq X_t$ whenever $s \leq t$. For each cell e of X let $S_e = \{s \in S : e \subseteq X_s\}$. If each subcomplex X_s is acyclic and each partially ordered set S_e is acyclic, then $H_*(X) \approx H_*(S)$. More precisely, there is a chain complex C, constructed canonically from the given data, which admits canonical weak equivalences $C(X) \leftarrow C \rightarrow C(S)$.*

PROOF. We will construct a double complex analogous to that used in §VII.4. Let $|S|^{(p)}$ be the set of p-simplices of $|S|$. For each simplex $\sigma = (s_0 < \cdots < s_p)$ in $|S|^{(p)}$, let $X_\sigma = X_{s_p}$. Then $X_\sigma \subseteq X_\tau$ if τ is a face of σ. In particular, we have inclusion maps

$$\partial_i : C(X_\sigma) \to C(X_{\partial_i \sigma}) \quad (i = 0, \ldots, p),$$

where $\partial_i \sigma = (s_0 < \cdots < \hat{s}_i < \cdots < s_p)$. Let

$$C_p = \bigoplus_{\sigma \in |S|^{(p)}} C(X_\sigma).$$

Letting $\partial = \sum(-1)^i\partial_i: C_p \to C_{p-1}$, we obtain a chain complex

$$\cdots \to C_p \xrightarrow{\partial} C_{p-1} \to \cdots$$

in the category of chain complexes, and hence a double complex. The acyclicity of each X_σ implies that one of the spectral sequences of the double complex collapses, yielding

$$H_*(C) \approx H_*(S),$$

where C is the total complex associated to the double complex. On the other hand, exactly as in §VII.4, the acyclicity of each S_e implies that the other spectral sequence also collapses and yields

$$H_*(C) \approx H_*(X).$$

It is easy to check that these isomorphisms are induced by canonical chain maps $C(X) \leftarrow C \to C(S)$, cf. exercise 1 of §VII.4. □

We will also need the following simple observation:

(12.2) Lemma. *If S is a partially ordered set with a largest or smallest element, then S is contractible.*

Proof. If $s \in S$ is the extreme element and $S' = S - \{s\}$, then $|S|$ is the cone over $|S'|$, with s as the cone vertex. □

Proof of Lemma 11.2. We have $Z = \bigcup_{H \in S} Z^H$, where S is the set of non-trivial finite subgroups of Γ. For each cell e of Z, the set S_e defined in 12.1 is simply the set of all non-trivial finite subgroups of the isotropy group Γ_e. Since Γ_e is a non-trivial finite group, S_e has a largest element and hence is contractible by 12.2. The hypotheses of 12.1 are therefore satisfied, and we obtain a complex C and weak equivalences $C(Z) \leftarrow C \to C(S)$. Since Γ acts on Z and permutes the subcomplexes Z^H according to the conjugation action of Γ on S, it is clear that the complex C inherits a Γ-action and that the weak equivalences above are compatible with the Γ-action. □

13 The p-Fractional Part of $\chi(\Gamma)$

Theorem 11.1 is not very useful in practice because the computation of $\chi_\Gamma(S)$ tends to be extremely tedious. If we fix a prime p, however, and replace S by the set of non-trivial finite subgroups which are p-groups, then the computation often simplifies drastically. Moreover, it turns out that this gives the p-fractional part of $\chi(\Gamma)$. In fact, we can even compute the latter by looking only at the *elementary abelian* p-subgroups of Γ, i.e., the subgroups isomorphic to a finite direct product of copies of $\mathbb{Z}_p$:

(13.1) Theorem. *Let Γ be a group of finite homological type and let $\mathscr{A}$ be the partially ordered set of non-trivial elementary abelian p-subgroups of Γ, where p is a fixed prime. Assume that the normalizer $N(A)$ is of finite homological type for every $A \in \mathscr{A}$. Then $\chi_\Gamma(\mathscr{A})$ is defined and*

$$\chi(\Gamma) \equiv \chi_\Gamma(\mathscr{A}) \bmod \mathbb{Z}_{(p)}.$$

(This is an improvement due to Quillen [1978] of the main result of Brown [1975b].)

The proof will use:

(13.2) Lemma. *Let Γ be a group such that* vcd $\Gamma < \infty$ *and let p be a prime. If Γ has a torsion-free normal subgroup Γ' of finite index such that $H_*(\Gamma', \mathbb{Z}_p)$ is finitely generated, then Γ has only finitely many p-subgroups, up to conjugacy.*

Proof. Let X be a finite dimensional, contractible, proper, admissible Γ-complex such that $X^H \neq \varnothing$ for every finite $H \subseteq \Gamma$. Let $Y = X/\Gamma'$ and let $G = \Gamma/\Gamma'$. By hypothesis $H_*(Y; \mathbb{Z}_p)$ is finitely generated, so we know that $H_*(Y^P; \mathbb{Z}_p)$ is finitely generated for every p-subgroup $P \subseteq G$ (§VII.10, Theorem 10.5 and exercise 2). In particular, Y^P has only finitely many connected components.

On the other hand, I claim that

$$Y^P = \coprod_{H \in \mathscr{C}} X^H/(\Gamma' \cap N(H)),$$

where $\mathscr{C}$ is a set of representatives for the Γ'-conjugacy classes of finite subgroups of Γ whose image in G is P. To see this, consider the inverse image $\widetilde{Y^P}$ of Y^P in X, and let $\widetilde{\mathscr{C}}$ be the set of finite subgroups of Γ whose image in G is P. For any $x \in \widetilde{Y^P}$, we know from §10 that the isotropy group Γ_x maps isomorphically to $G_y \subseteq G$, where $y \in Y$ is the image of x. Since $P \subseteq G_y$, it follows that there is a unique $H \subseteq \Gamma_x$ whose image in G is P. In other words, there is a unique $H \in \widetilde{\mathscr{C}}$ such that $x \in X^H$. Thus $\widetilde{Y^P} = \coprod_{H \in \widetilde{\mathscr{C}}} X^H$. Since Γ' permutes the X^H according to its conjugation action on $\widetilde{\mathscr{C}}$, it follows that

$$Y^P = \left(\coprod_{H \in \widetilde{\mathscr{C}}} X^H\right)\Big/\Gamma' = \coprod_{H \in \mathscr{C}} X^H/(\Gamma' \cap N(H)),$$

as claimed.

Since Y^P has only finitely many components and each X^H is non-empty, we conclude that $\mathscr{C}$ is finite, i.e., that the finite subgroups of Γ lying over P fall into finitely many conjugacy classes. Finally, there are only finitely many possibilities for P, so the lemma follows. $\square$

We also need:

(13.3) Lemma. *Let S and T be partially ordered sets and let $f_0, f_1: S \to T$ be order-preserving maps such that $f_0(s) \leq f_1(s)$ for all $s \in S$. Then $|f_0| \simeq |f_1|: |S| \to |T|$.*

[Note that this reduces to 12.2 if $S = T$ and either f_0 or f_1 is id_S while the other is a constant map.]

Proof. Heuristically, the desired homotopy moves $f_0(s)$ to $f_1(s)$ along the 1-simplex $\{f_0(s), f_1(s)\}$ of $|T|$. Unfortunately, this only defines the homotopy on vertices of $|S|$, so we instead proceed as follows. Let J be the ordered set $\{0, 1\}$ with $0 < 1$. Note that $|J|$ is the unit interval. Consider $S \times J$ with the product ordering: $(s, i) \leq (s', i')$ if and only if $s \leq s'$ and $i \leq i'$. Let $F: S \times J \to T$ be defined by $F(s, i) = f_i(s)$. The hypothesis that $f_0(s) \leq f_1(s)$ guarantees that F is order-preserving, so we have an induced map $|F|: |S \times J| \to |T|$. Now it is immediate from the definitions that $|S_1 \times S_2| = |S_1| \times |S_2|$ for any two partially ordered sets S_1, S_2, where the product on the right is the simplicial product defined in §VIII.11. Thus $|F|$ can be viewed as a map $|S| \times |J| \to |T|$, and this is the required homotopy. □

Proof of Theorem 13.1. By Lemma 13.2, Γ has only finitely many p-subgroups, up to conjugacy. It now follows exactly as in the first part of the proof of 11.1 that $\chi_\Gamma(\mathscr{A})$ is defined. We now wish to apply 11.5 to the Γ-complex $|\mathscr{A}|$, so we must verify that $|\mathscr{A}|^H$ is acyclic for each non-trivial p-subgroup $H \subseteq \Gamma$. Now $|\mathscr{A}|^H = |\mathscr{A}^H|$, where $\mathscr{A}^H$ is the set of non-trivial elementary abelian p-groups normalized by H, and I claim that this is in fact contractible. For let $C \subseteq H$ be a central subgroup of order p, and consider for any $A \in \mathscr{A}^H$ the chain of inequalities

$$A \geq A^C \leq C \cdot A^C \geq C$$

in $\mathscr{A}^H$. (Note that A^C, C, and $C \cdot A^C$ are indeed all in $\mathscr{A}^H$; in the case of A^C, this follows from VI.8.1.) This yields, via 13.3, a homotopy from the identity map of $\mathscr{A}^H$ to the constant map $A \mapsto C$, since the intermediate maps $A \mapsto A^C$ and $A \mapsto C \cdot A^C$ are order-preserving maps $\mathscr{A} \to \mathscr{A}$. Thus 11.5 is applicable and yields the desired congruence. □

In the next section we will show that $\chi_\Gamma(\mathscr{A})$ admits an explicit computation in terms of the normalizers of the elements $A \in \mathscr{A}$. For the moment, we just mention one case where this is obvious. Suppose that every $A \in \mathscr{A}$ is of rank ≤ 1. [This is equivalent to saying that the finite subgroups of Γ have p-periodic cohomology, cf. VI.9.7. It is also equivalent, as we will see in the next chapter, to saying that Γ itself has p-periodic cohomology in high dimensions.] In this case $|\mathscr{A}|$ is discrete and we obviously have

$$\chi_\Gamma(\mathscr{A}) = \sum_{P \in \mathscr{P}} \chi(N(P)),$$

where $\mathscr{P}$ is a set of representatives for the conjugacy classes of subgroups of Γ of order p. Thus

$$\chi(\Gamma) \equiv \sum_{P \in \mathscr{P}} \chi(N(P)) \bmod \mathbb{Z}_{(p)} \tag{13.4}$$

in this case.

For computational purposes, it is often more convenient to work with elements of order p and their centralizers, rather than subgroups of order p and their normalizers. It is easy to rewrite 13.4 in these terms:

(13.5) Corollary. *Let Γ be a group of finite homological type and assume that every elementary abelian p-subgroup of Γ has rank ≤ 1. Suppose further that the centralizer $C(\gamma)$ is of finite homological type for every $\gamma \in \Gamma$ of order p, and let $\mathscr{C}$ be a set of representatives for the conjugacy classes of elements of Γ of order p. Then $\mathscr{C}$ is finite and*

$$\chi(\Gamma) \equiv \frac{1}{p-1} \sum_{\gamma \in \mathscr{C}} \chi(C(\gamma)) \bmod \mathbb{Z}_{(p)}.$$

PROOF. For each $P \in \mathscr{P}$ choose a set $\mathscr{C}_P$ of representatives for the non-trivial elements of P under the conjugation action of $N(P)/C(P)$. Then $\coprod_{P \in \mathscr{P}} \mathscr{C}_P$ is a set of representatives for the conjugacy classes of elements of Γ of order p, so we may assume that $\mathscr{C}$ is equal to this union. Since $C(P) = C(\gamma)$ for every nontrivial $\gamma \in P$, the action of $N(P)/C(P)$ on $P - \{1\}$ is free. Thus

$$\operatorname{card}(\mathscr{C}_P) \cdot (N(P) : C(P)) = p - 1.$$

Consequently,

$$\begin{aligned} \sum_{\gamma \in \mathscr{C}} \chi(C(\gamma)) &= \sum_{P \in \mathscr{P}} \operatorname{card}(\mathscr{C}_P) \cdot \chi(C(P)) \\ &= \sum_{P \in \mathscr{P}} \operatorname{card}(\mathscr{C}_P) \cdot (N(P) : C(P)) \cdot \chi(N(P)) \\ &= (p-1) \sum_{P \in \mathscr{P}} \chi(N(P)). \end{aligned}$$

The corollary now follows from 13.4. □

We mention briefly one interesting application of 13.5 to number theory. (This application is due to Serre, and a detailed treatment is given in Brown [1974, 1975b].) Fix an odd prime p and let Γ be the symplectic group $Sp_{2n}(\mathbb{Z})$, where $2n = p - 1$.

On the one hand, we have Harder's formula (8.6) which gives $\chi(\Gamma)$ in terms of $\zeta(-1), \ldots, \zeta(1-2n)$, and hence in terms of the Bernoulli numbers $B_2, \ldots, B_{p-1}$. Using known facts about Bernoulli numbers ("von Staudt's Theorem") one concludes that $\chi(\Gamma) \equiv 0 \bmod \mathbb{Z}_{(p)}$ if and only if p divides the numerator of one of the numbers $B_2, \ldots, B_{p-3}$.

On the other hand, it is not hard to classify the elements of Γ of order p up to conjugacy and to compute their centralizers. One finds (i) that the hypotheses of 13.5 are satisfied; (ii) that the number of conjugacy classes of elements of order p is closely related to the class number h of the p-th cyclotomic number field; and (iii) that the centralizer of any element γ of order p is finite and of order $2p$, so that $\chi(C(\gamma)) = 1/2p$. It then follows from 13.5 that $\chi(\Gamma) \equiv 0 \bmod \mathbb{Z}_{(p)}$ if and only if $h/2p \equiv 0 \bmod \mathbb{Z}_{(p)}$, i.e., if and only if $p \mid h$. (The prime p is then said to be *irregular*.)

Combining the results of the last two paragraphs, we have: A prime p is irregular if and only if p divides the numerator of one of the Bernoulli numbers $B_2, \ldots, B_{p-3}$. This result is not new; indeed, it is known as *Kummer's criterion* for regularity. But our method of proof via Euler characteristics generalizes to rings of integers other than $\mathbb{Z}$ and yields number-theoretic results which were not previously known.

14 A Formula for $\chi_\Gamma(\mathscr{A})$

The formula to be given in this section is essentially due to Quillen [private communication]. It is based on:

(14.1) Proposition. *Let V be a vector space of dimension $r \geq 1$ over the field $\mathbb{F}_q$ with q elements, and let $S = S(V)$ be the partially ordered set of non-trivial proper subspaces of V. Then*

$$\tilde{H}_i(S) \approx \begin{cases} \mathbb{Z}^{n(r)} & \text{if } i = r - 2 \\ 0 & \text{if } i \neq r - 2, \end{cases}$$

where $n(r) = q^{r(r-1)/2}$.

(This is a special case of the so-called "Solomon–Tits theorem." The proof that we will give is due to Quillen [1973]. It will follow easily from the proof that $|S|$ in fact has the homotopy type of a bouquet of $n(r)$ spheres of dimension $r - 2$, but we will not need to know this.)

PROOF. The proposition is trivially true if $r = 1$, since S is then empty. Assume now that $r > 1$ and that the proposition is known for vector spaces of dimension $r - 1$. Let $L \subset V$ be a fixed 1-dimensional subspace and let $S' \subset S$ be the set of non-trivial subspaces $V' \subset V$ such that $V' + L \neq V$ (i.e., such that $V' + L \in S$). For any $V' \in S'$, we have relations

$$V' \leq V' + L \geq L$$

in S', and it follows easily that S' is contractible (cf. proof of 13.1). Thus

$$\tilde{H}_i(S) \approx H_i(S, S').$$

Let $\mathscr{H}$ be the set of hyperplanes H in V which are complementary to L (i.e., such that $V = L \oplus H$). Let $\bar{S}(H)$ (resp. $S(H)$) be the set of non-trivial subspaces (resp. non-trivial proper subspaces) of H. Note that every simplex σ of $|S|$ which is not in $|S'|$ has the form

$$V_0 < \cdots < V_n,$$

where $V_n \in \mathscr{H}$. Hence σ is a simplex of $|\bar{S}(H)|$ but not $|S(H)|$ for a unique $H \in \mathscr{H}$. Consequently, there is an excision isomorphism (already on the level of simplicial chain complexes)

$$H_*(S, S') \overset{\approx}{\leftarrow} \bigoplus_{H \in \mathscr{H}} H_*(\bar{S}(H), S(H)).$$

Now $\bar{S}(H)$ is contractible since it has a largest element, so

$$H_i(\bar{S}(H), S(H)) \approx \tilde{H}_{i-1}(S(H)) \approx \begin{cases} \mathbb{Z}^{n(r-1)} & \text{if } i = r-2 \\ 0 & \text{if } i \neq r-2 \end{cases}$$

by the induction hypothesis. Thus

$$\tilde{H}_i(S) \approx \begin{cases} \mathbb{Z}^{\operatorname{card}(\mathscr{H})\cdot n(r-1)} & \text{if } i = r-2 \\ 0 & \text{if } i \neq r-2. \end{cases}$$

To compute $\operatorname{card}(\mathscr{H})$, note that the elements of $\mathscr{H}$ correspond to splittings of the exact sequence

$$0 \to L \to V \to V/L \to 0,$$

so there is a bijection $\mathscr{H} \approx \operatorname{Hom}_{\mathbb{F}_q}(V/L, L)$. The latter being a vector space of dimension $r - 1$, it follows that $\operatorname{card}(\mathscr{H}) = q^{r-1}$. One now completes the proof by verifying that the function $n(r) = q^{r(r-1)/2}$ satisfies $n(r) = q^{r-1}n(r-1)$. □

(14.2) Theorem. *Let Γ be a group and p a prime such that $\chi_\Gamma(\mathscr{A})$ is defined, where $\mathscr{A}$ is the set of non-trivial elementary abelian p-subgroups of Γ. Let $\mathscr{A}_0$ be a set of representatives for $\mathscr{A}$* mod Γ. *Then*

$$\chi_\Gamma(\mathscr{A}) = \sum_{A \in \mathscr{A}_0} (-1)^{r(A)-1} p^{n(A)} \chi(N(A)),$$

where $r(A)$ is the rank of A and $n(A) = r(A)(r(A) - 1)/2$. Consequently,

$$\chi(\Gamma) \equiv \sum_{A \in \mathscr{A}_0} (-1)^{r(A)-1} p^{n(A)} \chi(N(A)) \mod \mathbb{Z}_{(p)}$$

under the hypotheses of 13.1.

Proof. For each $A \in \mathscr{A}$, let $\bar{S}(A)$ (resp. $S(A)$) be the set of non-trivial subgroups (resp. non-trivial proper subgroups) of A. Let $\mathscr{S}(A)$ be a set of representatives mod $N(A)$ for the simplices of $|\bar{S}(A)|$ which are not in $|S(A)|$. Then

$\coprod_{A \in \mathscr{A}_0} \mathscr{S}(A)$ is a set of representatives for the simplices of $|\mathscr{A}| \bmod \Gamma$. It therefore follows from the definition of $\chi_\Gamma(\mathscr{A})$ that

$$\chi_\Gamma(\mathscr{A}) = \sum_{A \in \mathscr{A}_0} \chi_{N(A)}(\bar{S}(A), S(A)),$$

where the relative equivariant Euler characteristic on the right is equal, by definition, to $\sum_{\sigma \in \mathscr{S}(A)} (-1)^{\dim \sigma} \chi(N(A)_\sigma)$.

I claim that $\chi_{N(A)}(\bar{S}(A), S(A)) = \chi(N(A)) \cdot \chi(\bar{S}(A), S(A))$. For let $C \subseteq N(A)$ be a torsion-free subgroup of finite index which centralizes A. Then C acts trivially on $\bar{S}(A)$, so we have a spectral sequence (cf. VII.7.2)

$$E^2_{pq} = H_p(C, \mathbb{Q}) \otimes_{\mathbb{Q}} H_q(\bar{S}(A), S(A); \mathbb{Q}) \Rightarrow H^C_{p+q}(\bar{S}(A), S(A); \mathbb{Q}).$$

Therefore $\tilde{\chi}_C(\bar{S}(A), S(A)) = \tilde{\chi}(C) \cdot \chi(\bar{S}(A), S(A))$. The claim now follows at once from 7.3a and b and from relative versions of 7.3b′ and c.

Thus

$$\chi_\Gamma(\mathscr{A}) = \sum_{A \in \mathscr{A}_0} \chi(N(A)) \cdot \chi(\bar{S}(A), S(A)).$$

Now A can be regarded as a vector space of dimension $r = r(A)$ over $\mathbb{F}_p$, so we can apply 14.1. Since $\bar{S}(A)$ is contractible, we conclude that $\chi(\bar{S}(A), S(A)) = (-1)^{r-1} p^{r(r-1)/2}$, whence the theorem. □

Remark. Let p^a be the maximal order of a p-subgroup of Γ. Then the denominators of the $\chi(N(A))$ which occur in the theorem involve p to at most the a-th power by 9.3. Since $r(r-1)/2 \geq 1$ if $r \geq 2$, it follows from 14.2 that

$$\chi(\Gamma) \equiv \sum_{P \in \mathscr{P}} \chi(N(P)) \mod p^{-(a-1)}\mathbb{Z}_{(p)},$$

where $\mathscr{P}$ is a set of representatives for the subgroups of Γ of order p. Thus we can compute the "leading term" of the partial fractions decomposition of $\chi(\Gamma)$ (i.e., the term involving p^a in the denominator) by considering only the subgroups of order p and their normalizers.

Exercises

1. Extract from the proof of 14.2 the following fact: Let S be a partially ordered Γ-set such that $\chi_\Gamma(S)$ is defined. For each $s \in S$, let $S_{\leq s}$ (resp. $S_{<s}$) be the set of $t \in S$ such that $t \leq s$ (resp. $t < s$). If each $S_{\leq s}$ is finite, then

$$\chi_\Gamma(S) = \sum_s \chi(\Gamma_s) \cdot \chi(S_{\leq s}, S_{<s}),$$

where s ranges over a set of representatives for $S \bmod \Gamma$.

2. Show that the hypothesis that each $S_{<s}$ is finite in exercise 1 can be replaced by the weaker hypothesis that $H_*(S_{<s})$ is finitely generated for each s. [Hint: Γ_s has a subgroup of finite index which acts trivially on $H_*(S_{\leq s}, S_{<s}; \mathbb{Z}_2)$.]

CHAPTER X
Farrell Cohomology Theory

1 Introduction

Let Γ be a group such that vcd $\Gamma < \infty$. If Γ is torsion-free, then we know from Chapter VIII that cd $\Gamma < \infty$, so that Γ has no high-dimensional cohomology. In the general case, one might hope to "explain" the high-dimensional cohomology of Γ in terms of the torsion in Γ. (This is analogous to the situation of Chapter IX, where we tried to explain the non-integrality of $\chi(\Gamma)$ in terms of the torsion in Γ.)

For this purpose it is convenient to use modified cohomology groups $\hat{H}^*(\Gamma, M)$, first introduced by Farrell [1977]. Farrell's theory generalizes the Tate cohomology theory for finite groups (Chapter VI), and it has the following two properties which make it appropriate for our present purposes: (i) $\hat{H}^i = H^i$ for $i >$ vcd Γ and (ii) $\hat{H}^* = 0$ if Γ is torsion-free.

The next three sections will be devoted to the foundations of the Farrell cohomology theory. Then in §§5 and 6 we will generalize to the Farrell theory some of the results of §§VI.8 and VI.9 on cohomological triviality and periodicity. Finally, §7 will contain the main results of the chapter, relating $\hat{H}^*(\Gamma)$ to the torsion in Γ.

2 Complete Resolutions

Let vcd $\Gamma = n < \infty$. By a *complete resolution* for Γ we mean an acyclic chain complex F of projective $\mathbb{Z}\Gamma$-modules, together with an ordinary projective resolution $\varepsilon: P \to \mathbb{Z}$ of $\mathbb{Z}$ over $\mathbb{Z}\Gamma$ such that F and P coincide in sufficiently high dimensions. (As in Chapter VI, F is *not* in general nonnegative.) If Γ is finite, for example, then any complete resolution in the sense of §VI.3 is also a complete resolution in the present sense. In §VI.3, however, we required F and P to agree in *all* non-negative dimensions, so the present notion of complete resolution is more general. (It might seem reasonable, by analogy with §VI.3, to require F and P to agree in dimensions $\geq n = \operatorname{vcd} \Gamma$. Indeed, we will see below that complete resolutions of this type always exist. Our reason for not requiring this is that we want a complete resolution for Γ to also be a complete resolution for any subgroup of Γ.)

If (F, P, ε) and (F', P', ε') are complete resolutions, then a chain map $\tau: F \to F'$ will be called a *map* of complete resolutions if there is an augmentation-preserving map $P \to P'$ which agrees with τ in sufficiently high dimensions.

(2.1) Proposition (a) *There exist complete resolutions (F, P, ε) such that F coincides with P in dimensions $\geq n =$* $\operatorname{vcd} \Gamma$. *If Γ is of type VFP, then F can be taken to be of finite type.*

(b) *If (F, P, ε) and (F', P', ε') are complete resolutions, then there is a unique homotopy class of maps from (F, P, ε) to (F', P', ε'), and these maps are homotopy equivalences.*

PROOF. Choose a torsion-free subgroup $\Gamma' \subseteq \Gamma$ of finite index. We will apply the relative homological algebra of §VI.2 to (Γ, Γ'). Thus we have a notion of *admissible* short exact sequence of Γ-modules and a corresponding notion of *relative injective* Γ-module. Let $\varepsilon: P \to \mathbb{Z}$ be a projective resolution of $\mathbb{Z}$ over $\mathbb{Z}\Gamma$ and let $K = \ker\{P_{n-1} \to P_{n-2}\}$. Then $(P_i)_{i \geq n}$ provides a projective resolution

$$\cdots \to P_{n+1} \to P_n \to K \to 0$$

of K. Since $\operatorname{cd} \Gamma' = n$, we know (VIII.2.1) that K is $\mathbb{Z}\Gamma'$-projective. Proposition VI.2.6 therefore gives us a relative injective resolution

$$0 \to K \to Q^0 \to Q^1 \to \cdots$$

with each Q^i projective. Splicing together $(P_i)_{i \geq n}$ and $(Q^i)_{i \geq 0}$, we obtain the desired complete resolution. Note that P and Q can be taken to be of finite type if Γ is of type VFP, whence (a).

The proof of (b) will require the following generalization of VI.3.2:

(2.2) Lemma. *If* $\operatorname{cd} \Gamma < \infty$ *then any acyclic chain complex of projective $\mathbb{Z}\Gamma$-modules is contractible.*

PROOF. Note first that any Γ-module M which is $\mathbb{Z}$-free has $\operatorname{proj\,dim}_{\mathbb{Z}\Gamma} M \leq n = \operatorname{cd} \Gamma$. For if P is a projective resolution of $\mathbb{Z}$ of length n, then $P \otimes M$ (with diagonal Γ-action) is a projective resolution of $\mathbb{Z} \otimes M = M$ of length n. [Why?] Suppose now that F is an acyclic complex of projectives and let Z_k for any k be the module of k-cycles. Then we have an exact sequence

$$0 \to Z_k \to F_k \to \cdots \to F_{k-n+1} \to Z_{k-n} \to 0$$

with each F_i projective. Since proj $\dim_{\mathbb{Z}\Gamma} Z_{k-n} \leq n$, it follows (VIII.2.1) that Z_k is projective. Hence $F_{k+1} \twoheadrightarrow Z_k$ splits for all k and F is contractible. □

Returning now to the proof of 2.1b, the lemma shows that any complete resolution F is Γ'-contractible, hence it is admissible. Also, each projective module F_i is relatively injective by VI.2.3. Given two complete resolutions as in (b), we use ordinary homological algebra to construct an augmentation-preserving map $P \to P'$; this defines a chain map $\tau: F \to F'$ in high dimensions. Since F is acyclic and admissible and F' is relatively injective, relative homological algebra (VI.2.4) now allows us to extend τ to all dimensions. Similarly, any two such maps are homotopic in high dimensions by ordinary homological algebra applied to P and P', and the homotopy can be extended to all dimensions by relative homological algebra. It is clear from the uniqueness that all maps of complete resolutions are homotopy equivalences. □

Various other chain maps can be constructed by the same technique. For instance:

(2.3) Proposition. *If (F, P, ε) is a complete resolution, then there is a unique homotopy class of chain maps $\tau: F \to P$ such that τ is the identity in sufficiently high dimensions.*

PROOF. F is acyclic and admissible and each P_i is relatively injective. □

In addition, one can construct a "diagonal map," which will be needed for the theory of cup products:

(2.4) Proposition. *Let (F, P, ε) be a complete resolution and let $\tau: P \to P \otimes P$ be a diagonal approximation (§V.1) with components $\tau_{pq}: P_{p+q} \to P_p \otimes P_q$. Let $F \hat{\otimes} F$ be the completed tensor product as in §VI.5. If m is an integer such that F and P agree in dimensions $\geq m - 1$, then there is a chain map $\Delta: F \to F \hat{\otimes} F$ whose (m, m)-component $\Delta_{mm}: F_{2m} \to F_m \otimes F_m$ is equal to τ_{mm}.*

PROOF. This is almost identical to the proof of the corresponding result for finite groups (end of §VI.5). As in that case, it suffices to define Δ in dimension $2m$, in such a way that $\partial\Delta | B_{2m} = 0$ and $\Delta_{mm} = \tau_{mm}$. This reduces to the construction of a family of maps $\alpha_p: F_{2m} \to F_{m+p} \otimes F_{m-p}$ ($p \in \mathbb{Z}$) satisfying (i) $(\partial'\alpha_p + \partial''\alpha_{p-1}) | B_{2m} = 0$ and (ii) $\alpha_0 = \tau_{mm}$. (Here ∂', ∂'', and B_{2m} are as in §VI.5.) Let $\alpha_0 = \tau_{mm}$ and $\alpha_1 = \tau_{m+1,m-1}$; this makes sense because $F = P$ in dimensions $\geq m - 1$. Since $\tau: P \to P \otimes P$ is a chain map, condition (i) (with $p = 1$) is satisfied. We can therefore define α_p inductively for all p, exactly as in §VI.5. □

Finally, recall that in the finite group case it is possible to construct a complete resolution by splicing together an ordinary resolution and its dual.

It turns out that there is an analogous (but more complicated) construction whenever Γ is of type VFP. For simplicity we will treat here only the case where Γ is a virtual duality group, referring to the exercises below for the general case.

Let Γ be a virtual duality group with "dualizing module" $D = H^n(\Gamma, \mathbb{Z}\Gamma)$ (cf. VIII.11.3). I claim that D is of type FP_∞ as a $\mathbb{Z}\Gamma$-module. For let $\Gamma' \subseteq \Gamma$ be a torsion-free subgroup of finite index and recall that D is of type FP as a $\mathbb{Z}\Gamma'$-module (§VIII.10, exercise 1); it now follows as in the proof of VIII.5.1 that D is of type FP_∞ over $\mathbb{Z}\Gamma$.

(2.5) Proposition. *Let Γ be a virtual duality group with dualizing module D, let $\varepsilon: P \to \mathbb{Z}$ be a finite type projective resolution of $\mathbb{Z}$ over $\mathbb{Z}\Gamma$, and let $\eta: Q \to D$ be a finite type projective resolution of D over $\mathbb{Z}\Gamma$. Then there is a complete resolution F such that the dual complex $\bar{F} = \mathcal{H}om_{\mathbb{Z}\Gamma}(F, \mathbb{Z}\Gamma)$ is the mapping cone of a chain map $\Sigma^{-n}Q \to \bar{P}$. In particular, $F_i = P_i$ for $i \geq n$ and $F_i = (Q_{n-1-i})^*$ for $i \leq -1$.*

Proof Consider the n-th suspension $\Sigma^n\bar{P}$ of the dual $\bar{P}$ of P. Then $H_i(\Sigma^n\bar{P}) = H_{i-n}(\bar{P}) = H^{n-i}(\Gamma, \mathbb{Z}\Gamma)$, which is 0 for $i \neq 0$ and D for $i = 0$. It follows easily that there is a weak equivalence $Q \to \Sigma^n\bar{P}$. [Start by lifting $\eta: Q_0 \to D$ to a map to the zero-cycles of $\Sigma^n\bar{P}$; now extend this to a chain map by the fundamental lemma I.7.4.] This map can also be viewed as a weak equivalence $f: \Sigma^{-n}Q \to \bar{P}$, and the mapping cone C of f is an acyclic complex of finitely generated projectives. Let F be the dual complex $\bar{C}$. Thus $F_i = (C_{-i})^* = (\bar{P}_{-i} \oplus (\Sigma^{-n}Q)_{-i-1})^* = P_i^{**} \oplus (Q_{n-i-1})^* = P_i \oplus (Q_{n-i-1})^*$; in particular, $F_i = P_i$ for $i \geq n$. Checking the definition of the differential in F, one finds that it agrees with that in P in dimensions $\geq n$, up to sign. The proposition will be proved, then, if we verify that F is acyclic. Let $\Gamma' \subseteq \Gamma$ be a torsion-free subgroup of finite index. Then we have $F = \mathcal{H}om_{\mathbb{Z}\Gamma}(C, \mathbb{Z}\Gamma) \approx \mathcal{H}om_{\mathbb{Z}\Gamma'}(C, \mathbb{Z}\Gamma')$. [This follows from the theory of induced and coinduced modules, exactly as in the proof of VI.3.4; alternatively, use Lemma VIII.7.4.] But C is Γ'-contractible (by 2.2 for instance), so its Γ'-dual $\mathcal{H}om_{\mathbb{Z}\Gamma'}(C, \mathbb{Z}\Gamma')$ is also Γ'-contractible and hence acyclic. □

Remark. Consider the complex $C = \bar{F}$ of this proof. It coincides with $\Sigma^{1-n}Q$ in dimensions ≥ 1, so $\Sigma^{n-1}C$ coincides with Q in dimensions $\geq n$. Thus $\Sigma^{n-1}C = \Sigma^{n-1}\bar{F}$ is what one would reasonably call a complete resolution of D.

Exercises

1. Let $\Gamma = \Gamma_1 \times \Gamma_2$, where $\operatorname{cd}\Gamma_1 < \infty$ and $\operatorname{vcd}\Gamma_2 < \infty$. Show that one can construct a complete resolution for Γ by taking the tensor product of a finite length projective resolution of $\mathbb{Z}$ over $\mathbb{Z}\Gamma_1$ and a complete resolution for Γ_2.

*2. A non-negative chain complex C is said to be of *type* FP_∞ if it admits a weak equivalence $P \to C$, where P is a non-negative chain complex of finitely generated projectives. (If C consists of a single module M concentrated in dimension 0, for example, this just means that M is of type FP_∞.) If Γ is a group and Γ' is a subgroup of finite index, prove that a chain complex of $\mathbb{Z}\Gamma$-modules is of type FP_∞ over $\mathbb{Z}\Gamma$ if and only if it is of type FP_∞ over $\mathbb{Z}\Gamma'$. [This is proved in Brown [1982]. All the essential ideas can be found in Brown [1975a], and one can even deduce the result from Theorem 2 of the latter by a "Shapiro's lemma" argument.]

*3. Suppose Γ is of type VFP and let $\bar{P}$ be as in 2.5. Let $\mathscr{D}$ be the subcomplex

$$0 \to \bar{P}^0 \to \cdots \to \bar{P}^{n-1} \to Z^n \to 0$$

of $\bar{P}$, where Z^n is the module of n-cocycles. Note that the inclusion $\mathscr{D} \hookrightarrow \bar{P}$ is a weak equivalence. Show that $\mathscr{D}$ is of type FP_∞ and deduce that 2.5 extends to groups of type VFP, with Q now taken to be a finite type "projective resolution" of $\Sigma^n\mathscr{D}$, i.e., a finite type complex of projectives such that there is a weak equivalence $Q \to \Sigma^n\mathscr{D}$.

3 Definition and Properties of $\hat{H}^*(\Gamma)$

We continue to assume that Γ is a group such that vcd $\Gamma = n < \infty$. We can then choose a complete resolution (F, P, ε) and set

$$\hat{H}^*(\Gamma, M) = H^*(\mathscr{H}om_\Gamma(F, M))$$

for any Γ-module M. By 2.1, $\hat{H}^*$ is well-defined up to canonical isomorphism. If Γ is finite, the present definition is obviously consistent with that of Chapter VI. For simplicity we will often write $\hat{H}^*(\Gamma)$ or even $\hat{H}^*$ instead of $\hat{H}^*(\Gamma, M)$. It will always be understood, however, that there is an arbitrary Γ-module of coefficients.

The cohomology theory $\hat{H}^*(\Gamma, —)$ will be called the *Farrell cohomology theory*. It has formal properties analogous to those of the Tate theory for finite groups (in which case $n = 0$):

(3.1) $\hat{H}^*(\Gamma) = 0$ if Γ is torsion-free.

For in this case we can take $F = 0$.

(3.2) $\hat{H}^*(\Gamma, -)$ has all the "usual" cohomological properties: long exact sequences, Shapiro's lemma, restriction and transfer maps, and cup products.

This is proved exactly as in ordinary cohomology theory, except for the construction of cup products. We will discuss the latter below, after stating two more properties.

(3.3) $\hat{H}^*(\Gamma, M) = 0$ if M is an induced module $\mathbb{Z}\Gamma \otimes_{\mathbb{Z}\Gamma'} M'$, where Γ' is a torsion-free subgroup of finite index and M' is an arbitrary Γ'-module. Consequently, the functors $\hat{H}^i(\Gamma, —)$ are effaceable and coeffaceable.

This follows from 3.1 and 3.2 (Shapiro's lemma). Note that 3.2 and 3.3 allow us to shift dimensions in both directions, as in VI.5.4. In view of the next property, this means that questions about Farrell cohomology can often be reduced to questions about (high-dimensional) ordinary cohomology.

(3.4) There is a canonical map $H^i \to \hat{H}^i$ which is an isomorphism for $i > n$ and an epimorphism for $i = n$. Moreover, the sequence

$$H^n(\Gamma') \xrightarrow{\text{tr}} H^n(\Gamma) \longrightarrow \hat{H}^n(\Gamma) \longrightarrow 0$$

is exact for any torsion-free subgroup $\Gamma' \subseteq \Gamma$ of finite index, where tr is the transfer map.

The map $H^i \to \hat{H}^i$ is induced by the chain map $F \to P$ of 2.3. Since the latter can be taken to be the identity in dimensions $\geq n$ (cf. 2.1), the first assertion of 3.4 is clear. To prove the second assertion, one could simply directly examine the n-coboundaries in $\mathcal{H}om_\Gamma(F, M)$, where F is assumed to be constructed from P as in the proof of 2.1. It is easier, however, to proceed as follows: Consider the "dimension-shifting" exact sequence

$$0 \to K \to \mathbb{Z}\Gamma \otimes_{\mathbb{Z}\Gamma'} M \to M \to 0.$$

This gives rise to a long exact sequence in ordinary cohomology which maps to the corresponding sequence in Farrell cohomology. Using Shapiro's lemma and definition (A) of the transfer map (§III.9), we deduce a diagram

$$\begin{array}{ccccccc}
H^n(\Gamma', M) & \xrightarrow{\text{tr}} & H^n(\Gamma, M) & \longrightarrow & H^{n+1}(\Gamma, K) & \longrightarrow & 0 \\
 & & \downarrow & & \downarrow \approx & & \\
0 & \longrightarrow & \hat{H}^n(\Gamma, M) & \xrightarrow{\approx} & \hat{H}^{n+1}(\Gamma, K) & \longrightarrow & 0
\end{array}$$

with exact rows. (The zero in the top row is $H^{n+1}(\Gamma', M)$, which vanishes because cd $\Gamma' = n$.) The second assertion of 3.4 follows at once.

Remarks

1. If Γ is finite, 3.4 reduces to the statement that $H^i = \hat{H}^i$ for $i > 0$ and that H^0 is the cokernel of the norm map $M \to M^G$.
2. It follows from 3.4 that the image of tr: $H^n(\Gamma') \to H^n(\Gamma)$ is independent of the choice of Γ'. This should not surprise the reader who has done exercise 6 of §VIII.2.

We can now explain the construction of cup products. Let (F, P, ε) be a complete resolution and choose $\Delta: F \to F \hat{\otimes} F$ as in 2.4. Using Δ we can construct cup products

$$\hat{H}^p(\Gamma, M) \otimes \hat{H}^q(\Gamma, N) \to \hat{H}^{p+q}(\Gamma, M \otimes N)$$

in the usual way, and these will be compatible with coboundary maps in long exact sequences. Moreover, we have by 2.4 an integer m, such that

$H^i \xrightarrow{\approx} \hat{H}^i$ for $i \geq m$ and such that the cup product $\hat{H}^m \otimes \hat{H}^m \to \hat{H}^{2m}$ agrees with the ordinary cup product $H^m \otimes H^m \to H^{2m}$. The standard dimension-shifting arguments (cf. proof of VI.5.8) now show that the cup product in $\hat{H}^*$ is compatible with that in H^*, in the sense that the diagram

$$\begin{array}{ccc} H^p \otimes H^q & \xrightarrow{\cup} & H^{p+q} \\ \downarrow & & \downarrow \\ \hat{H}^p \otimes \hat{H}^q & \xrightarrow{\cup} & \hat{H}^{p+q} \end{array}$$

commutes for all p, q. [Assuming this for (p, q), use the coeffaceability of H^* to deduce it for $(p + 1, q)$ and $(p, q + 1)$, and use the effaceability of $\hat{H}^*$ to deduce it for $(p - 1, q)$ and $(p, q - 1)$.]

It is now clear that our cup product is independent of the choice of F and Δ. For any two choices would give the same product in high dimensions by what we have just done, and hence they would agree in all dimensions by dimension-shifting. Finally, one uses dimension-shifting once again to establish the usual properties of the cup product (associativity, commutativity, unit); for these properties are known to hold in high dimensions. [Note: The unit for the cup product is the image in $\hat{H}^0(\Gamma, \mathbb{Z})$ of $1 \in H^0(\Gamma, \mathbb{Z}) = \mathbb{Z}$.]

Recall that the Tate group $\hat{H}^i$ for $i > 0$ could be interpreted in terms of homology. This resulted from the duality theory which allowed us to interpret the negative part of a complete resolution as the dual of an ordinary resolution. We can generalize this to Farrell cohomology, provided Γ is of type *VFP*. For simplicity we will assume that Γ is a virtual duality group, but the reader who has done exercises 2 and 3 of §2 should be able to treat the general case. [See also exercise 5 of §VIII.10.]

(3.5) Suppose that Γ is a virtual duality group with dualizing module D, and set $\tilde{H}_*(\Gamma, M) = H_*(\Gamma, D \otimes M)$. Then there is an isomorphism $\hat{H}^i \approx \tilde{H}_{n-1-i}$ for $i < -1$ and a monomorphism $\hat{H}^{-1} \to \tilde{H}_n$. Moreover, the sequence

$$0 \longrightarrow \hat{H}^{-1}(\Gamma) \longrightarrow \tilde{H}_n(\Gamma) \xrightarrow{\text{tr}} \tilde{H}_n(\Gamma')$$

is exact for any torsion-free subgroup $\Gamma' \subseteq \Gamma$ of finite index.

For let F be as in 2.5, and let $E = \Sigma^{n-1}\bar{F}$ (cf. remark at the end of §2). Then we have

$$\hat{H}^i(\Gamma, -) = H^i(\mathcal{H}om_\Gamma(F, -)) = H_{-i}(\bar{F} \otimes_\Gamma -) = H_{n-1-i}(E \otimes_\Gamma -).$$

Now the construction of E shows that there is a projective resolution Q of D which receives a chain map $E \to Q$ which is the identity in dimensions $\geq n$. Hence we have a map $H_j(E \otimes_\Gamma -) \to \operatorname{Tor}_j^\Gamma(D, -)$ which is an isomorphism for $j > n$ and a monomorphism for $j = n$. Since $\operatorname{Tor}_j^\Gamma(D, -) = \tilde{H}_j(\Gamma, -)$ by III.2.2, this proves the first assertion of 3.5. The second assertion follows by a dimension-shifting argument analogous to that used in 3.4; details are left to the reader.

We have not yet said anything about H^i for $0 \leq i \leq n-1$. This range of dimensions, which is vacuous if Γ is finite, is harder to understand than those treated in 3.4 and 3.5. We can say something about it, however, in view of the mapping cone construction in 2.5. For $\mathscr{H}om_\Gamma(F, M) = \bar{F} \otimes_\Gamma M$ is the mapping cone of a map $\Sigma^{-n}Q \otimes_\Gamma M \to \bar{P} \otimes_\Gamma M = \mathscr{H}om_\Gamma(P, M)$. We therefore have a long exact homology sequence, which takes the form

$$\cdots \to \tilde{H}_{n-i} \to H^i \to \hat{H}^i \to \tilde{H}_{n-1-i} \to H^{i+1} \to \cdots.$$

Since $H^{-1} = 0$ and $\tilde{H}_{-1} = 0$, we conclude:

(3.6) With the hypotheses and notation of 3.5, there is an exact sequence

$$0 \to \hat{H}^{-1} \to \tilde{H}_n \to H^0 \to \hat{H}^0 \to \tilde{H}_{n-1} \to H^1 \to \cdots \to \tilde{H}_0 \to H^n \to \hat{H}^n \to 0.$$

We can summarize 3.4–3.6 as follows. If Γ is a virtual duality group, then the Farrell cohomology functors $\hat{H}^i$ include: the cohomology functors H^i for $i > n$; the homology functors $\tilde{H}_i$ for $i > n$; a certain quotient of H^n; a certain subfunctor of $\tilde{H}_n$; and n additional functors $\hat{H}^0, \ldots, \hat{H}^{n-1}$, which are some sort of mixture of $\{H^i\}_{0 \leq i \leq n}$ and $\{\tilde{H}_i\}_{0 \leq i \leq n}$:

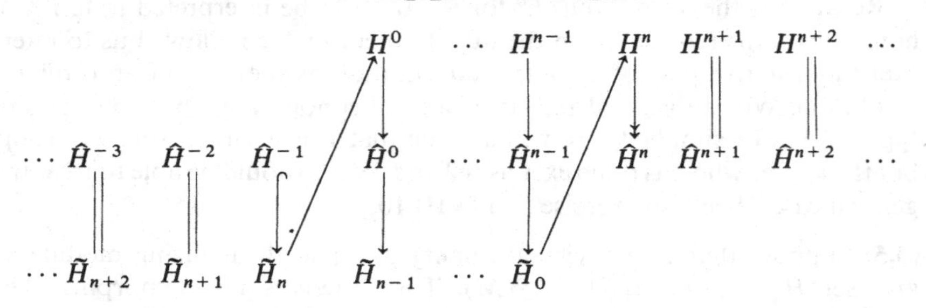

EXERCISES

1. Let $\Gamma = \mathbb{Z} \times \mathbb{Z}_2$. Construct an explicit complete resolution and use it to compute $\hat{H}^*(\Gamma, \mathbb{Z})$. [Cf. exercise 1 of §2.]

2. Prove that the groups $\hat{H}^*(\Gamma)$ are torsion-groups. More precisely, they are annihilated by the integer d of IX.9.1. [Hint: Transfer.]

   Remark. It is not known whether $\hat{H}^*(\Gamma)$ is annihilated by the integer m of IX.9.3. In view of the cup product structure on $\hat{H}^*(\Gamma)$, it would be enough to show that m annihilates the identity element $1 \in \hat{H}^0(\Gamma, \mathbb{Z})$.

3. (a) Define Farrell homology groups $\hat{H}_*(\Gamma)$.

   (b) If Γ is a virtual duality group with dualizing module D, show that $\hat{H}^i(\Gamma, M) \approx \hat{H}_{n-1-i}(\Gamma, D \otimes M)$ for all i. More precisely, show there is an element

$z \in \hat{H}_{n-1}(\Gamma, D)$ such that cap product with z is an isomorphism. [Hint: To construct an isomorphism $\psi: \hat{H}^i(\Gamma, M) \to \hat{H}_{n-1-i}(\Gamma, D \otimes M)$, use 3.5 and dimension-shifting; alternatively, take F and E as in the proof of 3.5 and show by homological algebra that there is a map $E \to F \otimes D$ which induces the desired isomorphism. Now let $z \in \hat{H}_{n-1}(\Gamma, D)$ be the image of $1 \in \hat{H}^0(\Gamma, \mathbb{Z})$ under ψ. Show that ψ is given by cap product with z (up to sign) as follows. The two composites

$$H^i(\Gamma, M) \to \hat{H}^i(\Gamma, M) \rightrightarrows \hat{H}_{n-1-i}(\Gamma, D \otimes M)$$

agree when $i = 0$ by naturality, hence they agree for all i by dimension-shifting. But then the two maps $\hat{H}^i(\Gamma, M) \rightrightarrows \hat{H}_{n-1-i}(\Gamma, D \otimes M)$ agree for large i and hence for all i.]

*(c) If D is $\mathbb{Z}$-free in (b), show that cap product with the same $z \in \hat{H}_{n-1}(\Gamma, D)$ gives isomorphisms $\hat{H}_i(\Gamma, M) \approx \hat{H}^{n-1-i}(\Gamma, \text{Hom}(D, M))$.

*(d) State and prove analogues of (b) and (c) for arbitrary groups Γ of type VFP.

*4. Let $\Gamma = \Gamma_1 \times \Gamma_2$ as in exercise 1 of §2. Derive, under suitable hypotheses, Künneth formulas relating $\hat{H}_*(\Gamma)$ to $H_*(\Gamma_1)$ and $\hat{H}_*(\Gamma_2)$ and similarly for cohomology.

5. Let $1 \to \Gamma' \to \Gamma \to \Gamma'' \to 1$ be a short exact sequence of groups of finite virtual cohomological dimension.

(a) If Γ'' is torsion-free, show that there is a Hochschild–Serre spectral sequence (derived as in ordinary cohomology) of the form

$$E_2^{pq} = H^p(\Gamma'', \hat{H}^q(\Gamma')) \Rightarrow \hat{H}^{p+q}(\Gamma).$$

*(b) If Γ' is torsion-free, show that there is a spectral sequence

$$E_2^{pq} = \hat{H}^p(\Gamma'', H^q(\Gamma')) \Rightarrow \hat{H}^{p+q}(\Gamma).$$

[Method 1: Use the ordinary Hochschild-Serre spectral sequence and dimension-shifting. Method 2: Use a complete resolution of the form $F = F'' \otimes C$, where C is a finite dimensional complex of Γ-modules which is a Γ'-projective resolution of $\mathbb{Z}$.]

*6. Define and study composition products $\hat{H}^* \otimes \hat{H}^* \to \hat{H}^*$ analogous to those of §VI.6. [The only tricky thing here is to prove the analogue of VI.6.1a. This is true, but the proof requires more work than in the case of finite groups.]

4 Equivariant Farrell Cohomology

Let Γ be a group of finite virtual cohomological dimension and let X be a *finite dimensional* Γ-CW-complex. As in Chapter VII, we then choose a complete resolution F for Γ and we set

$$\hat{H}^*_\Gamma(X; M) = H^*(\mathcal{H}om_\Gamma(F, C^*(X; M)))$$

for any Γ-module M. These *equivariant Farrell cohomology groups* have properties analogous to those of the ordinary equivariant cohomology groups (spectral sequences, etc.).

An important special case is that where X is taken to be proper and contractible (cf. VIII.11.1). We then obtain as in VII.7.10 a spectral sequence

$$E_1^{pq} = \prod_{\sigma \in \Sigma_p} \hat{H}^q(\Gamma_\sigma, M_\sigma) \Rightarrow \hat{H}^{p+q}(\Gamma, M), \tag{4.1}$$

where Σ_p is a set of representatives for the p-cells of X mod Γ. This spectral sequence relates the Farrell cohomology of Γ to the Tate cohomology of its finite subgroups. Note that the spectral sequence lives in the first and fourth quadrants, but there is no problem with convergence because $\dim X < \infty$. Indeed, the spectral sequence is concentrated in the vertical strip $0 \leq p \leq \dim X$, so $E_r = E_\infty$ as soon as $r > \dim X$.

The rest of this section will be devoted to establishing some properties of the spectral sequence 4.1 which will be needed in §6. For simplicity we will assume that X is an *ordered simplicial* complex with order preserving simplicial Γ-action. In addition we will assume that X^H is non-empty and connected for every finite subgroup $H \subseteq \Gamma$; we saw in §VIII.11 that such an X exists.

Note that the "order preserving" assumption implies that $M_\sigma = M$, i.e., that $\mathbb{Z}_\sigma = \mathbb{Z}$ with trivial Γ_σ-action, for each simplex σ. We may therefore suppress M from the notation in what follows.

Our first goal is to give a purely algebraic description of the left-hand edge $E_2^{0,q}$. For this we will need to compute the differential $d_1^{0,q}$. Recall that for any simplex σ of X and any $\gamma \in \Gamma$ we have a conjugation isomorphism

$$c(\gamma^{-1})^*\colon \hat{H}^*(\Gamma_\sigma) \xrightarrow{\approx} \hat{H}^*(\Gamma_{\gamma\sigma}),$$

denoted $u \mapsto \gamma u$ (cf. §III.8).

(4.2) Lemma. *Let X_p be the set of p-simplices of X. Then E_1^{pq} can be identified with the subgroup of $\prod_{\sigma \in X_p} \hat{H}^q(\Gamma_\sigma)$ consisting of those families $(u_\sigma)_{\sigma \in X_p}$ such that $\gamma u_\sigma = u_{\gamma\sigma}$ for all $\gamma \in \Gamma$, $\sigma \in X_p$. The differential d_1^{pq} is the restriction to this subgroup of the map*

$$d\colon \prod_{\sigma \in X_p} \hat{H}^q(\Gamma_\sigma) \to \prod_{\tau \in X_{p+1}} \hat{H}^q(\Gamma_\tau)$$

defined as follows: For any $\tau = (v_0, \ldots, v_{p+1}) \in X_{p+1}$ with $v_0 < \cdots < v_{p+1}$, let $\tau_i = (v_0, \ldots, \hat{v}_i, \ldots, v_{p+1})$, $i = 0, \ldots, p+1$, and let $\rho_i\colon \hat{H}^q(\Gamma_{\tau_i}) \to \hat{H}^q(\Gamma_\tau)$ be the restriction map. Then d is given by

$$(u_\sigma) \mapsto \left(\tau \mapsto \sum_{i=0}^{p+1} (-1)^i \rho_i(u_{\tau_i})\right).$$

(Note that $\Gamma_\tau \subseteq \Gamma_{\tau_i}$ because of the assumption that the Γ-action is order-preserving. Thus the definition of d makes sense.)

PROOF. Note first that

$$\prod_{\sigma \in X_p} \hat{H}^q(\Gamma_\sigma) = \prod_{\sigma \in \Sigma_p} \prod_{\gamma \in \Gamma/\Gamma_\sigma} \hat{H}^q(\Gamma_{\gamma\sigma}).$$

Since the conjugation action of Γ_σ on $\hat{H}^q(\Gamma_\sigma)$ is trivial (cf. III.8.3), it follows that there is a well-defined injection

$$\alpha: E_1^{pq} = \prod_{\sigma \in \Sigma_p} \hat{H}^q(\Gamma_\sigma) \to \prod_{\sigma \in X_p} \hat{H}^q(\Gamma_\sigma)$$

given by $(u_\sigma)_{\sigma \in \Sigma_p} \mapsto (\gamma u_\sigma)_{\sigma \in \Sigma_p, \gamma \in \Gamma/\Gamma_\sigma}$. The image of α is clearly equal to the set of families $(u_\sigma)_{\sigma \in X_p}$ satisfying $\gamma u_\sigma = u_{\gamma\sigma}$ for all $\sigma \in X_p$ and $\gamma \in \Gamma$, whence the first part of the proposition. The description of d_1 follows from the cohomology analogue of VII.8.1. For the convenience of the reader, however, we will give a direct proof.

Let F be a complete resolution for Γ. Then the injection

$$\alpha: E_1^{pq} \to \prod_{\sigma \in X_p} \hat{H}^q(\Gamma_\sigma)$$

is given in terms of cochains as follows: We have

$$\begin{aligned} E_1^{pq} &= \hat{H}^q(\Gamma, C^p(X; M)) \\ &= H^q(\mathscr{H}om_{\mathbb{Z}\Gamma}(F, \operatorname{Hom}(X_p, M))) \\ &= H^q(\operatorname{Hom}_\Gamma(X_p, \mathscr{H}om_{\mathbb{Z}}(F, M))), \end{aligned}$$

where the second equality comes from the ordering on X. An element $u \in E_1$ is therefore represented by a family $(c_\sigma)_{\sigma \in X_p}$ such that $c_\sigma \in \mathscr{H}om_{\mathbb{Z}}(F, M)$ and $\gamma c_\sigma = c_{\gamma\sigma}$ for all $\gamma \in \Gamma$ and $\sigma \in X_p$. Taking $\gamma \in \Gamma_\sigma$, it follows that $c_\sigma \in \mathscr{H}om_{\mathbb{Z}\Gamma_\sigma}(F, M)$. Hence c_σ represents an element $u_\sigma \in \hat{H}^q(\Gamma_\sigma, M)$, and one checks that $\alpha(u) = (u_\sigma)_{\sigma \in X_p}$. The differential d_1 is simply the map

$$\hat{H}^q(\Gamma, \delta): \hat{H}^q(\Gamma, C^p(X; M)) \to \hat{H}^q(\Gamma, C^{p+1}(X; M)),$$

where $\delta: C^p(X; M) \to C^{p+1}(X; M)$ is the coboundary operator. The lemma now follows at once from the definition of δ in terms of the face maps $X_{p+1} \to X_p$ and from the fact that the restriction map $\rho_i: \hat{H}^q(\Gamma_{\tau_i}) \to \hat{H}^q(\Gamma_\tau)$ is induced by the inclusion $\mathscr{H}om_{\mathbb{Z}\Gamma_{\tau_i}}(F, M) \hookrightarrow \mathscr{H}om_{\mathbb{Z}\Gamma_\tau}(F, M)$. □

In particular, we can now easily calculate $E_2^{0,q} = \ker d_1^{0,q}$, and we obtain:

(4.3) Lemma. *$E_2^{0,q}$ can be identified with the subgroup of $\prod_{v \in X_0} \hat{H}^q(\Gamma_v)$ consisting of those families $(u_v)_{v \in X_0}$ satisfying the following two conditions:*

(i) *$\gamma u_v = u_{\gamma v}$ for any $\gamma \in \Gamma$, $v \in X_0$.*
(ii) *if e is a 1-simplex of X with vertices v_0, v_1, then u_{v_0} and u_{v_1} restrict to the same element of $\hat{H}^q(\Gamma_e)$.*

Using now our hypothesis that X^H is non-empty and connected for H finite, we will deduce the desired algebraic description of $E_2^{0,q}$:

(4.4) Proposition. *Let $\mathfrak{F}$ be the set of finite subgroups of Γ. Then $E_2^{0,q}$ is isomorphic to the subgroup $\mathfrak{H}^q(\Gamma) \subseteq \prod_{H\in\mathfrak{F}} \hat{H}^q(H)$ consisting of those families $(u_H)_{H\in\mathfrak{F}}$ satisfying the following two conditions*:

(i) $\gamma u_H = u_{\gamma H\gamma^{-1}}$ *for all* $\gamma \in \Gamma$, $H \in \mathfrak{F}$.
(ii) *If* $H' \subseteq H$ *then* $\operatorname{res}^H_{H'} u_H = u_{H'}$.

(Note: Conditions (i) and (ii) could be combined into the single condition that the u_H are compatible with respect to "restriction maps" $\hat{H}^q(H_2) \to \hat{H}^q(H_1)$ induced by embeddings $H_1 \hookrightarrow H_2$ given by conjugation by elements of Γ.)

PROOF. Using the description of $E_2^{0,q}$ given in 4.3, we have an obvious map $\varphi\colon \mathfrak{H}^q(\Gamma) \to E_2^{0,q}$ given by $(u_H)_{H\in\mathfrak{F}} \mapsto (u_{\Gamma_v})_{v\in X_0}$. Since every $H \in \mathfrak{F}$ is contained in some Γ_v [because $X^H \neq \varnothing$], it follows from condition (ii) above that φ is injective. To prove that φ is surjective suppose $(u_v)_{v\in X_0}$ satisfies conditions (i) and (ii) of 4.3. Given $H \in \mathfrak{F}$, choose a vertex v such that $H \subseteq \Gamma_v$ and set $w_H = \operatorname{res}^{\Gamma_v}_H u_v$. Since X^H is connected, condition (ii) of 4.3 shows that w_H is independent of the choice of v. It is easy to check that the resulting family $(w_H)_{H\in\mathfrak{F}}$ is in $\mathfrak{H}^q(\Gamma)$ and that its image under φ is (u_v). □

Next we wish to introduce a multiplicative structure into the spectral sequence 4.1. To simplify the notation, we will assume that the coefficient module M is a commutative ring R with trivial Γ-action. The simplicial cochain complex $C^*(X) = C^*(X; R)$ then has a cup product

$$C^*(X) \otimes C^*(X) \to C^*(X)$$

which is strictly associative and is commutative (in the graded sense) up to homotopy. Moreover, the product and the homotopy are defined canonically in terms of the structure of X as an ordered simplicial complex. [See the exercise below for a review of these facts.] They are therefore compatible with the Γ-action.

Now choose a complete resolution F with a diagonal map $\Delta\colon F \to F \hat{\otimes} F$. We then have a product

$$\mathscr{H}om_\Gamma(F, C^*(X)) \otimes \mathscr{H}om_\Gamma(F, C^*(X)) \to \mathscr{H}om_\Gamma(F, C^*(X) \otimes C^*(X)) \\ \to \mathscr{H}om_\Gamma(F, C^*(X)),$$

where the first map is induced by Δ and the second map is induced by the cochain cup product $C^*(X) \otimes C^*(X) \to C^*(X)$. As we explained at the end of §VII.5, this product is compatible with the filtration defining the spectral sequence 4.1, hence there is an induced product

$$E_r \otimes E_r \to E_r$$

for $r \geq 1$. The following proposition summarizes the properties of this product which we will need:

(4.5) Proposition. (i) *The differential d_r is a derivation with respect to the product on E_r, i.e.,*

$$d_r(uv) = d_r(u) \cdot v + (-1)^{\deg u} u \cdot d_r(v).$$

(ii) *The product on $E_{r+1} = H(E_r)$ is obtained from the product on E_r by passage to homology.*

(iii) *The product on E_1 is the composite*

$$\hat{H}^q(\Gamma, C^p(X)) \otimes \hat{H}^{q'}(\Gamma, C^{p'}(X)) \to \hat{H}^{q+q'}(\Gamma, C^p(X) \otimes C^{p'}(X)) \to \hat{H}^{q+q'}(\Gamma, C^{p+p'}(X))$$

where the first map is the usual cup product in $\hat{H}^*(\Gamma, -)$ and the second map is induced by the cup product in $C^*(X)$.

(iv) *The product on E_r is associative for $r \geq 1$ and commutative for $r \geq 2$.*

(v) *The product on E_∞ is compatible with the usual product on $\hat{H}^*(\Gamma, R)$ under the identification of E_∞ with* Gr $\hat{H}^*(\Gamma, R)$.

(vi) *The isomorphism $E_2^{0,*} \approx \mathfrak{H}^*(\Gamma)$ of 4.4 is a ring isomorphism.*

Proof. (i) and (ii) follow from the fact that the product on E_r is induced by the product on $\mathscr{H}om_\Gamma(F, C^*(X))$. (iii) is immediate from the definitions. In view of (iii) and properties of the ordinary cup product, the product on E_1 is associative and is commutative up to homotopy (with respect to the differential d_1). (iv) now follows from (ii). The product on E_∞ is obviously compatible with the product on the abutment $H^*(\mathscr{H}om_\Gamma(F, C^*(X))$. But the identification of the abutment with $\hat{H}^*(\Gamma)$ is induced by a chain map $\mathscr{H}om_\Gamma(F, R) \to \mathscr{H}om_\Gamma(F, C^*(X))$ which is easily seen to be a ring homomorphism, whence (v). Finally, to prove (vi) it suffices to show that the map $\alpha: E_1^{0,*} \to \prod_{v \in X_0} \hat{H}^*(\Gamma_v)$ defined in the proof of 4.2 is a ring homomorphism. Consider the v-component of α,

$$E_1^{0,*} = \hat{H}^*(\Gamma, C^0(X)) \to \hat{H}^*(\Gamma_v).$$

It is given on the cochain level by a map

$$\mathscr{H}om_\Gamma(F, C^0(X)) \to \mathscr{H}om_{\Gamma_v}(F, R),$$

which in turn is induced by the ring homomorphism $C^0(X) = \mathrm{Hom}(X_0, R) \to R$ given by evaluation at v. This cochain map is easily seen to be a ring homomorphism, so (vi) is proved. □

As a simple application of this multiplicative structure we will prove a result due to Quillen [1971]. Fix a prime p and consider the ring $\mathfrak{H}^*(\Gamma, \mathbb{Z}_p)$ defined in 4.4. There is an obvious homomorphism

$$\rho: \hat{H}^*(\Gamma, \mathbb{Z}_p) \to \mathfrak{H}^*(\Gamma, \mathbb{Z}_p)$$

given by the restriction maps $\hat{H}^*(\Gamma) \to \hat{H}^*(H)$ $(H \in \mathfrak{F})$.

(4.6) Proposition. *The map ρ has the following two properties:*

(i) *Every element of* $\ker \rho$ *is nilpotent.*
(ii) *For any $u \in \mathfrak{H}^*(\Gamma, \mathbb{Z}_p)$ there is an integer $k \geq 0$ such that $u^{p^k} \in \operatorname{im} \rho$.*

(Thus ρ is an isomorphism "up to p-th powers." Following Quillen, we say that ρ is an *F-isomorphism.*)

PROOF. Note first that ρ is simply the edge homomorphism

$$\hat{H}^*(\Gamma) \twoheadrightarrow E_\infty^{0,*} \hookrightarrow E_2^{0,*} \approx \mathfrak{H}^*(\Gamma)$$

associated to the spectral sequence 4.1. This is easily seen by direct definition-checking (see also exercise 1 of §VII.7). Recall that the construction of the spectral sequence gives us a filtration

$$\hat{H}^*(\Gamma) = F^0\hat{H}^*(\Gamma) \supseteq F^1\hat{H}^*(\Gamma) \supseteq \cdots$$

which is compatible with the cup product (i.e., $F^sF^t \subseteq F^{s+t}$). Since the kernel of the edge homomorphism above is precisely $F^1\hat{H}^*(\Gamma)$, it follows that any $z \in \ker \rho$ satisfies $z^k \in F^k\hat{H}^*(\Gamma)$. But $F^k\hat{H}^*(\Gamma) = 0$ for $k > \dim X$, so z is nilpotent. Next let $u \in E_2^{0,*}$. I claim that $d_2(u^p) = 0$. This follows from anti-commutativity if p and $\deg u$ are both odd (in which case $u^2 = 0$); otherwise, it follows from (i) and (iv) of 4.5, which give

$$d_2(u^p) = pu^{p-1}d_2(u) = 0$$

since the coefficient module is $\mathbb{Z}_p$. Thus $u^p \in E_3^{0,*}$. Iterating this argument, we find that $u^{p^k} \in E_{k+2}^{0,*}$, hence $u^{p^k} \in E_\infty^{0,*} = \operatorname{im} \rho$ as soon as $k + 2 > \dim X$. □

Remark. Quillen actually proved a much stronger result. Namely, he proved the analogue of 4.6 with $\mathfrak{H}^*(\Gamma)$ defined in terms of the finite subgroups of Γ which are elementary abelian p-groups.

EXERCISE

Let K be an ordered simplicial complex and let $C(K)$ be its chain complex, with one basis element for every $(n + 1)$-tuple $(v_0, \ldots, v_n)$ of vertices such that $\{v_0, \ldots, v_n\}$ is an n-simplex with $v_0 < \cdots < v_n$. The *Alexander–Whitney* diagonal map $\Delta: C(K) \to C(K) \otimes C(K)$ is defined by

$$\Delta(v_0, \ldots, v_n) = \sum_{p=0}^{n} (v_0, \ldots, v_p) \otimes (v_p, \ldots, v_n).$$

Verify that Δ is a chain map and that the induced cochain cup product is associative and is commutative up to (canonical) homotopy. [Hint: The homotopy commutativity can be proved by an acyclic models argument, which is most easily carried out as follows: Because of the ordering on the vertices of K, a simplex of K can be regarded as a singular simplex in the geometric realization of K. Thus $C(K)$ is a subcomplex of the singular chain complex $C^{\text{sing}}(K)$. Now look at the standard proof via acyclic models that the Alexander–Whitney map on $C^{\text{sing}}(K)$ is naturally homotopy commutative, and observe that the homotopy can be taken to map $C(K)$ into $C(K) \otimes C(K)$.]

5 Cohomologically Trivial Modules

We continue to assume that $\operatorname{vcd} \Gamma = n < \infty$.

(5.1) Lemma. *If M is a Γ-module such that $\hat{H}^*(G, M) = 0$ for every finite subgroup $G \subseteq \Gamma$, then $\hat{H}^*(\Gamma, M) = 0$.*

PROOF. Consider the spectral sequence 4.1, and recall that X can be chosen so that $M_\sigma = M$ for all σ. Our hypothesis implies that $E_1 = 0$, hence the abutment $\hat{H}^*(\Gamma, M)$ is also zero. □

We will say that M is *cohomologically trivial* if $\hat{H}^*(\Gamma', M) = 0$ for every subgroup $\Gamma' \subseteq \Gamma$. Applying 5.1 to each Γ', we obtain:

(5.2) Proposition. *A Γ-module is cohomologically trivial if and only if it is cohomologically trivial as a G-module for every finite subgroup $G \subseteq \Gamma$.*

Thus the theory of cohomologically trivial modules is reduced to the case where Γ is finite. In particular, we can use this to prove the following generalization of Rim's theorems VI.8.10 and VI.8.12:

(5.3) Theorem. *The following conditions on a Γ-module M are equivalent:*

(i) *M is cohomologically trivial.*
(ii) $\operatorname{proj\,dim}_{\mathbb{Z}\Gamma} M \leq n + 1$.
(iii) $\operatorname{proj\,dim}_{\mathbb{Z}\Gamma} M < \infty$.

If these conditions hold and M is $\mathbb{Z}$-free, then $\operatorname{proj\,dim}_{\mathbb{Z}\Gamma} M \leq n$.

PROOF. We have (ii) ⇒ (iii) trivially, and (iii) ⇒ (i) exactly as in the proof of VI.8.12. To prove (i) ⇒ (ii), assume first that M is $\mathbb{Z}$-free, in which case we will prove that $\operatorname{proj\,dim}_{\mathbb{Z}\Gamma} M \leq n$. According to VIII.2.1, it will suffice to show that $\operatorname{Ext}_\Gamma^{n+1}(M, N) = 0$ for every Γ-module N. Moreover, a glance at the proof of VIII.2.1 shows that we only need to consider modules N which are $\mathbb{Z}$-free. In this case we know from VI.8.11 that the Γ-module $\operatorname{Hom}(M, N)$ is cohomologically trivial over every finite subgroup of Γ, hence $\operatorname{Hom}(M, N)$ is cohomologically trivial by 5.2. We therefore have

$$\begin{aligned} \operatorname{Ext}_\Gamma^{n+1}(M, N) &= H^{n+1}(\Gamma, \operatorname{Hom}(M, N)) && \text{by III.2.2} \\ &= \hat{H}^{n+1}(\Gamma, \operatorname{Hom}(M, N)) && \text{by 3.4} \\ &= 0. \end{aligned}$$

If M is now cohomologically trivial but not $\mathbb{Z}$-free, choose a short exact sequence $0 \to M' \to P \to M \to 0$ with P projective. Then M' is cohomologically trivial and $\mathbb{Z}$-free, so proj dim $M' \leq n$ by what we just proved. Therefore proj dim $M \leq n + 1$. □

EXERCISE

Show that the following conditions are equivalent:

(i) M is cohomologically trivial.
(ii) For every subgroup $\Gamma' \subseteq \Gamma$ there is an integer m such that $H^i(\Gamma', M) = 0$ for $i > m$.
(iii) $\hat{H}_*(\Gamma', M) = 0$ for every subgroup $\Gamma' \subseteq \Gamma$.
(iv) For every subgroup $\Gamma' \subseteq \Gamma$ there is an integer m such that $H_i(\Gamma', M) = 0$ for $i > m$.

6 Groups with Periodic Cohomology

A group Γ of finite virtual cohomological dimension is said to have *periodic cohomology* if for some $d \neq 0$ there is an element $u \in \hat{H}^d(\Gamma, \mathbb{Z})$ which is invertible in the ring $\hat{H}^*(\Gamma, \mathbb{Z})$. Cup product with u then gives a *periodicity isomorphism*

$$\hat{H}^i(\Gamma, M) \approx \hat{H}^{i+d}(\Gamma, M)$$

for any Γ-module M and any $i \in \mathbb{Z}$. Similarly, we say that Γ has *p-periodic cohomology* (where p is a prime) if the p-primary component $\hat{H}^*(\Gamma, \mathbb{Z})_{(p)}$, which is itself a ring (cf. exercise 2a of §VI.5), contains an invertible element of non-zero degree d. We then have

$$\hat{H}^i(\Gamma, M)_{(p)} \approx \hat{H}^{i+d}(\Gamma, M)_{(p)}.$$

As in exercise 2b of §VI.5 (see also exercise 2 of §3), we have

$$\hat{H}^*(\Gamma, \mathbb{Z}) \approx \prod_p \hat{H}^*(\Gamma, \mathbb{Z})_{(p)},$$

where p ranges over the primes such that Γ has p-torsion. Since this is a finite direct product of rings, it is clear that Γ has periodic cohomology if and only if Γ has p-periodic cohomology for every prime p. We will therefore restrict our attention to p-periodicity. Our goal, as in the previous section, will be to reduce to the case where Γ is finite, which we already understand by §VI.9. First of all, it is sometimes convenient to use $\hat{H}^*(\Gamma, \mathbb{Z}_p)$ instead of $\hat{H}^*(\Gamma, \mathbb{Z})_{(p)}$. This is justified by:

(6.1) Proposition. Γ *has p-periodic cohomology if and only if* $\hat{H}^*(\Gamma, \mathbb{Z}_p)$ *contains an invertible element of non-zero degree.*

The proof will be based on a "Bockstein" argument. Let $H(k) = \hat{H}^*(\Gamma, \mathbb{Z}_{p^k})$. For any k, l there is a canonical short exact sequence $0 \to \mathbb{Z}_{p^l} \to \mathbb{Z}_{p^{k+l}} \to \mathbb{Z}_{p^k} \to 0$, which yields a cohomology exact sequence

$$(6.2) \qquad \cdots \to H(l) \to H(k+l) \xrightarrow{\alpha_{k+l,k}} H(k) \xrightarrow{\beta_{k,l}} H(l) \to \cdots.$$

(6.3) Lemma. *If $k \geq l$ then $\beta_{k,l}$ is a derivation, i.e.,*

$$\beta_{k,l}(uv) = \beta_{k,l}(u) \cdot v + (-1)^{\deg u} u \cdot \beta_{k,l}(v).$$

(The right-hand side makes sense and is in $H(l)$ because we have a cup product $H(k) \otimes H(l) \to H(l)$ for $k \geq l$.)

PROOF. Choose a complete resolution F and a diagonal map $F \to F \hat{\otimes} F$. It is easy to see that $\beta_{k,l}$ can be computed as follows: Given $u \in H(k)$, choose a cochain $c \in \mathcal{H}om_\Gamma(F, \mathbb{Z})$ whose reduction mod p^k is a cocycle representing u; then δc is divisible by p^k, and $\delta c/p^k$ is a cocycle whose reduction mod p^l represents $\beta_{k,l}(u)$. The derivation property of $\beta_{k,l}$ now follows from the corresponding property of δ. □

(6.4) Lemma. *For any $u \in H(1)$ and any $k \geq 1$, $u^{p^k} \in \operatorname{im} \alpha_{k+1,1} = \ker \beta_{1,k}$.*

PROOF. I claim first that for any $u \in H(k)$, we have $u^p \in \ker \beta_{k,1} = \operatorname{im} \alpha_{k+1,k}$. This follows from the fact that $\beta_{k,1}$ is a derivation whose target $H(1)$ is annihilated by p (cf. proof of 4.6). Thus $\alpha_{k+1,1}$ is the composite of ring homomorphisms

$$H(k+1) \xrightarrow{\alpha_{k+1,k}} H(k) \to \cdots \to H(1),$$

each of which has the property that p-th powers are in the image. The lemma follows at once. □

Let $H(\infty) = \hat{H}^*(\Gamma, \mathbb{Z})_{(p)}$. Note that we still have an exact sequence of the form 6.2 if $l = \infty$ (and $k < \infty$); this is obtained by taking p-primary components in the cohomology exact sequence associated to

$$0 \to \mathbb{Z} \xrightarrow{p^k} \mathbb{Z} \to \mathbb{Z}_{p^k} \to 0.$$

(6.5) Lemma. *For sufficiently large k, $\operatorname{im} \alpha_{k,1} = \operatorname{im} \alpha_{\infty,1}$.*

PROOF. We know (exercise 2 of §3) that $H(\infty)$ is annihilated by p^k for some k. The sequence 6.2 with $l = \infty$ therefore yields an injection $\alpha_{\infty,k}\colon H(\infty) \hookrightarrow H(k)$ for large k. Now it is easy to see that $\beta_{1,k} = \alpha_{\infty,k}\beta_{1,\infty}$, so $\ker \beta_{1,k} = \ker \beta_{1,\infty}$ for large k. But $\ker \beta_{1,k} = \operatorname{im} \alpha_{k+1,1}$ and $\ker \beta_{1,\infty} = \operatorname{im} \alpha_{\infty,1}$. □

(6.6) Lemma. *The map $\alpha = \alpha_{\infty,1}\colon H(\infty) \to H(1)$ has the following two properties:*

(i) *Every element of* $\ker \alpha$ *is nilpotent.*
(ii) *For any $u \in H(1)$ there is an integer k such that $u^{p^k} \in \operatorname{im} \alpha$.*

(Thus α is an F-isomorphism in the sense of §4.)

PROOF. If $u \in \ker \alpha$, then 6.2 with $l = \infty$ and $k = 1$ shows that $u = pv$ for some $v \in H(\infty)$. Hence $u^k = p^k v^k = 0$ for large k, since $H(\infty)$ is annihilated by some p^k. This proves (i), and (ii) follows from 6.4 and 6.5. □

PROOF OF 6.1. The map $\alpha: H(\infty) \to H(1)$ is a ring homomorphism (taking 1 to 1). This follows from the fact that the map $\hat{H}^*(\Gamma, \mathbb{Z}) \to \hat{H}^*(\Gamma, \mathbb{Z}_p)$ is a ring homomorphism which maps all primary components to zero except $\hat{H}^*(\Gamma, \mathbb{Z})_{(p)}$. Thus α takes invertible elements to invertible elements, whence the "only if" part of the proposition. Conversely, suppose $H(1)$ contains an invertible element u of non-zero degree. Raising u to a power if necessary, we can assume (6.6(ii)) that $u, u^{-1} \in \operatorname{im} \alpha$. Let $u = \alpha(\tilde{u})$ and $u^{-1} = \alpha(\tilde{v})$. Then $\alpha(\tilde{u}\tilde{v}) = 1$, so 6.6(i) implies that $\tilde{u}\tilde{v} = 1 - x$ with x nilpotent. But then $1 - x$ is invertible, with inverse $1 + x + x^2 + \cdots$, so $\tilde{u}$ is invertible and Γ has p-periodic cohomology. □

We now give the main result of this section:

(6.7) Theorem. *The following conditions are equivalent:*

(i) Γ *has p-periodic cohomology.*
(ii) *There exist integers i and d with* $d \neq 0$ *such that* $\hat{H}^i(\Gamma, M)_{(p)} \approx \hat{H}^{i+d}(\Gamma, M)_{(p)}$ *for all* Γ*-modules M.*
(iii) *Every finite subgroup of* Γ *has p-periodic cohomology.*
(iv) *Every elementary abelian p-subgroup of* Γ *has rank* ≤ 1.

PROOF. (i) $\Rightarrow$ (ii) trivially. If (ii) holds, then (ii) holds for every finite subgroup of Γ by Shapiro's lemma. But this implies (iii) by the analogue of VI.9.1 for p-periodicity. Since (iii) $\Leftrightarrow$ (iv) by VI.9.7, it remains to prove (iii) $\Rightarrow$ (i).

Let $\mathfrak{F}$ and $\mathfrak{H}^*(\Gamma)$ be as in 4.4, with $\mathbb{Z}_p$ as coefficient module. If (iii) holds, then I claim that we can find a family $u = (u_H)_{H \in \mathfrak{F}}$ of invertible elements $u_H \in \hat{H}^d(H) = \hat{H}^d(H, \mathbb{Z}_p)$ such that $u \in \mathfrak{H}^d(\Gamma)$. To see this, note first that the elements of $\mathfrak{F}$ fall into finitely many isomorphism classes; for if $\Gamma' \subseteq \Gamma$ is a torsion-free normal subgroup of finite index, then every $H \in \mathfrak{F}$ is isomorphic to a subgroup of the finite group Γ/Γ'. Now choose for each $H \in \mathfrak{F}$ an invertible element $v_H \in \hat{H}^*(H)$ of degree $d_H \neq 0$. By what we have just observed, we can certainly assume that only finitely many distinct degrees d_H occur. Raising each v_H to a suitable power, if necessary, we can then arrange that all the d_H are equal to a single integer d.

Now let $c: H_1 \hookrightarrow H_2$ be an embedding of finite subgroups given by conjugation by some element of Γ. Then v_{H_1} and $c^* v_{H_2}$ are two generators of $\hat{H}^d(H_1) \approx \hat{H}^0(H_1)$, hence $c^* v_{H_2} = \lambda v_{H_1}$ for some non-zero scalar $\lambda \in \mathbb{Z}_p$. Since $\lambda^{p-1} = 1$, it follows that $v_{H_1}^{p-1} = (c^* v_{H_2})^{p-1} = c^*(v_{H_2}^{p-1})$. Thus if we set $u_H = v_H^{p-1}$, we have $(u_H) \in \mathfrak{H}^*(\Gamma)$. Clearly (u_H^{-1}) is also in $\mathfrak{H}^*(\Gamma)$, so $\mathfrak{H}^*(\Gamma)$ contains an invertible element of non-zero degree. In view of the F-isomorphism $\hat{H}^*(\Gamma) \to \mathfrak{H}^*(\Gamma)$ of 4.6, it follows exactly as in the proof of 6.1 that $\hat{H}^*(\Gamma) = \hat{H}^*(\Gamma, \mathbb{Z}_p)$ contains an invertible element of non-zero degree. □

Remark. A weaker version of the implication (iii) $\Rightarrow$ (i) was proved by Venkov [1965], but not in the language of Farrell cohomology—he spoke instead of the ordinary cohomology being periodic in sufficiently high dimensions. Restated in terms of Farrell cohomology, his result said: if $\hat{H}^d(\Gamma)$ contains an element u whose restriction to $\hat{H}^*(H)$ is invertible for every $H \in \mathfrak{F}$, then u is invertible. (His proof, like ours, was based on the multiplicative structure in the spectral sequence 4.1.) The missing ingredient which was needed to go from Venkov's result to Theorem 6.7 was Quillen's observation 4.6.

EXERCISE

If 6.7(ii) holds and the isomorphism is natural, show that $\hat{H}^*(\Gamma, \mathbb{Z})_{(p)}$ contains an invertible element of degree d. [Hint: By dimension-shifting, 6.7(ii) holds for all i, with isomorphisms that are compatible with connecting homomorphisms. Show that the composite $H^i \to \hat{H}^i_{(p)} \to \hat{H}^{i+d}_{(p)}$ is necessarily given (up to sign) by cup product with the image of $1 \in H^0$.]

7 $\hat{H}^*(\Gamma)$ and the Ordered Set of Finite Subgroups of Γ

In this section we will prove analogues for Farrell cohomology of Theorems IX.11.1 and IX.13.1.

If Γ operates on a partially ordered set S such that $|S|$ is finite dimensional, then we set

$$\hat{H}^*_\Gamma(S) = \hat{H}^*_\Gamma(|S|).$$

As usual, it is understood here that there is an arbitrary Γ-module of coefficients. Recall that for any finite dimensional Γ-complex X there is a canonical map

$$\hat{H}^*(\Gamma) = \hat{H}^*_\Gamma(\text{pt.}) \to \hat{H}^*_\Gamma(X),$$

induced by the map $X \to \text{pt.}$

(7.1) Theorem. *Let Γ be a group such that* vcd $\Gamma < \infty$, *and let S be the set of non-trivial finite subgroups of Γ. Then the canonical map*

$$\hat{H}^*(\Gamma) \to \hat{H}^*_\Gamma(S)$$

is an isomorphism.

PROOF. As in the proof of IX.11.1, let X be a finite dimensional, contractible, admissible, proper Γ-complex with contractible fixed-point sets X^H $(H \in S)$, and let $X_0 = \bigcup_{H \in S} X^H$. Then we have

$$\begin{aligned}
\hat{H}^*(\Gamma) &\approx \hat{H}^*_\Gamma(X) && \text{because } X \text{ is contractible}\\
&\approx \hat{H}^*_\Gamma(X_0) && \text{because } \Gamma \text{ acts freely in } X - X_0 \text{ (cf. VII.10.1)}\\
&\approx \hat{H}^*_\Gamma(S) && \text{by IX.11.2.}
\end{aligned}$$

To see that the composite isomorphism is given by the canonical map $\hat{H}^*(\Gamma) \to \hat{H}^*_\Gamma(S)$, consider the diagram.

$$(*) \qquad \begin{array}{c} \hat{H}^*(\Gamma) \\ \swarrow^{\approx} \quad \downarrow \quad \searrow \\ \hat{H}^*_\Gamma(X) \xrightarrow{\approx} \hat{H}^*_\Gamma(X_0) \approx \hat{H}^*_\Gamma(S), \end{array}$$

where all maps coming from $\hat{H}^*(\Gamma)$ are the canonical ones. The left-hand triangle obviously commutes. For the right-hand triangle, we must go back to IX.11.2 and note that, in the notation of the latter, the square

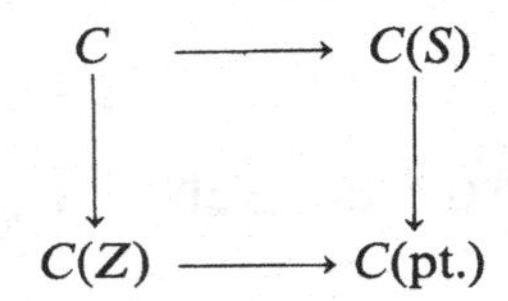

$$\begin{array}{ccc} C & \longrightarrow & C(S) \\ \downarrow & & \downarrow \\ C(Z) & \longrightarrow & C(\text{pt.}) \end{array}$$

commutes. This is easily verified. Thus $(*)$ commutes, whence the theorem. □

As in Chapter IX, there is a p-local version of 7.1 which is more useful. Its proof will be based on the following analogue of IX.11.5:

(7.2) Proposition. *Let Z be a finite dimensional admissible Γ-complex and let p be a prime such that Z^H is acyclic for every non-trivial finite p-subgroup of Γ. Then the canonical map*

$$\hat{H}^*(\Gamma)_{(p)} \to \hat{H}^*_\Gamma(Z)_{(p)}$$

is an isomorphism.

PROOF. Assume first that every finite subgroup of Γ is a p-group. We can then apply the method of proof of 7.1 to Γ acting on $X \times Z$ (with X as above), and we obtain an isomorphism $\hat{H}^*_\Gamma(X \times Z) \approx \hat{H}^*_\Gamma(S)$ such that the triangle

$$\begin{array}{c} \hat{H}^*(\Gamma) \\ \swarrow \qquad \searrow^{\approx} \\ \hat{H}^*_\Gamma(X \times Z) \approx \hat{H}^*_\Gamma(S) \end{array}$$

commutes. On the other hand, the projection $X \times Z \to Z$ yields

$$\begin{array}{ccc} & \hat{H}^*(\Gamma) & \\ \swarrow & & \searrow \\ \hat{H}^*_\Gamma(Z) & \xrightarrow{\approx} & \hat{H}^*_\Gamma(X \times Z). \end{array}$$

We therefore have

$$\hat{H}^*(\Gamma) \xrightarrow{\approx} \hat{H}^*_\Gamma(Z)$$

in this case.

In the general case, choose a subgroup $\Gamma' \subseteq \Gamma$ of finite index such that $(\Gamma : \Gamma')$ is relatively prime to p and every finite subgroup of Γ' is a p-group (cf. proof of IX.9.2). Using restriction and transfer maps in the usual way (cf. §III.10), we have a natural embedding of $\hat{H}^*_\Gamma(W)_{(p)}$ (for any W) as a direct summand of $\hat{H}^*_{\Gamma'}(W)$. In particular, taking $W = Z$ and $W = \text{pt.}$, we see that the canonical map

$$\hat{H}^*(\Gamma)_{(p)} \to \hat{H}^*_\Gamma(Z)_{(p)}$$

is a direct summand of the canonical map

$$\hat{H}^*(\Gamma') \to \hat{H}^*_{\Gamma'}(Z).$$

The latter being an isomorphism by the previous paragraph, it follows that the former is also an isomorphism. □

We can now prove the main result of this section:

(7.3) Theorem. *Let Γ be a group such that* vcd $\Gamma < \infty$, *let p be a prime, and let $\mathscr{A}$ be the ordered set of non-trivial elementary abelian p-subgroups of Γ. Then*

$$\hat{H}^*(\Gamma)_{(p)} \xrightarrow{\approx} \hat{H}^*_\Gamma(\mathscr{A})_{(p)}.$$

PROOF. Apply 7.2 with $Z = |\mathscr{A}|$ (cf. proof of IX.13.1). □

One special case where $\hat{H}^*_\Gamma(\mathscr{A})$ can be easily understood is that where $\mathscr{A}$ is discrete. In that case 7.3 yields:

(7.4) Corollary. *Suppose that every elementary abelian p-subgroup of Γ has rank ≤ 1. Then*

$$\hat{H}^*(\Gamma)_{(p)} \xrightarrow{\approx} \prod_{P \in \mathscr{P}} \hat{H}^*(N(P))_{(p)},$$

where $\mathscr{P}$ is a set of representatives for the conjugacy classes of subgroups of Γ of order p.

For a concrete example where 7.4 leads to explicit cohomology calculations, see Brown [1976]. (These calculations, however, which involve $\Gamma = SL_3(\mathbb{Z})$, have been superseded by the calculations of Soulé [1978].) For an example of 7.3 with $\dim \mathscr{A} > 0$, see Brown [1979]. This time $\Gamma = SL_3(\mathbb{Z}[1/2])$, $p = 2$, and $\dim \mathscr{A} = 1$.

Remark. Since $\hat{H}^i(\Gamma) = H^i(\Gamma)$ for $i > \operatorname{vcd} \Gamma$, the canonical map

$$H^i(\Gamma)_{(p)} \to \prod_{P \in \mathscr{P}} H^i(N(P))_{(p)}$$

is an isomorphism for $i > \operatorname{vcd} \Gamma$ under the hypotheses of 7.4. It is natural to ask, by analogy with similar situations in algebraic topology, whether the map is surjective for $i = \operatorname{vcd} \Gamma$. It can be shown that this question, whose answer is not known, is closely related to the question asked in §VIII.11 as to whether one can always find a proper, contractible Γ-complex of dimension equal to $\operatorname{vcd} \Gamma$.

Exercise

Show that the components of the map in 7.4 are restriction maps

$$\hat{H}^*(\Gamma)_{(p)} \to \hat{H}^*(N(P))_{(p)}.$$

References

G. Almkvist [1973], Endomorphisms of finitely generated projective modules over a commutative ring, *Ark. Mat.* **11** (1973), 263–301.

R. C. Alperin and P. B. Shalen [1982], Linear groups of finite cohomological dimension, *Invent. Math.* **66** (1982), 89–98.

A. Ash [1977], Deformation retracts with lowest possible dimension of arithmetic quotients of self-adjoint homogeneous cones, *Math. Ann.* **225** (1977), 69–76.

M. F. Atiyah and I. G. Macdonald [1969], *Introduction to commutative algebra*, Addison-Wesley, Reading, Mass., 1969.

H. Bass [1976], Euler characteristics and characters of discrete groups, *Invent. Math.* **35** (1976), 155–196.

H. Bass [1979], Traces and Euler characteristics, Homological group theory (C. T. C. Wall, ed.), *London Math. Soc. Lecture Notes 36*, Cambridge University Press, Cambridge, 1979, 105–136.

R. Bieri [1976], Homological dimension of discrete groups, Queen Mary College Mathematics Notes, London, 1976.

R. Bieri and B. Eckmann [1973], Groups with homological duality generalizing Poincaré duality, *Invent. Math.* **20** (1973), 103–124.

R. Bieri and B. Eckmann [1974], Finiteness properties of duality groups, *Comment. Math. Helv.* **49** (1974), 74–83.

A. Borel et al. [1960], Seminar on transformation groups, *Ann. of Math. Studies* **46**, Princeton, 1960.

A. Borel and J-P. Serre [1974], Corners and arithmetic groups, *Comment. Math. Helv.* **48** (1974), 244–297.

A. Borel and J-P. Serre [1976], Cohomologie d'immeubles et de groupes S-arithmétiques, *Topology* **15** (1976), 211–232.

R. Brauer [1926], Über Zusammenhänge zwischen arithmetischen und invariantentheoretischen Eigenschaften von Gruppen linearer Substitutionen, *Sitzungsber. Preuss. Akad. Wiss.* (1926), 410–416. Collected papers, vol. 1, 5–11.

G. E. Bredon [1972], *Introduction to compact transformation groups*, Academic Press, New York, 1972.

K. S. Brown [1974], Euler characteristics of discrete groups and G-spaces, *Invent. Math.* **27** (1974), 229–264.

K. S. Brown [1975a], Homological criteria for finiteness, *Comment. Math. Helv.* **50** (1975), 129–135.
K. S. Brown [1975b], Euler characteristics of groups: The p-fractional part, *Invent. Math.* **29** (1975), 1–5.
K. S. Brown [1976], High dimensional cohomology of discrete groups, *Proc. Nat. Acad. Sci. U.S.A.* **73** (1976), 1795–1797.
K. S. Brown [1979], Groups of virtually finite dimension, Homological group theory (C. T. C. Wall, ed)., *London Math. Soc. Lecture Notes 36*, Cambridge University Press, Cambridge, 1979, 27–70.
K. S. Brown [1982], Complete Euler characteristics and fixed-point theory, *J. Pure Appl. Algebra* **24** (1982), 103–121.
K. S. Brown and P. J. Kahn [1977], Homotopy dimension and simple cohomological dimension of spaces, *Comment. Math. Helv.* **52** (1977), 111–127.
W. Burnside [1911], *Theory of groups of finite order* (*2nd edition*), Cambridge University Press, Cambridge, 1911. (Reprinted by Dover, New York, 1955.)
H. Cartan et al. [1954/55], Algèbres d'Eilenberg-MacLane et homotopie, Séminaire Henri Cartan 1954/55, École Normale Supérieure, Paris.
H. Cartan and S. Eilenberg [1956], *Homological algebra*, Princeton University Press, Princeton, 1956.
I. M. Chiswell [1976a], Exact sequences associated with a graph of groups, *J. Pure Appl. Algebra* **8** (1976), 63–74.
I. M. Chiswell [1976b], Euler characteristics of groups, *Math. Z.* **147** (1976), 1–11.
D. E. Cohen [1972], Groups with free subgroups of finite index, Conference on group theory, *Lecture Notes in Math. 319*, Springer-Verlag, Berlin-Heidelberg-New York, 1972, 26–44.
D. E. Cohen [1978], Combinatorial group theory: A topological approach, Queen Mary College Mathematics Notes, London, 1978.
J. Cohen [1972], Poincaré 2-complexes—I, *Topology* **11** (1972), 417–419.
H. S. M. Coxeter and W. O. J. Moser [1980], *Generators and relations for discrete groups*, 4th ed., Springer-Verlag, Berlin-Heidelberg-New York, 1980.
A. Dold [1972], *Lectures on algebraic topology*, Springer-Verlag, Berlin-Heidelberg-New York, 1972.
M. J. Dunwoody [1972], Relation modules, *Bull. London Math. Soc.* **4** (1972), 151–155.
M. J. Dunwoody [1979], Accessibility and groups of cohomological dimension one, *Proc. London Math. Soc.* (*3*) **38** (1979), 193–215.
E. Dyer and A. T. Vasquez [1973], Some small aspherical spaces, *J. Austral. Math. Soc.* **16** (1973), 332–352.
B. Eckmann [1953], Cohomology of groups and transfer, *Ann. of Math.* **58** (1953), 481–493.
B. Eckmann and H. Müller [1980], Poincaré duality groups of dimension two, *Comment. Math. Helv.* **55** (1980), 510–520.
S. Eilenberg [1947], Homology of spaces with operators, I, *Trans. Amer. Math. Soc.* **61** (1947), 378–417; errata, **62** (1947), 548.
S. Eilenberg and T. Ganea [1957], On the Lusternik-Schnirelmann category of abstract groups, *Ann. of Math.* **65** (1957), 517–518.
S. Eilenberg and S. MacLane [1945], Relations between homology and homotopy groups of spaces, *Ann. of Math.* (*2*) **46** (1945), 480–509.
S. Eilenberg and S. MacLane [1947], Cohomology theory in abstract groups, II, Group extensions with a non-abelian kernel, *Ann. of Math* (*2*) **48** (1947), 326–341.
S. Eilenberg and N. Steenrod [1952], *Foundations of algebraic topology*, Princeton University Press, Princeton, 1952.
F. T. Farrell [1975], Poincaré duality and groups of type (FP), *Comment. Math. Helv.* **50** (1975), 187–195.
F. T. Farrell [1977], An extension of Tate cohomology to a class of infinite groups, *J. Pure Appl. Algebra* **10** (1977), 153–161.

J. Fischer, A. Karrass, and D. Solitar [1972], On one-relator groups having elements of finite order, *Proc. Amer. Math. Soc.* **33** (1972), 297–301.
R. Godement [1958], *Topologie algébrique et théorie des faisceaux*, Hermann, Paris, 1958.
O. Goldman [1961], Determinants in projective modules, *Nagoya Math. J.* **18** (1961), 27–36.
D. H. Gottlieb [1965], A certain subgroup of the fundamental group, *Amer. J. Math.* **87** (1965), 840–856.
A. Grothendieck [1957], Sur quelques points d'algèbre homologique, *Tôhoku Math. J.* (*2*) **9** (1957), 119–221.
K. W. Gruenberg [1970], Cohomological topics in group theory, *Lecture Notes in Math. 143*, Springer-Verlag, Berlin-Heidelberg-New York, 1970.
M. Hall, Jr. [1959], *The theory of groups*, MacMillan, New York, 1959.
G. Harder [1971], A Gauss-Bonnet formula for discrete arithmetically defined groups, *Ann. Sci. École Norm. Sup* (*4*) **4** (1971), 409–455.
A. Hattori [1965], Rank element of a projective module, *Nagoya J. Math.* **25** (1965), 113–120.
G. Hochschild [1965], *The structure of Lie groups*, Holden-Day, San Francisco, 1965.
G. Hochschild and J-P. Serre [1953], Cohomology of group extensions, *Trans. Amer. Math. Soc.* **74** (1953), 110–134.
H. Hopf [1942], Fundamentalgruppe und zweite Bettische Gruppe, *Comment. Math. Helv.* **14** (1942), 257–309.
J. F. P. Hudson [1969], *Piecewise linear topology*, Benjamin, New York, 1969.
N. J. S. Hughes [1951], The use of bilinear mappings in the classification of groups of class 2, *Proc. Amer. Math. Soc.* **2** (1951), 742–747.
W. Hurewicz [1936], Beiträge zur Topologie der Deformationen. IV. Asphärische Räume, *Nederl. Akad. Wetensch. Proc.* **39** (1936), 215–224.
S. Illman [1978], Smooth equivariant triangulations of G-manifolds for G a finite group, *Math. Ann.* **233** (1978), 199–220.
S. Jackowski [1978], The Euler class and periodicity of group cohomology, *Comment. Math. Helv.* **53** (1978), 643–650.
A. Karrass, A. Pietrowski, and D. Solitar [1973], Finitely generated groups with a free subgroup of finite index, *J. Austral. Math. Soc.* **16** (1973), 458–466.
R. C. Kirby and L. C. Siebenmann [1969], On the triangulation of manifolds and the Hauptvermutung, *Bull. Amer. Math. Soc.* **75** (1969), 742–749.
J. Lehner [1964], *Discontinuous groups and automorphic functions*, Mathematical surveys no. 8, Amer. Math. Soc., Providence, 1964.
A. T. Lundell and S. Weingram [1969], *The topology of CW complexes*, Van Nostrand Reinhold, New York, 1969.
R. C. Lyndon [1950], Cohomology theory of groups with a single defining relation, *Ann. of Math.* **52** (1950), 650–665.
R. C. Lyndon and P. E. Schupp [1977], *Combinatorial group theory*, Springer-Verlag, Berlin-Heidelberg-New York, 1977.
S. MacLane [1949], Cohomology theory in abstract groups, III, Operator homomorphisms of kernels, *Ann. of Math.* (*2*) **50** (1949), 736–761.
S. MacLane [1963], *Homology*, Springer-Verlag, Berlin, 1963.
S. MacLane [1978], Origins of the cohomology of groups, *Enseign. Math.* (*2*) **24** (1978), 1–29.
S. MacLane [1979], Historical note, *J. Algebra* **60** (1979), 319–320.
I. Madsen, C. B. Thomas, and C. T. C. Wall [1976], The topological space form problem—II: Existence of free actions, *Topology* **15** (1976), 375–382.
A. I. Malcev [1949], On a class of homogeneous spaces, *Izv. Akad. Nauk SSSR Ser. Mat.* **13** (1949), 9–32. (English translation: *Amer. Math. Soc. Transl. No.* **39** (1951).)

W. S. Massey [1967], *Algebraic topology: an introduction*, Harcourt, Brace & World, New York, 1967. (Reprinted by Springer-Verlag.)

C. Miller [1952], The second homology group of a group; relations among commutators, *Proc. Amer. Math. Soc.* **3** (1952), 588–595.

J. Milnor [1957a], Groups which act on S^n without fixed points, *Amer. J. Math.* **79** (1957), 623–630.

J. Milnor [1957b], The geometric realization of a semi-simplicial complex, *Ann. of Math.* (*2*) **65** (1957), 357–362.

J. Milnor [1963], Morse theory, *Ann. of Math. Studies* **51**, Princeton University Press, Princeton, 1963.

J. Milnor [1971], Introduction to algebraic K-theory, *Ann. of Math. Studies* **72**, Princeton University Press, Princeton, 1971.

M. S. Montgomery [1969], Left and right inverses in group algebras, *Bull. Amer. Math. Soc.* **75** (1969), 539–540.

J. Munkres [1966], Elementary differential topology, revised ed., *Ann. of Math. Studies* **54**, Princeton University Press, Princeton, 1966.

C. D. Papakyriakopoulos [1957], On Dehn's lemma and the asphericity of knots, *Ann. of Math.* **66** (1957), 1–26.

D. S. Passman [1977], *The algebraic structure of group rings*, Wiley, New York, 1977.

H. Poincaré [1904], Cinquième complément à l'Analysis situs, *Rend. Circ. Mat. Palermo* **18** (1904), 45–110. (*Oeuvres*, Gauthier-Villars (1953), t. VI, pp. 435–498.)

D. Quillen [1971], The spectrum of an equivariant cohomology ring, I, II, *Ann. of Math.* **94** (1971), 549–572 and 573–602.

D. Quillen [1973], Finite generation of the groups K_i of rings of algebraic integers, Algebraic K-theory I, *Lecture Notes in Math. 341*, Springer-Verlag, Berlin-Heidelberg-New York, 1973, 179–198.

D. Quillen [1978], Homotopy properties of the poset of non-trivial p-subgroups of a group, *Advances in Math.* **28** (1978), 101–128.

H. R. Schneebeli [1978], On virtual properties of group extensions, *Math. Z.* **159** (1978), 159–167.

O. Schreier [1926], Über die Erweiterungen von Gruppen, I, *Monatsh. Math. u. Phys.* **34** (1926), 165–180.

H. Schubert [1968], *Topology*, Allyn and Bacon, Boston, 1968.

I. Schur [1902], Neuer Beweis eines Satzes über endliche Gruppen, *Sitzungsber. Preuss. Akad. Wiss.* (1902), 1013–1019. Gesammelte Abhandlungen, I, 79–85.

I. Schur [1904], Über die Darstellung der endlichen Gruppen durch gebrochene lineare Substitutionen, *J. Reine Angew. Math.* **127** (1904), 20–50. Gesammelte Abhandlungen, I, 86–116.

G. P. Scott [1974], An embedding theorem for groups with a free subgroup of finite index, *Bull. London Math. Soc.* **6** (1974), 304–306.

G. P. Scott and C. T. C. Wall [1979], Topological methods in group theory, Homological group theory (C. T. C. Wall, ed.), *London Math. Soc. Lecture Notes 36*, Cambridge University Press, Cambridge, 1979, 137–203.

J-P. Serre [1965], Algèbre locale; multiplicités (2nd edition), *Lecture Notes in Math. 11*, Springer-Verlag, Berlin-Heidelberg-New York, 1965.

J-P. Serre [1968], *Corps locaux*, Hermann, Paris, 1968.

J-P. Serre [1971], Cohomologie des groupes discrets, *Ann of Math. Studies* **70** (1971), 77–169.

J-P. Serre [1977a], Arbres, amalgames, SL_2, *Astérisque* **46** (1977).

J-P. Serre [1977b], *Linear representations of finite groups*, Springer-Verlag, New York, 1977.

J-P. Serre [1979], Arithmetic groups, Homological group theory (C. T. C. Wall, ed.), *London Math. Soc. Lecture Notes 36*, Cambridge University Press, Cambridge, 1979, 105–136.

C. L. Siegel [1971], *Topics in complex function theory*, vol. II, Wiley-Interscience, New York, 1971.

C. Soulé [1978], The cohomology of $SL_3(\mathbb{Z})$, *Topology* **17** (1978), 1–22.

E. H. Spanier [1966], *Algebraic Topology*, McGraw-Hill, New York, 1966.

J. R. Stallings [1963], A finitely presented group whose 3-dimensional integral homology is not finitely generated, *Amer. J. Math.* **85** (1963), 541–543.

J. R. Stallings [1965a], Homology and central series of groups, *J. Algebra* **2** (1965), 170–181.

J. R. Stallings [1965b], Centerless groups—an algebraic formulation of Gottlieb's theorem, *Topology* **4** (1965), 129–134.

J. R. Stallings [1968], On torsion-free groups with infinitely many ends, *Ann. of Math.* **88** (1968), 312–334.

J. R. Stallings [unpublished], An extension theorem for Euler characteristics of groups, unpublished.

U. Stammbach [1966], Anwendungen der Homologietheorie der Gruppen auf Zentralreihen und auf Invarianten von Präsentierungen, *Math. Z.* **94** (1966), 157–177.

U. Stammbach [1973], Homology in group theory, *Lecture Notes in Math. 359*, Springer-Verlag, Berlin-Heidelberg-New York, 1973.

R. Strebel [1976], A homological finiteness criterion, *Math. Z.* **151** (1976), 263–275.

R. Strebel [1977], A remark on subgroups of infinite index in Poincaré duality groups, *Comment. Math. Helv.* **52** (1977), 317–324.

U. Stuhler [1980], Homological properties of certain arithmetic groups in the function field case, *Invent. Math.* **57** (1980), 263–281.

M. Suzuki [1955], On finite groups with cyclic Sylow subgroups for all odd primes, *Amer. J. Math.* **77** (1955), 657–691.

R. G. Swan [1960a], A new method in fixed point theory, *Comment. Math. Helv.* **34** (1960), 1–16.

R. G. Swan [1960b], Induced representations and projective modules, *Ann. of Math.* **71** (1960), 552–578.

R. G. Swan [1969], Groups of cohomological dimension one, *J. Algebra* **12** (1969), 585–601.

O. Teichmüller [1940], Über die sogenannte nichtcommutative Galoische Theorie und die Relation $\xi_{\lambda,\mu,\nu}\xi_{\lambda,\mu\nu,\pi}\xi^{\lambda}_{\mu,\nu,\pi} = \xi_{\lambda,\mu,\nu\pi}\xi_{\lambda\mu,\nu,\pi}$, *Deutsche Math.* **5** (1940) 138–149.

B. B. Venkov [1965], On homologies of groups of units in division algebras, *Trudy Mat. Inst. Steklov* **80** (1965), 66–89. [English translation: *Proc. Steklov Inst. Math.* **80** (1965), 73–100.]

G. Voronoi [1907], Nouvelles applications des paramètres continus à la théorie des formes quadratiques, I, *J. Reine Angew. Math.* **133** (1907), 97–178.

C. T. C. Wall [1961], Rational Euler characteristics, *Proc. Cambridge Philos. Soc.* **57** (1961), 182–183.

C. T. C. Wall [1965], Finiteness conditions for CW-complexes, *Ann. of Math.* **81** (1965), 56–69.

C. T. C. Wall [1966], Finiteness conditions for CW-complexes II, *Proc. Royal Soc. A* **295** (1966), 129–139.

C. T. C. Wall (ed.) [1979], List of problems, Homological group theory (C. T. C. Wall, ed.), *London Math. Soc. Lecture Notes 36*, Cambridge University Press, Cambridge, 1979, 369–394.

A. Weil [1952], Sur les théorèmes de de Rham, *Comment. Math. Helv.* **26** (1952), 119–145.

J. E. West [1977], Mapping Hilbert cube manifolds to ANR's: A solution of a conjecture of Borsuk, *Ann. of Math.* **106** (1977), 1–18.

J. H. C. Whitehead [1949], Combinatorial homotopy, I, II, *Bull. Amer. Math. Soc.* **55** (1949), 213–245, 453–496.

J. A. Wolf [1974], *Spaces of constant curvature* (3rd edition), Publish or Perish, Boston, 1974.

H. J. Zassenhaus [1958], *The theory of groups*, 2nd ed., New York, Chelsea, 1958.

Notation Index

Subject Index

Graduate Texts in Mathematics

continued from page ii

119 ROTMAN. An Introduction to Algebraic Topology.
120 ZIEMER. Weakly Differentiable Functions: Sobolev Spaces and Functions of Bounded Variation.
121 LANG. Cyclotomic Fields I and II. Combined 2nd ed.
122 REMMERT. Theory of Complex Functions. *Readings in Mathematics*
123 EBBINGHAUS/HERMES et al. Numbers. *Readings in Mathematics*
124 DUBROVIN/FOMENKO/NOVIKOV. Modern Geometry—Methods and Applications. Part III.
125 BERENSTEIN/GAY. Complex Variables: An Introduction.
126 BOREL. Linear Algebraic Groups.
127 MASSEY. A Basic Course in Algebraic Topology.
128 RAUCH. Partial Differential Equations.
129 FULTON/HARRIS. Representation Theory: A First Course. *Readings in Mathematics*
130 DODSON/POSTON. Tensor Geometry.
131 LAM. A First Course in Noncommutative Rings.
132 BEARDON. Iteration of Rational Functions.
133 HARRIS. Algebraic Geometry: A First Course.
134 ROMAN. Coding and Information Theory.
135 ROMAN. Advanced Linear Algebra.
136 ADKINS/WEINTRAUB. Algebra: An Approach via Module Theory.
137 AXLER/BOURDON/RAMEY. Harmonic Function Theory.
138 COHEN. A Course in Computational Algebraic Number Theory.
139 BREDON. Topology and Geometry.
140 AUBIN. Optima and Equilibria. An Introduction to Nonlinear Analysis.
141 BECKER/WEISPFENNING/KREDEL. Gröbner Bases. A Computational Approach to Commutative Algebra.
142 LANG. Real and Functional Analysis. 3rd ed.
143 DOOB. Measure Theory.
144 DENNIS/FARB. Noncommutative Algebra.
145 VICK. Homology Theory. An Introduction to Algebraic Topology. 2nd ed.
146 BRIDGES. Computability: A Mathematical Sketchbook.
147 ROSENBERG. Algebraic K-Theory and Its Applications.
148 ROTMAN. An Introduction to the Theory of Groups. 4th ed.
149 RATCLIFFE. Foundations of Hyperbolic Manifolds.
150 EISENBUD. Commutative Algebra with a View Toward Algebraic Geometry.
151 SILVERMAN. Advanced Topics in the Arithmetic of Elliptic Curves.
152 ZIEGLER. Lectures on Polytopes.
153 FULTON. Algebraic Topology: A First Course.
154 BROWN/PEARCY. An Introduction to Analysis.
155 KASSEL. Quantum Groups.
156 KECHRIS. Classical Descriptive Set Theory.
157 MALLIAVIN. Integration and Probability.
158 ROMAN. Field Theory.
159 CONWAY. Functions of One Complex Variable II.
160 LANG. Differential and Riemannian Manifolds.
161 BORWEIN/ERDÉLYI. Polynomials and Polynomial Inequalities.
162 ALPERIN/BELL. Groups and Representations.
163 DIXON/MORTIMER. Permutation Groups.
164 NATHANSON. Additive Number Theory: The Classical Bases.
165 NATHANSON. Additive Number Theory: Inverse Problems and the Geometry of Sumsets.
166 SHARPE. Differential Geometry: Cartan's Generalization of Klein's Erlangen Programme.